Slurry Erosion

Slurry Erosion: Flow Phenomenon, Complexities, and Protection Methods presents the necessary fundamentals to understand slurry flows and erosion phenomenon. It describes the complexities in the critical assessment on minimizing erosion and the process of selecting an appropriate protection technique.

Addressing new technologies for modeling and achieving reduction in slurry erosion with optimized design, the book highlights how CFD simulations and AI techniques are used to predict slurry erosion. It explores the application of slurry erosion in various industries, such as thermal power plants, hydropower plants, mining, chemical and food processing, and offshore structures. The chapters cover various problems in slurry erosion, including head loss, erosion–corrosion, erosion–abrasion, leakage flow, puncture point failure, and leaching.

This resource is intended to support researchers and graduate students studying slurry erosion, surface engineering, tribology, erosion and corrosion prevention.

Dr. Jashanpreet Singh is an Assistant Professor in the Department of Mechanical Engineering and University Center for Research and Development at Chandigarh University, Mohali, India. He earned his PhD in the field of slurry erosion and thermal spray coatings from Thapar Institute of Engineering and Technology, Patiala, in 2019 and has three years of teaching experience and two years of industry experience. Dr. Singh has authored over 100 research publications in various international journals, books, and conferences, and he serves as a reviewer for various journals. He has authored one book and several book chapters with CRC Press. He received the Trusted Reviewer Status in 2020 and the Outstanding Reviewer Award in 2021 from the Institute of Physics (IOP) for *Surface Topography: Metrology and Properties.* Dr. Singh is a life member of AICTSD, IAENG, and NITTSD.

Dr. Satish Kumar is both a Dean (Research and Consultancy) and a Professor in the Department of Mechanical Engineering at the National Institute of Technology, Jamshedpur, India. He is currently the Faculty in Charge of Admission, Controller of Examination, Chairman of Institute Publication Cell, and Chairman of Institute Newsletter at NIT Jamshedpur. He received his bachelor's and master's in mechanical engineering from SVNIT Surat in 2002 and 2004, respectively. Dr. Kumar completed his Ph.D. under the supervision of Professor B. K. Gandhi at the Department of Mechanical and Industrial Engineering, IIT Roorkee, in 2014. He has authored and co-authored over 150 research articles in the field of multiphase flows and coatings and completed five sponsored research projects funded by different agencies, with three research projects currently ongoing. Dr. Kumar is a life member of ISTE, ASME, NSFMFP, IAENG, Institution of Engineers, and NITTSD.

Advances in Surface Engineering and Tribology

Series Editors: Hitesh Vasudev, Jashanpreet Singh, and Jeyaprakash Natarajan

Advances in Surface Engineering and Tribology explores the interdisciplinary aspects of surface science and engineering, as well as tribology discussions pertaining to the examination of friction, wear, and lubrication. It covers topics ranging from surface modification techniques, nanomaterials and their applications in nanotechnology, catalytic and photocatalytic processes, synthesis of materials, colloids and surfaces, thin film deposition processes, methods for characterizing surfaces, and surface engineering for diverse industrial purposes. The volumes comprehensively explore the complex mechanisms that regulate surface phenomena, presenting both established principles and emerging developments in the field of surface science and engineering. The series offers a forum for professionals, engineers, and researchers in the energy, automotive, aerospace, and biomedical sectors to delve into innovative developments in these interrelated disciplines.

Should you have an idea for this series, please get in contact with Kyra Lindholm (kyra.lindholm@taylorandfrancis.com).

Thermal Spray Coatings for High-Temperature Conditions
Edited by Gaurav Prashar, Hitesh Vasudev, and Lalit Thakur

Microwave Claddings
Engineering Applications
Edited by Sarbjeet Kaushal

Fabrication of Wear and Corrosion Resistant Materials
Surface Processing Techniques
Edited by Gurbhej Singh and Hitesh Vasudev

Metal Oxides
Advances in Sustainability and Nanotechnology
Edited by Virat Khanna, Rohit Jasrotia, Tejeet Singh, and Gulab Singh

Multifunctional and Smart Materials
Advances in Sustainable Engineering Applications
Edited by Hitesh Vasudev and Gurbhej Singh

Microwave Processing of Metallic Materials
Revolutionizing Surface Engineering for Green Manufacturing
Edited by Hitesh Vasudev, Gurbhej Singh, and Amit Bansal

Slurry Erosion
Flow Phenomenon, Complexities, and Protection Methods
Edited by Jashanpreet Singh and Satish Kumar

For more information about this series, please visit: www.routledge.com/Advanced-Materials-Processing-and-Manufacturing/book-series/CRCAMPM

Slurry Erosion

Flow Phenomenon, Complexities, and Protection Methods

Edited by
Jashanpreet Singh and Satish Kumar

CRC Press
Taylor & Francis Group
Boca Raton London

CRC Press is an imprint of the
Taylor & Francis Group, an **informa** business

Designed cover image: Jashanpreet Singh and Satish Kumar

First edition published 2026
by CRC Press
2385 NW Executive Center Drive, Suite 320, Boca Raton FL 33431

and by CRC Press
4 Park Square, Milton Park, Abingdon, Oxon, OX14 4RN

CRC Press is an imprint of Taylor & Francis Group, LLC

ISBN: 978-1-032-91060-4 (hbk)
ISBN: 978-1-032-91062-8 (pbk)
ISBN: 978-1-003-56115-6 (ebk)

DOI: 10.1201/9781003561156

Typeset in Times
by Apex CoVantage, LLC

This book is dedicated to our families, friends, and teachers.

Contents

About the Editors ix
List of Contributors xi
Preface xiii

SECTION I Fundamentals of Slurry Flow, Erosion Phenomena, and Flow Complexities

Chapter 1 Slurries and Their Properties 3

Jashanpreet Singh and Moniya

Chapter 2 Investigation of Slurry Erosion, Erosion–Abrasion, and Erosion–Corrosion Using Different Testers and Test Loops 24

Jibrin Muhammad Yelwa and Haruna Musa

Chapter 3 Head Losses Associated with Slurry Flows 49

Jashanpreet Singh, Jatinder Pal Singh, Virinder Kumar, and Miroslaw Szala

Chapter 4 Numerical and Computational Analysis of Slurry Flows in Mineral Beneficiation and Processing 68

Sukhwinder Singh Sekhon, Virinder Kumar, Parminder Singh, and Ashif Equbal

Chapter 5 Assessment of Slurry Erosion in Pumps and Hydraulic Turbines 80

Virinder Kumar and Jashanpreet Singh

SECTION II Protection Methods against Slurry Flows and Erosion

Chapter 6 Erosion Wear Behavior of Composite Materials for Slurry and Solid Particle Erosion Applications 105

Praveen Kumar Verma, Hitesh Vasudev, Sunitha K, and Yanhai Cheng

Chapter 7 Fabrication of Metallic Materials for Erosion Resistance 120

Sunitha K, Praveen Kumar Verma, Hitesh Vasudev, and Che Hua Yang

Chapter 8 Surface Engineering for Protection against Slurry Erosion ... 136
Praveen Kumar Verma, Hitesh Vasudev, and Sule Emmanuel Oshioriameh

Chapter 9 Coating Methods for Protection against Slurry Erosion ... 152
Moniya, Jashanpreet Singh, Anuj Kumar Sehgal, and Mohamed Aboughaly

Chapter 10 Internal Coatings to Prevent Slurry Erosion in Pipelines ... 166
Nayana Acharya, Raghavendra Sagar, and Nacer Badi

Chapter 11 Surface Treatment Methods for Protection against Slurry Erosion in Marine Environments ... 182
Praveen Kumar Verma, Hitesh Vasudev, and Bhupinder Singh

Chapter 12 Post-Processing of Surfaces for Improving Their Performance against Slurry Erosion ... 197
Praveen Kumar Verma, Hitesh Vasudev, and Jürgen Markus Lackner

Chapter 13 Artificial Intelligence in Slurry Erosion and Protection Techniques ... 211
Jyoti Bhattacharjee, Subhasis Roy, and Hyun Suk Jung

SECTION III Case Studies on Slurry Flows and Erosion

Chapter 14 Modeling and Analysis of Flow-Induced Erosion in Pipeline Systems ... 231
Asisha Ranjan Pradhan, Satish Kumar, and Shivam Kumar

Chapter 15 Case Study on Erosive Wear from Mineral Slurry Handling ... 243
Praveen Kumar Verma, Hitesh Vasudev, and Yanhai Cheng

Chapter 16 Case Studies on Optimizing Industrial Slurry Systems ... 261
Jyoti Bhattacharjee, Subhasis Roy, and Hyun Suk Jung

Index ... 279

About the Editors

Dr. Jashanpreet Singh
Designation: Assistant Professor
University Center for Research and Development,
Department of Mechanical Engineering,
Chandigarh University, Mohali, Punjab, India

Dr. Jashanpreet Singh is currently working as Assistant Professor in the University Center for Research and Development at Chandigarh University, Mohali, India. He obtained his PhD in the field of slurry erosion and thermal spray coatings from Thapar Institute of Engineering and Technology, Patiala, in 2019. He has teaching experience of more than three years and industry experience of more than two years. His areas of research are tribology–solid particle erosion, power plant engineering, ash conveying, thermal spray coatings, CFD simulation, composites, and material characterization. He has contributed comprehensively in thermal spray coatings and tribology in reputed journals for various publishers. He has authored over 100 research publications, of which 58 publications are indexed in SCIE/ESCI (Web of Science) and 93 in the SCOPUS database in various international journals, books, and conferences of repute. His *h*-index is 31 as per the SCOPUS database. He has authored one book with CRC Press, edited two books with CRC Press, and book chapters in different CRC Press books. He serves as a reviewer for various journals. He is currently an editor of Elsevier, CRC Press, Wiley-Scrivener, and Bentham Science books. He serves as a guest editor of several SCOPUS indexed journals. He received Trusted Reviewer Status from the Institute of Physics in 2020. He was awarded an Outstanding Reviewer Award 2021 by the Institute of Physics for the journal *Surface Topography: Metrology and Properties*. He is a life member of AICTSD, IAENG, and NITTSD.

Prof. (Dr.) Satish Kumar
Designation: Dean (Research and Consultancy)
Professor, Department of Mechanical Engineering,
National Institute of Technology, Jamshedpur, Jharkhand, India

Dr. Satish Kumar is presently working as a professor in the Department of Mechanical Engineering, National Institute of Technology Jamshedpur, India. Prof. Satish Kumar is also working as Dean (Research & Consultancy), and Nodal Officer (Research Innovation Park). He is the Chairman & Nodal officer of NIRF activities at the institute. He also held the position of Faculty in Charge of Admission, Controller of Examination, from 2020 to 2023; Chairman of Institute Publication Cell; and Chairman of Institute Newsletter from 2019 to 2023 at NIT Jamshedpur. He has more than 21 years of teaching and research experience. Dr. Satish Kumar obtained his bachelor's and master's from Department of Mechanical Engineering, SVNIT Surat, in 2002 and 2004. He completed his Ph.D. under the supervision of Professor B. K. Gandhi, Department of Mechanical and Industrial Engineering, IIT Roorkee, in the area of solid–liquid flow in 2014. His research interests are rheology of slurry flow, computational fluid dynamics, erosion wear, and flow phenomena of solid–liquid flow in pumps and pipelines. He has authored (and co-authored) over 176 research articles, including 90 in SCI Index journals in the field of multiphase flow. He has been the editor of the many reputed journals and books. He has supervised 16 Ph.D. and 55 M.Tech theses in the area of multiphase flow. He has organized 2 national conferences, 2 international conferences, 2 Science and Engineering Research Board (SERB) sponsored high-end Karyshala workshops,

2 ATAL-AICTE sponsored faculty development programs, and more than 16 faculty development programs/short-term courses. He has completed seven sponsored research projects funded by different agencies such as SERB, DST, and UGC successfully. Currently, he has four research projects ongoing. He is a life member of ISTE, ASME, NSFMFP, IAENG, Institution of Engineers, NITTSD, and others.

Contributors

Mohamed Aboughaly
Faculty of Energy Systems and Nuclear Science
University of Ontario Institute of Technology
Canada

Nayana Acharya
Department of Physics
Bhuvanendra Degree College
India

Nacer Badi
Department of Physics
University of Tabuk
Saudi Arabia

Jyoti Bhattacharjee
Department of Chemical Engineering
University of Calcutta
India

Yanhai Cheng
School of Mechanical and Electrical Engineering
China University of Mining and Technology
China

Ashif Equbal
Ikuta Sanki Kogyo
Japan

Hyun Suk Jung
Advanced Materials Science and Engineering
SKKU
South Korea

Sunitha K
Research and Development Cell
Lovely Professional University
India

Satish Kumar
Department of Mechanical Engineering
National Institute of Technology
India

Shivam Kumar
WA School of Mines: Minerals, Energy and Chemical Engineering
Curtin University
Australia

Virinder Kumar
University Centre for Research and Development
Chandigarh University
India

Jürgen Markus Lackner
JOANNEUM RESEARCH Forschungsgesellschaft mbH
Institute for Sensors, Photonics and Manufacturing Technologies, Research Group Laser and Plasma Processing,
Austria

Moniya
Department of Mechanical Engineering
Chandigarh University
India

Haruna Musa
Department of Chemistry
Bayero University
Nigeria

Sule Emmanuel Oshioriameh
Industrial Engineering and Management (Engineering Management Program)
American University of Beirut
Lebanon

Asisha Ranjan Pradhan
Department of Mechanical & Aerospace Engineering
Indian Institute of Technology
India

Subhasis Roy
Department of Chemical Engineering
University of Calcutta
India

Raghavendra Sagar
Department of Physics
Mangalore Institute of Technology & Engineering
India

Agus Saptoro
Chemical and Energy Engineering
Curtin University Malaysia
Malaysia

Anuj Kumar Sehgal
Department of Aerospace Engineering
Chandigarh University
India

Sukhwinder Singh Sekhon
ICAR-Central Institute of Post-Harvest Engineering & Technology
India

Bhupinder Singh
Department of Chemical and Environmental Engineering
University of Nottingham
United Kingdom

Jashanpreet Singh
University Center for Research and Development
Chandigarh University
India

Jatinder Pal Singh
Department of Mechanical Engineering
Guru Nanak Dev University
India
jatinder.mech@gndu.ac.in

Parminder Singh
University Centre for Research and Development
Chandigarh University
India

Miroslaw Szala
Department of Materials Engineering
Lublin University of Technology
Poland

Hitesh Vasudev
Research and Development Cell
Lovely Professional University
India

Praveen Kumar Verma
Research and Development Cell
Lovely Professional University
India

Che Hua Yang
School of Mechanical and Electrical Engineering
National Taipei University of Technology
Taiwan

Jibrin Muhammad Yelwa
Department of Chemistry
Bayero University
Nigeria

Preface

Slurry Erosion: Flow Phenomenon, Complexities and Protection Methods book describes the fundamentals and mechanism of slurry erosion caused by the erosive particles immersed in liquid. It provides the basic fundamental to understand the slurry flows and erosion phenomenon, complexities in minimizing erosion, critical assessment on minimizing the erosion, and selection of adequate protection technique.

This book presents the overview about slurry flow phenomenon and its applications inside different industries, and complexities that arise due to the flow of solid–liquid or gas-solid flows. This book aims to provide the understanding and designing of slurry flow and conveying systems in order to reduce adverse effects by the flow the erosive particles. This book also presents the details of investigations of different properties of bulk solid particles and slurries as well as their advanced characterization.

This book highlight the basics of slurry flows and erosion, application of slurry erosion in various industries (like thermal power plants, hydropower plants, mining, chemical and food processing industries, and offshore structures), various parameters involved in slurry flow, existing testing facilities, computational fluid dynamics (CFD) simulation of slurry flows, and artificial intelligence (AI) in slurry erosion field. The other problems which are associated to slurry erosion such as head loss, erosion–corrosion, erosion–abrasion, leakage flow, and puncture point failure are also covered in this book. In this book, the case studies related to the CFD, wear, corrosion head loss analysis of slurry flow phenomenon in different applications and the adverse effects caused due to slurry flow have been discussed. This book covers the literature of mechanistic flow models, head loss models, cavitation models, and erosion models.

This reference book is directed toward professors, students, practitioners, and researchers of higher education institutions working in the fields that deal with the application of solid particle erosion. Also, the engineers can study the different parameters of specific erosive slurry flow on-site. This book enables the design engineers to optimize the design by selecting the tested and proved newly developed methodologies in a small dataset. The literature based on scientific principles can be helpful for the interpretation of field data, selection of machinery, selection of materials, and protection of materials.

Section I

Fundamentals of Slurry Flow, Erosion Phenomena, and Flow Complexities

1 Slurries and Their Properties

Jashanpreet Singh and Moniya

1.1 INTRODUCTION

Slurries are the mixtures of solid particles suspended in a liquid medium. They are found in natural processes like river sediments and volcanic flows and also in engineered processes, like mineral processing, ceramics manufacturing, chemical industries, wastewater treatment, and energy production. The individual behavior of slurries is due to the synergistic effect of fluid dynamics and particle interactions which can determine their flow, stability, and processing properties.

The study of slurries is important since their characteristics, including viscosity, density, particle size distribution, and concentration, have direct effects on the efficiency, safety, and cost of operation. An example is the transportation of coal slurries in pipelines, management of cement slurries in construction, or management of tailings in mining, which all require well characterized slurry systems. Furthermore, slurries are generally not homogeneous. They may have various behaviors of both Newtonian and highly non-Newtonian flow, which depends on the interactions between particles and fluids and external conditions.

This chapter presents the basic ideas of slurries, including their classification according to the composition, concentration, flow characteristics, and areas of use. The chapter classifies slurries into several types, including dilute, concentrated, stable, unstable, and Newtonian and non-Newtonian, which gives a systematic way of explaining their behavior. Moreover, it preconditions the investigation of the technological applicability of each type of slurry, as well as the challenges and opportunities that they offer in the practice.

1.2 TYPES OF SLURRIES

1.2.1 Based on Composition and Phase

Many engineers prefer to organize slurries according to the number of distinct phases they observe during the experiments. Solid granular particles are suspended in a liquid, typically normal water, or in a two-phase suspension in a particular concentration. The concentration is frequently encountered in mining haul lines, ceramic slip tanks, and metallurgical pour pots due to its simplicity and dependability. Occasionally, the concentration becomes more complex when a third component like gas is introduced into the mixture, which results in a three-phase slurry. This situation is frequently seen in froth flotation cells, where a multitude of air bubbles assist in the separation of valuable minerals from the rock. Everything is disrupted by the mere presence of gas, including the design, tuning, and monitoring of pipelines and separators, as well as the stability of foam and fluid performance.

1.2.2 Based on Behavior

Slurries can also be classified based on the behavior of the particulates when the mixture remains stationary for a small time period. Slurries may be categorized based on their flow behavior, which is dictated by the rheological interactions between the particles and the liquid medium. Newtonian slurries have a constant viscosity regardless of shear rate, usually at low solid concentrations. Non-Newtonian slurries, in contrast, are characterized by variable viscosity, including shear-thinning

DOI: 10.1201/9781003561156-2

(pseudoplastic), shear-thickening (dilatant), Bingham plastic (yield stress to begin to flow), thixotropy (reduction in viscosity with time under shear), and the uncommon rheopecty (increase in viscosity with shear time). This type of behavior is important, as it directly influences the efficiency of pumping, transport, and processing of slurries in natural and industrial systems.

Slurries are frequently classified as either settling or non-settling depending upon the capacity of the solid particles to be suspended in the liquid media. Settling slurries have coarse or heavy particles that will tend to separate and settle due to the effect of gravity when the agitation or flow velocity is inadequate; such as sand–water and coal–water mixtures. This type of slurry is frequently observed in ash conveying, mining, or when transporting extremely dense ores. Non-settling slurries are mixture of fine particles and liquid which have colloidal size and are held in suspension by Brownian motion or repulsive interparticle forces, as in paint formulations or clay suspensions. Chemical processing industries and pharmaceutical laboratories rely on this category to ensure that all components are well mixed. This difference should be considered in the design of transport systems, as settling slurries need greater flow velocities or agitation to resist sedimentation, whereas non-settling slurries tend to be stable but can have complicated rheological behavior. Also, the selection of pump type is entirely determined by the rate at which particulates settle. It steers the slurry towards a centrifugal impeller or a slow-gear positive-displacement rig.

1.2.3 Based on Terminology in Industry

The terminology used to characterize slurries in industrial practice frequently exhibits significant variation, even within the specialized field of slurry handling and processing. Terms such as mud, paste, slime, sludge, slip, pulp, dispersion, broth, tailings, silt, and froth are commonly used. The most common are cement slurry, drilling slurry or drilling mud, coal-water slurry, mineral slurry, and chemical slurry, which are used to indicate their main area of application. Cement and grout slurries find application in the construction industry to provide binding and structural stability, whereas drilling slurries or muds are needed in the oil and gas industry to provide lubrication, cooling, and cuttings transportation. Ore slurries and tailings slurries are also extensively used in the mining and mineral processing industries, and concentration and flowability are important in pipeline transport and waste management. On the same note, coal-water slurries are used as alternative fuels in power generation. By selecting the appropriate term, engineers can attempt to predict the necessary pumping capacity, conduit diameter, and other equipment to maintain the flow of operations.

1.2.4 Based on Particle Size

Sorting solids according to their particulate size provides engineers with a rapid understanding of the slurry's gritty texture. Pump design, piping selection, and perhaps even routine maintenance schedules can be altered by this singular measurement. Clay and sediment particles are tightly packed at a size of less than 2 microns, while silt is dispersed between 2 and 50 microns. The fine sand, which is located at the broadest end of the scale, ranges from 50 to 100 microns, while the medium sand continues to progress up to 500 microns. Coarse sand is situated in the 500-to-2000-micron range, with gravel larger than 2 millimeters following its path. In certain dredging scenarios, chunks larger than 64 millimeters are introduced into the mixture, rendering soft rubber hoses virtually useless. Ceramic or hardened linings are the standard due to the fact that larger granules consume components at a rate that exceeds expectations.

1.2.5 Application-Based Classifications of Slurry Pumps

The category into which a specific slurry pump is classified is frequently determined by its applications, which provides engineers with a convenient shortcut when numerous equipment options are available. When specifications are implemented, a slurry pump is typically identified as a coarse-wear pump that is capable of handling fluids that are laden with grit or clay. A gravel pump with thicker impellers is designed to accommodate particulates that are slightly larger, such as pea gravel or shattered stone, and transport them with ease. Dredge pumps effortlessly navigate through debris the size of fist-sized boulders in dredging machines that traverse riverbeds. Aerated mixtures float like shaving foam, and froth blowers intervene to move low-density slurries while simultaneously maintaining bubble control. In both the CIP (Carbon-in-Pulp) and CIL (Carbon-in-Leach) methods, the slurry is gently handled by carbon transfer pumps with the main aim of reducing abrasive wear, whilst being used to transport activated carbon effectively through the gold recovery circuit even in the most aggressive mineral stream environment. In the meantime, sump pumps serve as guardians in the collection pits, and their cantilevered shaft causes fluids with grit sediments to rise and maintains important parts of the pump in places of safe operation out of the wear zone, e.g. keeping seals in place.

1.3 PROPERTIES OF SLURRIES

Slurry properties are essential in determining the efficacy of their handling, transport, and processing. Key characteristics that influence flow behavior and storage requirements include the type of slurry, viscosity, and static settling tendency. Slurries are affected by the applied forces during pumping, which are influenced by rheological properties that include shear behavior and yield stress. The solid concentration has an impact on both density and attrition potential, while the pH value is crucial for evaluating chemical compatibility and corrosion risk. In industrial applications, the behavior of the slurry is collectively determined by these factors. The typical property ranges for a variety of prevalent slurry types are summarized in Table 1.1.

1.3.1 Settling Properties

The static settled concentration can be estimated using a gravitational settling method to assess the settling nature of slurry. The static resolved concentration of various slurries is determined by the role of specific gravity. Specific gravity is the ratio of the measured density of a substance to the density of a reference substance. The pycnometer method can be used to determine the specific gravity

TABLE 1.1
Properties of Solid Particles Processed through Hydraulic Machinery

Materials	Specific Gravity (g/cm^3)	Hardness (Hv)	Shape	pH value	References
Sand	2.667	360	Conical-knife edges	7.71–7.16 for Cw = 30–60%	Singh [1]
Quartz	2.650	1100	Blocky and conical	–	Clark and Llewellyn [2]
Alumina	3.940	1800	Angular	–	Desale et al. [3]
SiC	3.220	2500	Angular	–	Desale et al. [3]
Fly ash	2.259	–	Spherical	6.32–6.09 for Cw=30–60%	Singh [1]
Bottom ash	2.081	–	Blocky	7.5–7.34 for Cw=30–60%	Singh [1]
Ore	4.620	–	Conical	8.93–9.03 for Cw=20–50%	Singh et al. [4]
Coal	1.460 deviation: ±2%	–	Conical	6.38–5.93 Cw=10–60%	Singh et al. [5]
Tailings	2.890	–	Knife-edge structure	6.99–7.30 Cw=25–55%	Lindgren [6] and Verma et al. [7]

of slurries, which necessitates the following weight measurements and empirical correlations [1]. Pipelines, elbows, tees, and other fittings are substantially affected by the settling properties. Factors such as the specific gravity of the various slurries and their settling characteristics must also be taken into account in slurry transport systems. The density of a substance is compared to a prescribed reference density, which is known as specific gravity. The empirical correlation, which employs the pycnometer method, can be employed to determine specific gravity by measuring weight:

$$Specific\ gravity = \frac{Ws}{\left(W_w - W_{sw}\right) + W_s} \quad (1.1)$$

In this equation, W_s, W_w, and W_{sw} are the weight of solid particles, weight of water, and weight of solid-water slurry, respectively, in a jar to attain a fixed level of volume.

1.3.2 Acidity and Alkalinity

The slurry's pH level is also essential. Erosion–corrosion is influenced by the acidity of the sediment. AISI 4330 steel's erosion–corrosion characteristics in a saline–sand slurry were investigated by Ballesteros et al. [8]. The results indicated that the volume loss of AISI 4330 steel was greater when the saline–sand was pH = 3 than when it was pH = 7. The following correlation [9] is typically used to calculate the erosion–corrosion rate (EC):

$$EC = \frac{WL}{A \times T} \quad (1.2)$$

In this equation, WL, A, and T represents the iron loss weight (mg), elbow surface area (m^2), and time (min), respectively.

The total volume loss (TVL) from materials can be calculated by using the following relationship [10]:

$$TVL = ER + CR + EC \quad (1.3)$$

where the volume of reductions attributed to pure is represented by *ER, CR*, and *EC*, the synergistic interactions among erosion and corrosion, pure electrochemical degradation, and mechanical injury, respectively. In spite of this, accelerated wear resulted in the formation of a weak FeZn10 scale, which exacerbates corrosion. Therefore, erosion and corrosion's synergistic effect is classified into two categories [11]:

- Accelerated wear due to corrosion
- Accelerated corrosion due to wear

For the calculation of the synergistic effects between erosion and corrosion, the following expressions can be used [12]:

$$EC = \Delta K_E + \Delta K_C \quad (1.4)$$

The total metal wastage rate caused by corrosion (i.e., K_C) is given by:

$$K_C = CR + \Delta K_C \quad (1.5)$$

Similarly, the total metal wastage rate caused by erosion (i.e., K_E) is given by:

$$K_E = ER + \Delta K_E \quad (1.6)$$

where ΔK_E and ΔK_C represent corrosion-accelerated erosion and erosion-accelerated corrosion. For active metals, $K_C = CR$ and $K_E = K_{EO}$, whereas for passive metals, $K_C = \Delta K_C$ and $K_E = M.\ ER$ can be calculated by standard erosion wear correlations according to applications.

1.3.3 Shape of Particles

One of the most important factors affecting erosion wear in slurry erosion is the form of the particles. Usually hydraulic machines handle variously shaped particles. Most hydraulically transported solids are alumina, mining/tailing [12], coal [5], fly ash, bottom ash, sand/silt [13], silicon carbide [3], quartz [2], glass beads [14], concrete [6], and ores [4]. Figure 1.1 shows the different solids moved by hydraulic equipment. When it comes to hydraulic equipment, particulate wear is greatly affected by particle shape. A number of researchers have offered models that may be used to compute the shape factor, which is used to define the particle shape. Bouwman et al. [15] reported an empirical association that Riley (1941) suggested as a means of determining the form factor (SF):

$$SF = \frac{4\pi A}{P^2} \tag{1.7}$$

In this equation, A and P denote the area and perimeter of particles. Rathod et al. [16] proposed a formula to estimate the particle's perimeter:

$$P = \frac{\pi}{2}\left[\frac{3}{2}\left(L_P + W_P\right) - \left(\frac{L_P}{W_P}\right)\right]^{\frac{1}{2}} \tag{1.8}$$

where L_P is the measured length, and W_P is the width of the particle. According to Levy and Chik [17], the wear rate produced by angular particles is about four times higher than that of spherical particles (wear caused by spherical particles).

There is a significant difference in the form factor between angular and spherical particles. As the erodent's hardness and form factor vary, so does the erosion wear mechanism of the target surface [20]. Ahlert [21] reported that SF = 1 for completely spherical particles, SF = 0.53 for partially spherical particles, and SF = 0.20 for angular/conical particles. Bahadur and Badruddin [22] assessed erosion-induced wear at jet speeds ranging from 40 to 65 m/s. They

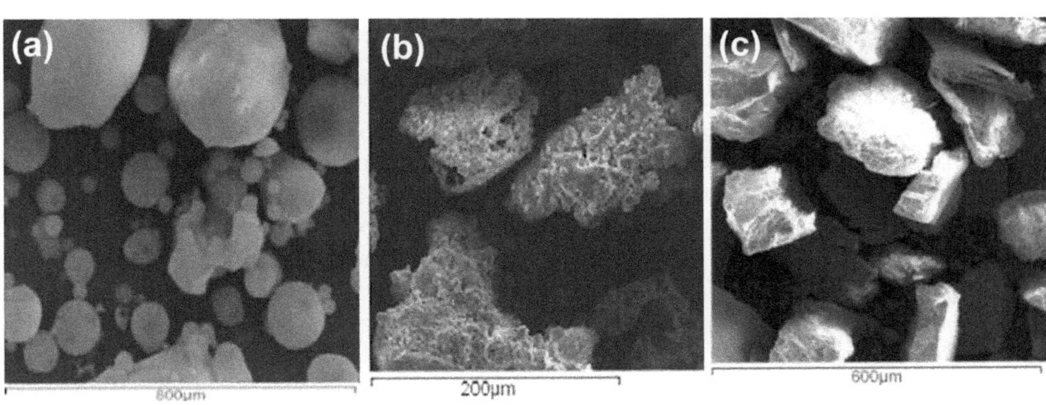

FIGURE 1.1 Various solids transported through hydraulic machinery: (a) fly ash [18], (b) bottom ash [18], and (c) sand [19].

used several erodents to investigate the impact of particle shape and size, which ranged from 10–500 µm. The results showed that the particle circularity factors were within a certain range (i.e., 0.70–0.97).

Zou and Yu [23] introduced another factor called "sphericity of particles", which can be evaluated by using the following correlation:

$$\psi = \frac{A_S}{A_P} \tag{1.9}$$

In this equation, A_S and A_P represent the areas of the sphere and particle, respectively.

Additionally, the following factor to determine the sphericity of fault rock particle sizes was proposed by Heilbronner and Keulen [24]:

$$RD = \frac{4 \times A_p}{\pi \times D^2_{\max fret}} \tag{1.10}$$

where $D_{maxfret}$ stands for the long axis of rectangular particles with defective properties. They also reported the Paris factor, as given here:

$$PF = 2\left(\frac{P_p - P_{ch}}{P_{chA}}\right) \tag{1.11}$$

The Paris factor is used to analyze the shape of grains.

Desale et al. [25] developed a correlation by modifying the empirical correlation proposed by Riley (1941):

$$SF = \left(SF_{Avg} \times SF_{Max} \times SF_{Min}\right) \tag{1.12}$$

To evaluate the effect particle shape in terms of size fraction, they examined the erosion wear of AISI 1020 and AISI 1045 steel. The erosion wear performance of both ductile materials was examined using alumina, quartz, and silicon carbides as erodents. The erodents were considered to be the same size and had a solid particle velocity of 3 m/s. Their results showed that the drop in the form factor of the erodent raised the erosion wear. With a dry sand rubber wheel, Woldman et al. [26] looked at erosion wear. For the particle size distribution (PSD) range of 80 to 500 µm, they discovered throughout the investigation that the CF of sand particles falls from 0.56 to 0.65. They also covered the Krumbein chart, which uses physical appearance to estimate the form factor (up to 1 decimal), as depicted in Figure 1.2.

Thakur et al. [27] performed an experimental study to examine the effect of solid concentration (varied as 10, 20, and 30 weight percent) on SS-304 steel's erosion wear at 0.5 m/s. They also discovered that surface property determines the erosion wear of a material. Walker and Hambe [28] used a Coriolis tester to investigate the erosion wear of white cast iron utilizing particle form. Using silicon carbide, sand, and alumina as erodent materials, they conducted extensive Coriolis tester erosion wear studies. As shown, they created a correlation for the prediction of erosion wear with regard to particle form.

$$Er = k \times CF^{-x} \tag{1.13}$$

The shape of the erodent revealed a significant effect on the erosion wear of white cast iron at various velocities and impact angles, they found. As shown in Figure 1.3, they also discovered that sand, alumina, and SiC had respective circularity factor (CF) values of 0.73, 0.71, and 0.64. Table 1.2 summarizes the form factor of many particle materials.

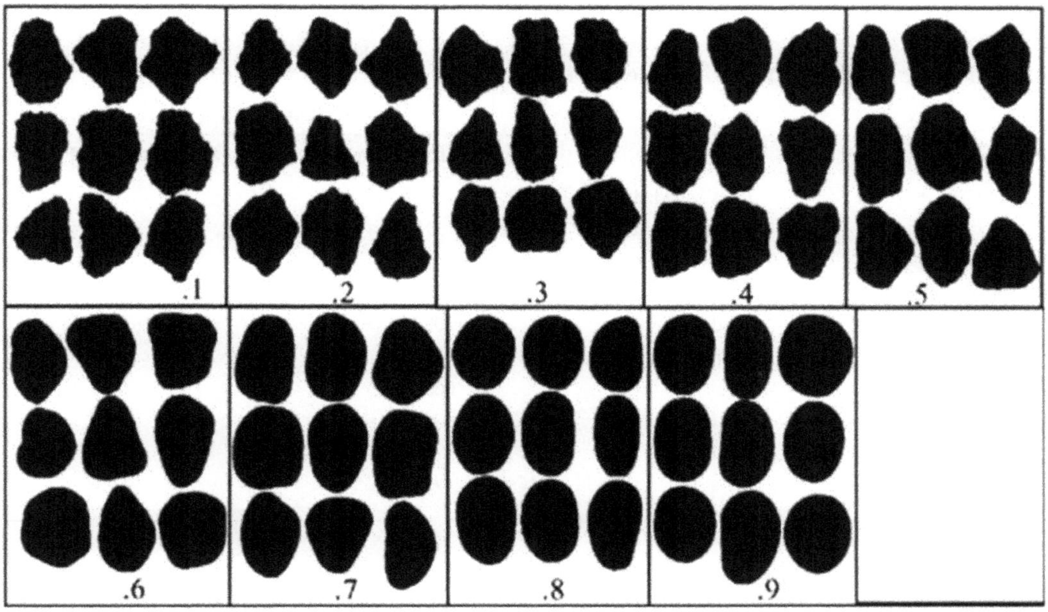

FIGURE 1.2 Krumbein chart [26].

(By permission of Elsevier)

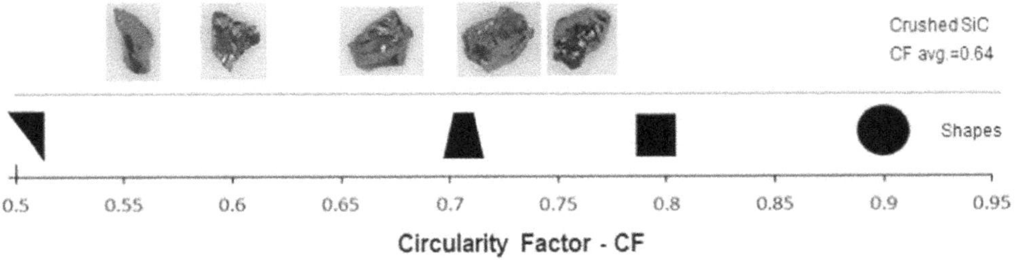

FIGURE 1.3 Circularity factors of sand, alumina, and SiC [28].

(By permission of Elsevier)

1.3.4 Density of Particles (ρ)

One of the most basic properties that affect the behavior of a slurry is the density of particles (ρ). It is the weight of solid particles per cubic meter (kg/m^3) or gram per cubic centimeter (g/cm^3). The density of the particles has a direct impact on the settling rate, suspension stability, and density of the slurry mixture. Dense materials like minerals and metal oxides are more likely to be deposited by gravity than the lighter particles like organic matter. When a larger density collides with

TABLE 1.2
Shape Factor of Different Particulate Materials

Authors	Particles	Equi-Sized Particles: Range of Size Fraction	Equi-Sized Particles: Shape Factor (SF) in μm
Ahlert [21]	Fully rounded	NA	1
	Semi-rounded	NA	0.53
	Angular/conical	NA	0.20
Woldman et al. [26]	Sand	50–800	0.56–0.65
Walker and Hambe [28]	Sand	1200–2000	0.73
	Alumina	600–800	0.71
	SiC	600–800	0.64
Singh et al. [13]	Bottom ash	<300	0.712
	Sand	<300	0.637
	Fly ash	<300	0.908

the target surface, the transfer of kinetic energy is higher as compared to a lower density of particles. Laboratory trials offer a hard yardstick. For instance, Stack and Pungwiwat [29] performed experimentation on aluminum-oxide pebbles and silicon-carbide bits at a metallic plate. The denser alumina plowed in harder and resulted in a greater mass loss. A separate run by Desale et al. [25] showed almost exactly the same findings. The density difference ($\Delta\rho$) between the particle density and liquid density is commonly the force behind sedimentation, classification, and separation processes in industrial applications. A correct understanding of the particle density is required to calculate slurry transport, design equipment, and optimize the process because it determines the rheology, power needed to pump, and extent of wear on pipelines and equipment.

1.3.5 Particle Size *(d)*

In addition, the size of the particles significantly influences the degradation rate. The particles' larger size implies that a significant segment of the target surface is in contact with the particle surface during the impact. Additionally, particulates of greater size generate a greater amount of surface wear due to their high kinetic energy in motion. In general, particles are of varying sizes when they are in powdered form. Therefore, particle size distribution or the range of size fraction is the metric for particle size. To evaluate the size fraction distribution of the particulates, the British standard sieve (BSS) number is used. Using the arithmetic mean diameter (d_i), as represented in the following equation [30], the BSS number of powdered particulates can be determined:

$$d_p = \sum\nolimits_{i=1}^{n} fi \times \left(N_i\right) \tag{1.14}$$

In this equation, N_i is the BSS number of the particle size range, and f_i is the solid fraction. Singh et al. [31] remarked that the average fineness number (AFN) is a more accurate indicator of particle size. By obtaining the arithmetic weighted mean diameter of particles and sifting the solids in several particle size ranges, AFN may be determined. An empirical expression may be used to get the average fineness number:

$$AFN = \frac{\Sigma\left(Mi \times \left(m_i \times g\right)\right)}{\Sigma\left(m_i \times g\right)} \tag{1.15}$$

where g, $\Sigma(m_i)$, and M_i denote the gravitational constant, sum of overall mass (i.e. each sieve), and British sieve standard multiplier, respectively.

Clark and Hartwitch [32] conducted an experiment on the erosion wear of AISI 1020 and AISI 1045 steel to investigate the impact of silica sand particle size. They observed that the size fraction of eroded materials is a determining factor in the erosion wear of materials. Tilly [33] found that the erosion ratio remained constant when the particle size increased above 1000 μm. Numerous researchers [12, 34, 35] have found the significance of particle size in slurry erosion. They also stated that erosion wear is correlated with particle size in a power-law fashion.

Singh et al. [36] provided the erosion wear reliance on weighted mean diameter described in earlier studies:

$$E_w = k \times (weighted\ mean\ diameter)^n \tag{1.16}$$

The value of the diameter exponent (n) varies for different particulates for different target materials. Generally, the value of the particle diameter exponent (n) lies in the range of 0.3 to 2.0. It has been noted in the literature that the erosion induced by certain slurries is reduced by the addition of a finer size fraction of particle [30]. Changes in particle–particle, particle–target, and flow phenomena are influenced by the mixing of particles of varying sizes.

1.3.6 Hardness of Particles (H_P)

Another crucial characteristic of particles that greatly influences wear rate is their hardness [37, 38]. Considering terms, notably fracture toughness (K_c), particle radius (r_P), particle density (ρ_P), and hardness of particles (H_P), Wiederhorn and Hockey [39] suggested a correlation for the prediction of erosion wear:

$$E_T = k \times V_P^{2.8} \times r_P^{3.9} \times K_C^{-1.8} \times H_P^{0.48} \times \rho_P^{1.4} \tag{1.17}$$

Wada and Watanabe [40] reported an empirical association based on the hardness factor, as shown:

$$E \prec \left(\frac{H_T}{H_P} \right)^k \tag{1.18}$$

The exponent of the hardness ratio is denoted by the symbol k in the previous equation. The hardness ratio is defined as the ratio of the material's hardness (H_T) to the particle hardness (H_P) [41]. Elkholy [41] used four distinct materials with varying degrees of hardness, ranging from 150 to 400 HB (Brinell hardness). The wear studies were conducted by adjusting the particle size of silica sand while keeping the other parameters constant, and the mass loss was determined. The researcher said that the value of silica sand was 710 HB. The transition point of the hardness ratio was reported as follows:

$$\frac{H_T}{H_P} = 1.9 \tag{1.19}$$

The exponent k fluctuates as the hardness ratio varies from the aforementioned number. Levy and Chik [17] investigated the impact of particle hardness on the erosion wear of AISI 1020 carbon steel. They discovered that the erosion wear of AISI 1020 steel was enhanced until a hardness value of 700 kgf/mm^2 (particles) was reached, at which point no further rise in erosion wear was seen. Table 1.3 provides a summary of the hardness of several solid particles.

The erosion wear of various materials is affected by the movement of slurry particles. In contrast, Desale et al. [3] observed that erosion wear increased in the following sequence of solid particles:

- Alumina > silicon carbide > quartz (same order of hardness also)

Singh [1] also conducted a comparison between erodent fly ash, bottom ash, and sand. An order was also reported:

- Sand > bottom ash > fly ash (on the basis of shape, hardness, and density)

Particle hardness is clearly important for particulate wear, but other particulate characteristics also matter. Therefore, it may be said that the particulate wear depends on the hardness of target material up to a given value.

1.3.7 Solid Concentration of Slurry

The solid concentration of particles (vol. % or wt. %) also influences the wear of the materials [41]. The following is the relationship between the erosion wear performance of the target surface and solid concentration:

$$E = k(Concentration)^p \tag{1.20}$$

where E is erosion wear, C is solid concentration, and p is the concentration's exponent. The value of the exponent p lies in the range 0.9 to 1.3 [42]. In Table 1.2, the value of the exponent p for various types of slurries and materials is reported from the work of many researchers. Erosion wear of materials typically increases as the concentration of solids increases [30, 43–45]. Dasgupta et al. [46] determined the erosion wear examination of a steel specimen using a multisized slurry of sand. They discovered that the erosion wear rate reduced when the solid concentration was raised from 30% to 50%. Nevertheless, the erosion rate reduces as the solid concentration increases, even at low concentrations, as shown in Figure 1.4 [47]. Similar results were reported by other researchers [48, 49]. The shielding effect of rebounding particles is the cause of this phenomena. They said that the erosion rate typically rises until it reaches a critical value in conjunction with the solid concentration during the early phase. The particle–particle interactions between hitting and rebounding particles become essential when this critical value is exceeded. The particle–particle interactions of impacting and rebounding particles result in a momentum loss and a change in their orientations at solid concentrations (<1 wt. %). The particle–target interaction is reduced as the particle–particle contact increases, resulting in a low erosion rate [50]. The properties of the target materials determine the rebounding of particles. Therefore, this phenomenon is restricted to a certain set of materials. Additionally, the flow characteristics of slurry are influenced by its solid content. The Reynolds number (*Re*) and Stokes (*St*) number are influenced by an increase in solid concentration. Bong [51, 52] observed that the viscosity of slurry rises as the solid concentration of slurry grows, whereas *Re* decreases as the solid concentration of slurry increases, as shown in Figure 1.5. In pipeline flow scenarios, the flows are classified as laminar when *Re* is less than 2100, transitional when *Re* is between 2100 and 4000, and turbulent when *Re* exceeds 4000. Shehadeh et al. [9] investigated the erosion–corrosion properties of carbon steel using a sand–water slurry that included a solid concentration of 3 to 9 g/l. Additionally, they said that the erosion–corrosion rate rises as the concentration and value of *Re* increase, as shown in Figure 1.6.

The fluid's mobility of solid particles is also influenced by the change in concentration. Study of the Stokes number may provide a more comprehensive understanding of this occurrence. Dabirian et al. [53] found that the value of *St* rises as the particle size and solid concentration increase. Further, Bartosik [54] mentioned that the shear stress between particles and the target wall is greatly

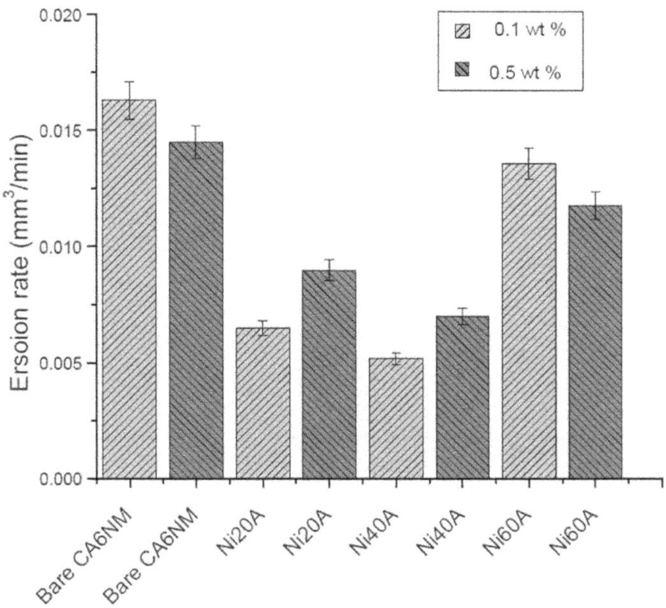

FIGURE 1.4 Erosion rate vs. solid concentration [47].

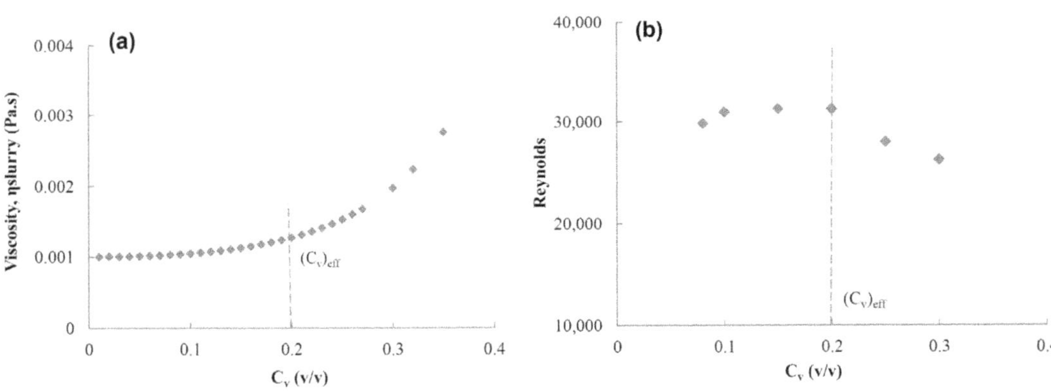

FIGURE 1.5 Effect of solid concentration on (a) viscosity and (b) Reynolds number [52].

influenced by the solid concentration and particle size. It is evident from the aforementioned literature that the concentration of the slurry influences several factors that have not yet been thoroughly investigated. The wear of erosion rises as the solid concentration of equi-sized and multisized dilute slurries increases [55]. A few studies have also observed that the erosion rate lowers as the concentration of the slurry increases [1, 46–50, 56]. The settling flow generated by slurry particles also influences erosion wear [1, 5]. According to refs. [1, 5, 50], the particle–particle and particle–wall interaction phenomena are affected by the change in solid concentration.

1.3.8 Flow Velocity

The velocity at which particles strike a surface is a major contributor to the wear. Granular particles of slurry achieve a great kinetic energy (KE) at higher velocities [20]. When they strike the surface

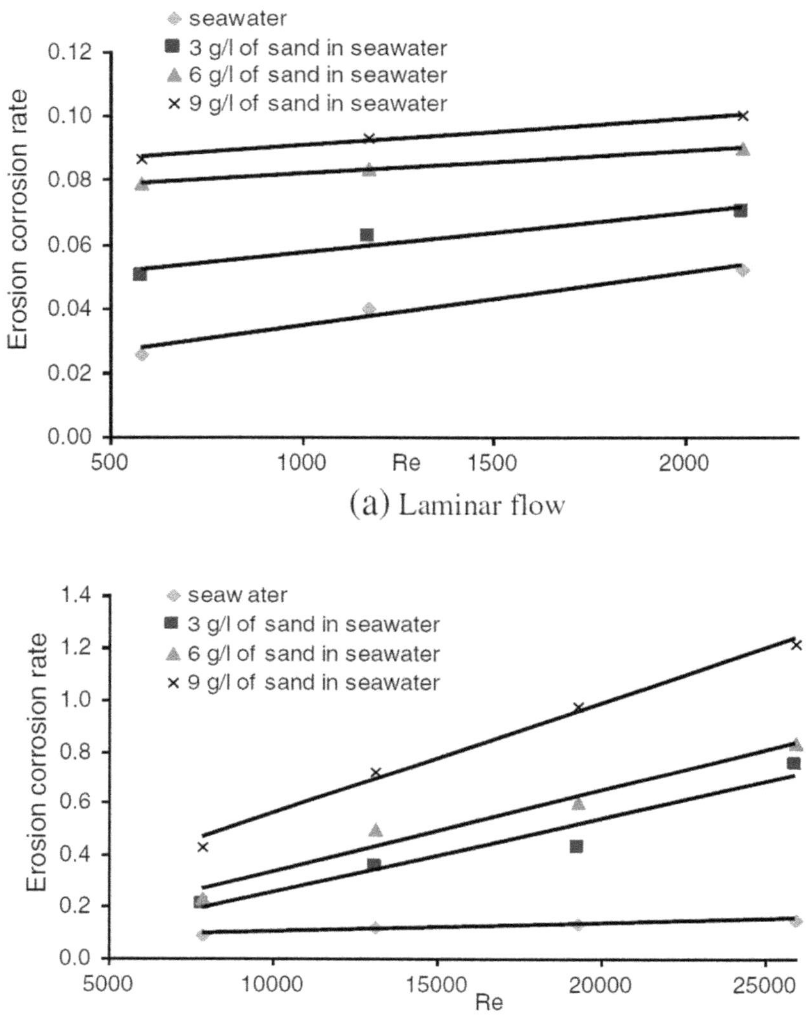

(a) Laminar flow

(b) Turbulent flow

FIGURE 1.6 Erosion–corrosion rate (mg/m^2/min) vs. Reynolds number [9].

of the target, the KE transfer is converted into impact energy [31, 34, 57, 58]. Therefore, high slurry erosion occurs. However, erosion generally increases with the square of particle velocity; that is, granular particles at a very high speed result in higher slurry erosion:

$$E = k(velocity)^m \tag{1.22}$$

where the velocity exponent is denoted by *m*. Finnie [59] found the value of the velocity exponent (*m*) is equal to 2. Numerous researchers [12, 32, 60, 61] have reported various values for the velocity exponent. The abrasive particle velocity was determined to be the most significant factor in slurry erosion, with an exponent value of *m* ranging from 0.2 to 4.0. Researchers [62–66] also mentioned that erosion rises in proportion to the impingement velocity of slurry particles. Padhy and Saini [67] examined the impact of velocity on the wear of erosion on turbine buckets.

The erosion wear investigation was conducted using sand with a particle size of 90–180 μm and a jet velocity of 5000–10000 rev/min. The authors stated that the erosion wear velocity is governed by the power law relationship. According to refs. [1, 5, 53, 54], the flow phenomena and dimensionless numbers are influenced by the solid concentration (*Re* and *St*). Ultimately, it may be concluded that the concealed elements of solid concentration require more investigation. Other researchers have also noted a similar kind of observation [12, 20, 31, 32, 36, 60, 61]. Singh et al. [31] presented the molecular aspects that underlie the varying values of the velocity exponent (m). It was said that the velocity exponent is significant under three distinct conditions:

Condition 1: The erosion rate rises in a linear manner as the velocity of particles increases when $m = 1$.

Condition 2: When $m > 1$, the erosion rate increases in a non-linear manner, with the particle-wall phenomena predominating over the particle–particle phenomenon.

Condition 3: The erosion rate is directly proportional to the kinetic energy of hitting particles when $m < 1$, and the particle–particle phenomena is more dominant than the particle-wall phenomenon. Rebounding particles emerge from the specimen's surface in this circumstance.

On the basis of two classes, low velocity and high solid concentration and high velocity and low solid concentration, several studies explored the impact of velocity on erosion. Numerous researchers [12, 32, 34, 58, 61] have changed the concentration within the range of 10–70 wt.% at a velocity of 0.05–10 m/s. A few researchers [35, 68–70] focused on erosion tests at high velocities and mass flow rates and low concentrations. In 1998, Wood and Wheeler [35] fabricated a jet tester facility to generate a high velocity and mass flow rate of sand particles. The erosion rate is a function of the kinetic energy of the particles that strike the surface, according to their research. They said that the erosion rate is marginally influenced by the sand flow (concentration) and is significantly dependent on the kinetic energy of the sand particles that strike the ground. A similar kind of observation was made in a previous study [68].

1.3.9 Temperature Produced during Flow

Due to the mechanical interaction between solid particles and the surface during transport of slurry through hydraulic components such as the impellers, pipelines, casings, and blades, the temperature rises. Several studies have shown that temperature may affect erosion wear behavior significantly [71–73]. In particular, higher slurry temperatures have been related to higher erosion rates brought about by more particle mobility and particle impact energy. The slurry suspension is also sensitive to temperature, and the temperature will influence slurry viscosity as well. With an increase in temperature, the viscosity decreases and leads to a decrease in adhesive forces between particles. This reduction promotes the particles moving more freely, so their kinetic energy rises, thereby leading to higher erosive potential [1]. In addition, Singh et al. noticed the predominant forces in terms of behavior of particles are dependent on the slurry conditions; that is, settling forces dominate at low solid concentrations, turbulent forces at high flows and concentrations, and drag forces at high viscosities. It was interestingly observed that some researchers [74, 75] witnessed a drop in erosion wear with the rise of temperature of the target surface. This is due to the increase of ductility and elongation of the material, which increases its potential to absorb the impacts of particles by elastic rebound rather than cracking or chipping. Sarlin et al. [76] also showed higher temperatures might lead to the growth of cavitation bubbles, which in turn could dissipate kinetic energy during the collision of particles more effectively and hence lower erosion. It is clear that temperature has a complex influence on slurry erosion, which is dependent upon a string of interlinked local parameters such as particle velocity, slurry composition, and material properties. A fundamental understanding of this relationship is important for maximization of durability and performance of components

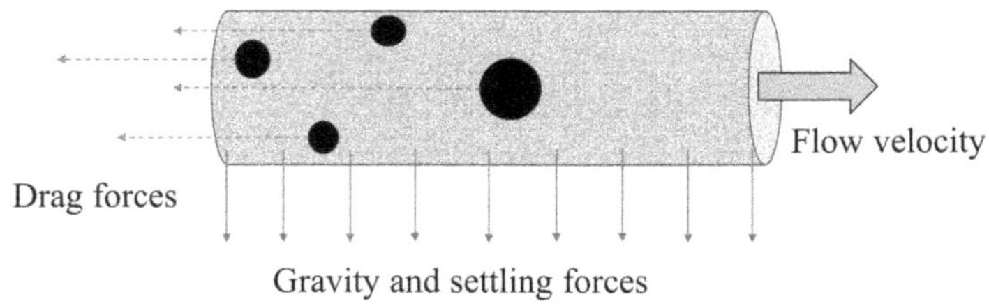

FIGURE 1.7 Schematic diagram of different forces acting on slurry suspended particles. **(Created by authors of this chapter)**

in slurry handling systems. The temperature of hydraulic components (including impellers, pipes, casing, and blades) rises as a result of the mechanical effects of solid particles during the transit of a variety of slurries. According to some experts, erosion wear is exacerbated by a rise in slurry temperature [71, 72]. High-temperature erosion wear tests are appropriate for the examination of such issues [73]. Figure 1.7 illustrates a schematic illustration of the many forces that influence the suspended particles of slurry. Singh et al. [5] observed that the settling force is the dominant force at lower solid concentrations; turbulence forces are the dominant force at high solid concentrations, velocities, and flow rates; and drag forces are the dominant force at high viscosity. The viscosity of a slurry suspension is significantly influenced by temperature. The viscosity of the slurry is reduced as the temperature of the slurry increases. The decrease in viscosity is a consequence of the decrease in adhesion forces that operate between particles, which leads to a high kinetic energy of the particles. Researchers [74, 75] observed that erosion wear may be mitigated by raising the temperature of the target surface. It is attributed to a rise in the ductility of the materials, which in turn increases the elongation property of the target material and boosts the rebounding nature of the particles. In a few instances, it has been observed that erosion diminishes as the temperature rises. Sarlin et al. [76] found that the kinetic energy of impacting particles is reduced as a result of increased cavitation bubbles at higher temperatures.

1.4 SLURRY EROSION

Several industries, for example, mining, manufacturing, maritime, and power generation, report damage to the surface of their hydraulic parts. The main sources of wear to materials are abrasion, adhesion, erosion, cavitation, and a combination of corrosion and erosion [77, 78]. The process of erosion mainly depends on how the particles in the slurry come into contact with the surfaces, and factors involved include the slurry's movement speed, thickness, its angle with the surface, and what it will erode [12, 37, 79]. Cavitation happens when bubbles collapse on a surface surrounded by moving fluids, whereas corrosion–erosion develops when the effects of both chemical attack and erosion happen together in harsh areas. Because of this, the lifespan of equipment and the efficiency of the plant both decline, so it is necessary to find ways to handle wear. Taking care of adhesive wear is possible with lubrication [55], while corrosion prevention can involve galvanization or the use of alternate materials [80, 81]. In most cases, erosion is more serious when sand is flowing through a narrow space [82].

Thermal power plants face issues of erosion–corrosion in the ash handling, boiler, heat exchanger, and ESP areas because they deal with the movement of coal and different kinds of ash [5, 83–86]. When slurries are abrasive, pipelines, pumps, and fittings frequently get damaged inside. Mining conditions include traveling high-speed sand and slurry particles that result in extensive deterioration on pumps, impellers, valves, and joints [87, 88]. Types of material transported in slurry are coal, oil, sand, and tailings from metals. Other types of hydraulic equipment, such as turbines, pipes, nozzles, and those used offshore, are affected by silt and cavitation [89–91]. A few examples of erosion in hydraulic systems can be found in Figure 1.8 [92, 93].

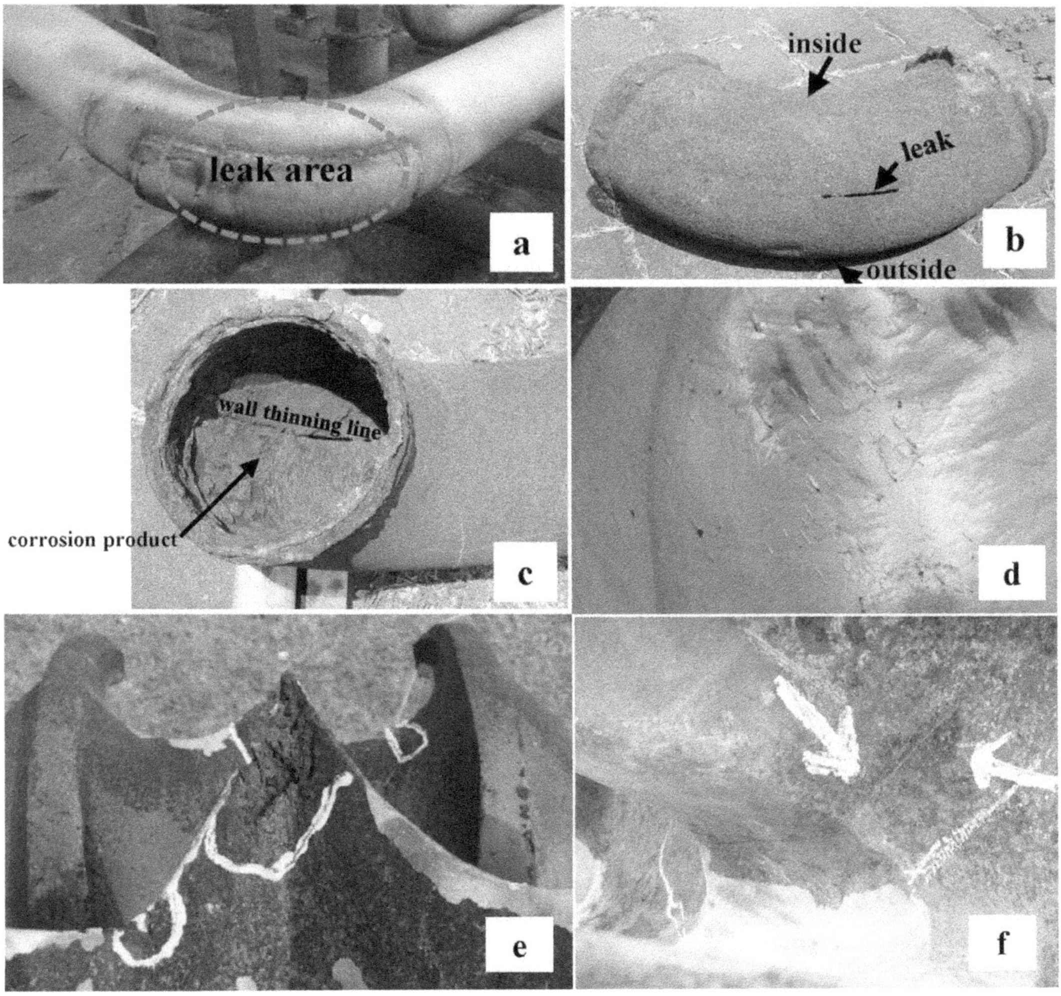

FIGURE 1.8 Wear on the surface of different machinery: (a) erosion on pipe bend, (b) eroded pipe bend from outside, (c) corrosion inside pipe bend [93], (d) erosion on recovery welding in impeller of Pelton turbine, (e) cavitation-erosion at bucket of Pelton turbine, and (f) impeller neck with crack [92].

1.5 MECHANISMS INVOLVED IN SLURRY EROSION

Surface deterioration in materials is often characterized by abrasion, corrosion, and erosion. Electrochemical reactions that may cause pitting and crater formation [57] are included in corrosion. Chip formation and detachment, followed by crack and fracture development, are mainly responsible for abrasion [94], as shown in Figure 1.9. Erosion is more complex; not yet completely understood; and dependent on, for example, particle characteristics, material properties, and process condition. According to Finnie [59], brittle materials break by cutting, but ductile materials will deform plastically. Erosion is reported by Levy [95] and others [96, 97] in ductile materials due to deformation and microcracking with the generation and formation of small craters, but in brittle materials, rapid crack initiation and propagation occur, as shown in Figure 1.9(c).

Erosion mechanisms can occur over one another. In order to quantify which of these processes was dominant, Sundararajan et al. [98] proposed an erosion efficiency (η) such that if $\eta > 1$, there

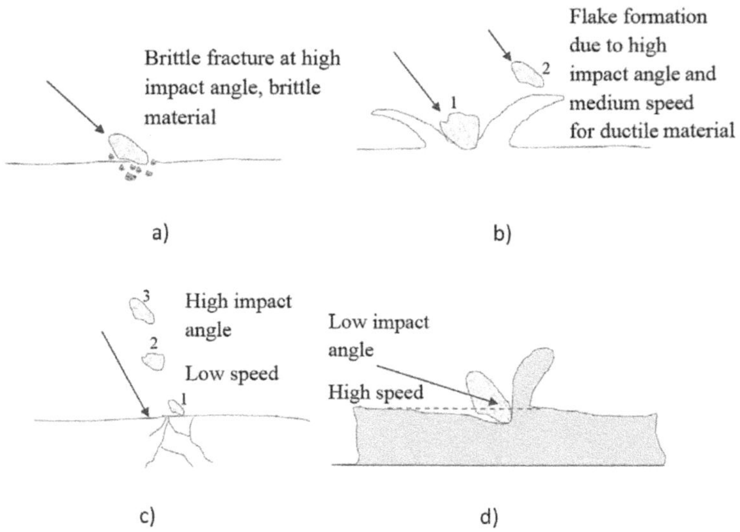

FIGURE 1.9 Mechanisms involved in particulate wear: (a) brittle fracture, (b) plastic deformation, (c) fatigue erosion, and (d) abrasive/cutting erosion [94].

(By permission of Elsevier)

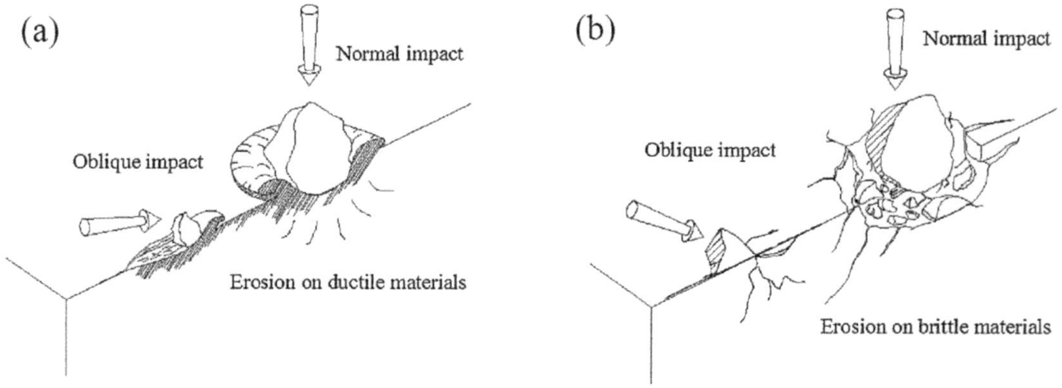

FIGURE 1.10 A schematic illustration of erosion in (a) ductile and (b) brittle materials [97].

(By permission of Elsevier)

was brittle erosion and if $\eta < 1$, there was plastic deformation. In addition, Clark and Wong [99] quantified the erosion through deformation (ED) and cutting (EC) energy contributions. An erosion identifier (ξ) was introduced by Grewal et al. [42]:

- Plowing: $\xi < 1$
- In this case (microcutting), $\xi = 1$
- For $\xi > 1$, the erosion is called brittle

Erosion behavior is also particle shape dependent. According to Singh et al. [13], spherical fly ash caused plastic deformation, and bottom ash induced plowing, while angular sand caused cracking on SS 316L. Hardness is not only a factor in determining cavitation erosion, brought about by bubble collapse in slurries, the surface toughness, roughness, and irregularities are important. Erosion is strongly dependent upon the impact angle. At low angles and low velocities, ridges are formed without material removal (plowing). When velocity increases, cutting takes over. Also, at high velocity and a high angle, erosion involves fracturing and metal cutting [100].

1.6 CONCLUSION

Slurries have different physicochemical and mineralogical properties, which play a crucial role in flow phenomenon and erosion. The main properties that play vital roles in transportation, erosion, and disposal are pH, shape factor, size distribution, solid concentration, hardness of particles, viscosity, and density. Slurry erosion is a prevalent phenomenon in a variety of sectors and applications, and it has the potential to have a significant impact. Due to the development of wear-resistant materials, the impact of slurry erosion has been reduced in numerous sectors. Consistent research and development in this field will be necessary to enhance the dependability and durability of apparatus in the presence of slurry erosion. This chapter provided a concise summary of the current state of the art in assessing the various properties of slurries or particles in terms of their transportation complexities. Consequently, it is imperative to optimize the impact of influencing parameters in order to regulate the deterioration phenomenon.

REFERENCES

[1] Singh J. Investigation on slurry erosion of different pumping materials and coatings. Thapar Institute of Engineering and Technology, Patiala, India. 2019.

[2] Clark HM, Llewellyn RJ. Assessment of the erosion resistance of steels used for slurry handling and transport in mineral processing applications. Wear. 2001;250–251:32–44.

[3] Desale GR, Gandhi BK, Jain SC. Slurry erosion of ductile materials under normal impact condition. Wear. 2008;264:322–330.

[4] Singh M, Kumar S, Kumar S, et al. Characterization of Iron-ore suspension at in-situ conditions. Mater Today Proc [Internet]. 2018;5:17845–17851. Available from: https://doi.org/10.1016/j.matpr.2018.06.110.

[5] Singh JP, Kumar S, Mohapatra SK. Head loss investigations inside 90° pipe bend for conveying of fine coal-water slurry suspension. Int J Coal Prep Util [Internet]. 2017;40:1–17. Available from: https://doi.org/10.1080/19392699.2017.1415892.

[6] Lindgren M, Perolainen J. Slurry pot investigation of the influence of erodent characteristics on the erosion resistance of austenitic and duplex stainless steel grades. Wear [Internet]. 2014;319:38–48. Available from: http://dx.doi.org/10.1016/j.wear.2014.07.006.

[7] Verma P, Kumar S, Lal K, et al. Prediction of pressure drop for the flow of zinc ore-water suspension in horizontal pipeline. IOP Conf Ser Mater Sci Eng. 2018;455.

[8] Peña Ballesteros DY, Mendez Camacho YE, Barreto Hernandez LV. Evaluation of the synergistic effect of erosion-corrosion on AISI 4330 steel in saline-sand multiphase flow by electrochemical and gravimetric techniques. Int J Electrochem. 2016;2016:1–7.

[9] Shehadeh M, Anany M, Saqr KM, et al. Experimental investigation of erosion–corrosion phenomena in a steel fitting due to plain and slurry seawater flow. Int J Mech Mater Eng. 2014;9:1–8.

[10] Karafyllias G, Galloway A, Humphries E. The effect of low pH in erosion–corrosion resistance of high chromium cast irons and stainless steels. Wear [Internet]. 2019;420–421:79–86. Available from: https://doi.org/10.1016/j.wear.2018.11.021.

[11] Yu Z, Chen M, Li F, et al. Synergistic effect of corrosion and wear of the 316 stainless steel in molten zinc alloy at 460 °C. Corros Sci [Internet]. 2020;165:108411. Available from: https://doi.org/10.1016/j.corsci.2019.108411.

[12] Gandhi BK, Singh SN, Seshadri V. Study of the parametric dependence of erosion wear for the parallel flow of solid-liquid mixtures. Tribol Int. 1999;32:275–282.

[13] Singh J, Kumar S, Mohapatra S. Study on role of particle shape in erosion wear of austenitic steel using image processing analysis technique. Proc Inst Mech Eng Part J J Eng Tribol [Internet]. 2019;233:712–725. Available from: http://journals.sagepub.com/doi/10.1177/1350650118794698.

[14] Arabnejad H, Mansouri A, Shirazi SA, et al. Development of mechanistic erosion equation for solid particles. Wear [Internet]. 2015;332–333:1044–1050. Available from: http://dx.doi.org/10.1016/j.wear.2015.01.031.
[15] Bouwman AM, Bosma JC, Vonk P, et al. Which shape factor(s) best describe granules? Powder Technol. 2004;146:66–72.
[16] Rathod A, Sapate SG, Khatirkar RK. Shape factor analysis of abrasive particles used in slurry abrasion testing. Int J Mech Ind Eng. 2012;2:35–39.
[17] Levy AV, Chik P. The effects of erodent composition and shape on the erosion of steel. Wear. 1983;89:151–162.
[18] Singh J. Slurry erosion performance analysis and characterization of high-velocity oxy-fuel sprayed Ni and Co hardsurfacing alloy coatings. J King Saud Univ—Eng Sci [Internet]. 2023;35:415–429. Available from: https://doi.org/10.1016/j.jksues.2021.06.009.
[19] Singh J. Wear performance analysis and characterization of HVOF deposited Ni-20Cr2O3, Ni-30Al2O3, and Al2O3-13TiO2 coatings. Appl Surf Sci Adv [Internet]. 2021;6:100161. Available from: https://doi.org/10.1016/j.apsadv.2021.100161.
[20] Hawthorne HM, Xie Y, Yick SK. A study of single particle-target surface interactions along a specimen in the Coriolis slurry erosion tester. Wear. 2002;253:403–410.
[21] Ahlert KR. Effects of Particle Impingement Angle and Surface Waiting on Solid Partilce Erosion of AISI 1018 Steel. 1994.
[22] Bahadur S, Badruddin R. Erodent particle characterization and the effect of particle size and shape on erosion. Wear. 1990;138:189–208.
[23] Zou RP, Yu AB. Evaluation of the packing characteristics of mono-sized non-spherical particles. Powder Technol. 1996;88:71–79.
[24] Heilbronner R, Keulen N. Grain size and grain shape analysis of fault rocks. Tectonophysics. 2006;427:199–216.
[25] Desale GR, Gandhi BK, Jain SC. Effect of erodent properties on erosion wear of ductile type materials. Wear. 2006;261:914–921.
[26] Woldman M, van der Heide E, Schipper DJ, et al. Investigating the influence of sand particle properties on abrasive wear behaviour. Wear [Internet]. 2012;294–295:419–426. Available from: http://dx.doi.org/10.1016/j.wear.2012.07.017.
[27] Thakur L, Arora N, Jayaganthan R, et al. An investigation on erosion behavior of HVOF sprayed WC-CoCr coatings. Appl Surf Sci [Internet]. 2011;258:1225–1234. Available from: http://dx.doi.org/10.1016/j.apsusc.2011.09.079.
[28] Walker CI, Hambe M. Influence of particle shape on slurry wear of white iron. Wear [Internet]. 2015;332–333:1021–1027. Available from: http://dx.doi.org/10.1016/j.wear.2014.12.029.
[29] Stack MM, Pungwiwat N. Slurry erosion of metallics, polymers, and ceramics: Particle size effects. Mater Sci Technol. 1999;15:337–344.
[30] Gupta R, Singh SN, Seshadri V. Study on the uneven wear rate in a slurry pipeline on the basis of mesurements in a pot tester. Wear. 1995;184:169–178.
[31] Singh J, Kumar S, Mohapatra SK. Erosion tribo-performance of HVOF deposited Stellite-6 and Colmonoy-88 micron layers on SS-316L. Tribol Int [Internet]. 2018;0–1. Available from: https://doi.org/10.1016/j.triboint.2018.06.004.
[32] Clark HMI, Hartwich RB. A re-examination of the "particle size effect" in slurry erosion. Wear. 2001;248:147–161.
[33] Tilly GP. A two stage mechanism of ductile erosion. Wear. 1973;23:87–96.
[34] Rajahram SS, Harvey TJ, Wood RJK. Electrochemical investigation of erosion–corrosion using a slurry pot erosion tester. Tribol Int [Internet]. 2011;44:232–240. Available from: http://dx.doi.org/10.1016/j.triboint.2010.10.008.
[35] Wood RJK, Wheeler DW. Design and performance of a high velocity air-sand jet impingement erosion facility. Wear. 1998;220:95–112.
[36] Singh J, Kumar S, Mohapatra SK. Tribological analysis of WC-10Co-4Cr and Ni-20Cr2O3 coating on stainless steel 304. Wear [Internet]. 2017;376–377:1105–1111. Available from: http://dx.doi.org/10.1016/j.wear.2017.01.032.
[37] Gandhi BK, Singh SN, Seshadri V. Performance characteristics of centrifugal slurry pumps. J Fluid Eng. 2001;123:271–280.
[38] Clark HMI. The influence of the flow field in slurry erosion. Wear. 1992;152:223–240.

[39] Wiederhorn SM, Hockey BJ. Effect of material parameters on the erosion resistance of brittle materials. J Mater Sci. 1983;18:766–780.
[40] Wada S, Watanabe N. Solid particle erosion of brittle materials (part 3). J Ceram Assoc Japan. 1987;95:573–578.
[41] Elkholy A. Prediction of abrasion wear for slurry pump materials. Wear. 1983;84:39–49.
[42] Grewal HS, Agrawal A, Singh H. Design and development of high-velocity slurry erosion test rig using CFD. J Mater Eng Perform. 2013;22:152–161.
[43] Desale GR, Gandhi BK, Jain SC. Development of correlations for predicting the slurry erosion of ductile materials. J Tribol. 2011;133:1–10.
[44] Prasad BK, Jha AK, Modi OP, et al. Effect of sand concentration in the medium and travel distance and speed on the slurry wear response of a zinc-based alloy alumina particle composite. Tribol Lett. 2004;17:301–309.
[45] Singh J. Analysis on suitability of HVOF sprayed Ni-20Al, Ni-20Cr and Al-20Ti coatings in coal-ash slurry conditions using artificial neural network model. Ind Lubr Tribol. 2019;71:972–982.
[46] Dasgupta R, Prasad BK, Jha AK, et al. Effects of sand concentration on slurry erosion of steels. Mater Transections. 1998;39:1185–1190.
[47] Grewal HS, Arora HS, Agrawal A, et al. Slurry erosion of thermal spray coatings: Effect of sand concentration. Procedia Eng [Internet]. 2013;68:484–490. Available from: http://dx.doi.org/10.1016/j.proeng.2013.12.210.
[48] Chandel S, Singh SN, Seshadri V. Experimental study of erosion wear in a centrifugal slurry pump using Coriolis wear test rig. Part Sci Technol. 2012;30:179–195.
[49] Hawthorne HM. Some Coriolis slurry erosion test developments. Tribol Int. 2002;35:625–630.
[50] Kleis I, Kulu P. Influence of particle concentration. Solid Part Eros. Springer-Verlag London Limited. 2008. pp. 24–27.
[51] Bong E. Solid-liquid mass transfer in agitated vessels with high solids concentration. 2013;144. Available from: https://researchbank.rmit.edu.au/view/rmit:160881.
[52] Bong EY, Eshtiaghi N, Wu J, et al. Optimum solids concentration for solids suspension and solid-liquid mass transfer in agitated vessels. Chem Eng Res Des [Internet]. 2015;100:148–156. Available from: http://dx.doi.org/10.1016/j.cherd.2015.05.021.
[53] Dabirian R, Mohan R, Shoham O, et al. Critical sand deposition velocity for gas-liquid stratified flow in horizontal pipes. J Nat Gas Sci Eng [Internet]. 2016;33:527–537. Available from: http://dx.doi.org/10.1016/j.jngse.2016.05.008.
[54] Bartosik A. Influence of coarse-dispersive solid phase on the "particles-wall" shear stress in turbulent slurry flow with high solid concentration. Arch Mech Eng. 2010;57:45–68.
[55] Qian ZD, Wang Y, Guo ZW, et al. Analysis of silt abrasion of the impeller ring in a centrifugal pump with J-grooves. IOP Conf Ser Mater Sci Eng. 2016;129:012021.
[56] Turenne S, Châtigny Y, Simard D, et al. The effect of abrasive particle size on the slurry erosion resistance of particulate-reinforced aluminium alloy. Wear. 1990;141:147–158.
[57] Rajahram SS, Harvey TJ, Wood RJK. Erosion–corrosion resistance of engineering materials in various test conditions. Wear. 2009;267:244–254.
[58] Kumar R, Bhandari S, Goyal A. Performance analysis of CA6NM hydro-turbine steel under accelerated conditions by using an indigenously developed slurry erosion tester. Trans Indian Inst Met. 2018;71:753–761.
[59] Finne I. Erosion of surfaces. Wear. 1960;3:87–103.
[60] Deng T, Li J, Chaudhry AR, et al. Comparison between weight loss of bends in a pneumatic conveyor and erosion rate obtained in a centrifugal erosion tester for the same materials. Wear. 2005;258:402–411.
[61] Divakar M, Agarwal VK, Singh SN. Effect of the material surface hardness on the erosion of AISI316. Wear. 2005;259:110–117.
[62] Islam MA, Alam T, Farhat ZN, et al. Effect of microstructure on the erosion behavior of carbon steel. Wear [Internet]. 2015;332–333:1080–1089. Available from: http://dx.doi.org/10.1016/j.wear.2014.12.004.
[63] Manisekaran T, Kamaraj M, Sharrif SM, et al. Slurry erosion studies on surface modified 13Cr-4Ni steels: Effect of angle of impingement and particle size. J Mater Eng Perform. 2007;16:567–572.
[64] Singh J, Singh S. Neural network prediction of slurry erosion of heavy-duty pump impeller/casing materials 18Cr-8Ni, 16Cr-10Ni-2Mo, super duplex 24Cr-6Ni-3Mo-N, and grey cast iron. Wear 2021; 476:203741.
[65] Singh J, Singh JP. Performance analysis of erosion resistant Mo2C reinforced WC-CoCr coating for pump impeller with Taguchi's method. Ind Lubr Tribol. 2021;(in press).

[66] Singh J. Tribo-performance analysis of HVOF sprayed 86WC-10Co4Cr & Ni-Cr2O3 on AISI 316L steel using DOE-ANN methodology. Ind Lubr Tribol. 2021;(in press).
[67] Padhy MK, Saini RP. Effect of size and concentration of silt particles on erosion of Pelton turbine buckets. Energy [Internet]. 2009;34:1477–1483. Available from: http://dx.doi.org/10.1016/j.energy.2009.06.015.
[68] Grewal HS, Agrawal A, Singh H. Design and development of high-velocity slurry erosion test rig using CFD. J Mater Eng Perform. 2013;22:152–161.
[69] Haugen K, Kvernvold O, Ronold A, et al. Sand erosion of wear-resistant materials: Erosion in choke valves. Wear. 1995;186–187:179–188.
[70] Feng Z, Tzeng Y, Field JE. Solid particle impact of CVD diamond films. Thin Solid Films. 1992;212:35–42.
[71] Young JP, Ruff AW. Particle erosion measurements on metals. Am Soc Mech Eng. 1976;121–125.
[72] Levy A, Man YF. The effect of temperature on the erosion–corrosion of 9Cr-1Mo steel. Wear. 1986;111:161–172.
[73] Hayashi N, Kagimoto Y, Notomi A, et al. Development of new testing method by centrifugal erosion tester at elevated temperature. Wear. 2005;258:443–457.
[74] Smeltzer CE, Gulden ME, Compton WA. Mechanisms of metal removal by impacting dust particles. J Fluids Eng Trans ASME. 1970;92:639–652.
[75] Humphrey JAC. Fundamentals of fluid motion in erosion by solid particle impact. Int J Heat Fluid Flow. 1990;11:170–195.
[76] Sarlin E, Lindgren M, Suihkonen R, et al. High-temperature slurry erosion of vinylester matrix composites—the effect of test parameters. Wear. 2015;328–329:488–497.
[77] Parent LL, Li DY. Wear of hydrotransport lines in Athabasca oil sands. Wear. 2013;301:477–482.
[78] Walker CI, Robbie P. Comparison of some laboratory wear tests and field wear in slurry pumps. Wear [Internet]. 2013;302:1026–1034. Available from: http://dx.doi.org/10.1016/j.wear.2012.11.053.
[79] Roco MC, Addie GR. Erosion wear in slurry pumps and pipes. Powder Technol. 1987;50:35–46.
[80] Li X, Lü T. Analysis of corrosion failure of petrochemical pipe elbow. J Cent South Univ Technol. 2005;12:119–123.
[81] Lee DI, Lim HC. Erosion–corrosion damages of water-pump impeller. Int J Automot Technol [Internet]. 2009;10:629–634. Available from: https://doi.org/10.1007/s12239-009-0074-5.
[82] Postlethwaite J, Tinker EB, Hawrylak MW. Erosion-corrosion in slurry pipelines. Corrosion. 1974;285–289.
[83] Li ZM, Sun FZ, Ma L, et al. Low-pressure economizer increases fly ash collection efficiency in ESP. Appl Therm Eng [Internet]. 2016;93:509–517. Available from: http://dx.doi.org/10.1016/j.applthermaleng.2015.10.006.
[84] Singh J, Kumar S, Singh G. Taguchi's approach for optimization of tribo-resistance parameters forss304. Mater Today Proc [Internet]. 2018;5:5031–5038. Available from: https://doi.org/10.1016/j.matpr.2017.12.081.
[85] Linak WP, Miller CA, Seames WS, et al. On trimodal particle size distributions in fly ash from pulverized-coal combustion. Proceed Comb Ins. 2002;29:441–447.
[86] Mudgal D, Ahuja L, Bhatia D, et al. High temperature corrosion behaviour of superalloys under actual waste incinerator environment. Eng Fail Anal [Internet]. 2016;63:160–171. Available from: http://dx.doi.org/10.1016/j.engfailanal.2016.02.016.
[87] Rabinovich E, Kalman H. Threshold velocities of particle-fluid flows in horizontal pipes and ducts: Literature review. Rev Chem Eng. 2011;27:215–239.
[88] Szala M, Łukasik D. Cavitation wear of pump impellers. J Technol Exploit Mech Eng. 2016;2:40–44.
[89] Mansouri A, Shirazi SA, Mclaury BS. Experimental and numerical investigation of the effect of viscosity and particle size on the erosion damage caused by solid particles. Proc FEDSM2014. Chicago, Illinois, USA. 2014. pp. 1–10.
[90] Mansouri A, Arabnejad H, Karimi S, et al. Improved CFD modeling and validation of erosion damage due to fine sand particles. Wear [Internet]. 2015;338–339:339–350. Available from: http://dx.doi.org/10.1016/j.wear.2015.07.011.
[91] Vieira RE, Mansouri A, McLaury BS, et al. Experimental and computational study of erosion in elbows due to sand particles in air flow. Powder Technol [Internet]. 2016;288:339–353. Available from: http://dx.doi.org/10.1016/j.powtec.2015.11.028.
[92] Chávez JC, Valencia JA, Jaramillo GA, et al. Failure analysis of a Pelton impeller. Eng Fail Anal [Internet]. 2015;48:297–307. Available from: http://dx.doi.org/10.1016/j.engfailanal.2014.08.012.

[93] Kusmono, K. Analysis of a failed pipe elbow in geothermal production facility. Case Stud Eng Fail Anal [Internet]. 2017;9:71–77. Available from: https://doi.org/10.1016/j.csefa.2017.08.001.
[94] Chitrakar S, Neopane HP, Dahlhaug OG. Study of the simultaneous effects of secondary flow and sediment erosion in Francis turbines. Renew Energy [Internet]. 2016;97:881–891. Available from: http://dx.doi.org/10.1016/j.renene.2016.06.007.
[95] Levy AV. Solid particle erosion and erosion-corrosion of materials. ASM Int. Ohio: ASM International. 1996.
[96] Bellman R, Levy A. Erosion mechanism in ductile metals. Wear. 1981;70:1–27.
[97] Wang YF, Yang ZG. Finite element model of erosive wear on ductile and brittle materials. Wear. 2008;265:871–878.
[98] Sundararajan G, Roy M, Venkataraman B. Erosion efficiency-a new parameter to characterize the dominant erosion micromechanism. Wear. 1990;140:369–381.
[99] Clark HMI, Wong KK. Impact angle, particle energy and mass loss in erosion by dilute slurries. Wear. 1995;186–187:454–464.
[100] Islam MA, Farhat ZN. Effect of impact angle and velocity on erosion of API X42 pipeline steel under high abrasive feed rate. Wear [Internet]. 2014;311:180–190. Available from: http://dx.doi.org/10.1016/j.wear.2014.01.005.

2 Investigation of Slurry Erosion, Erosion–Abrasion, and Erosion–Corrosion Using Different Testers and Test Loops

Jibrin Muhammad Yelwa and Haruna Musa

2.1 INTRODUCTION

Slurries, which are mixtures of liquid with suspended solid particles, serve a central function in several industries, facilitating the transport, processing, or handling of particles that could not have been handled by conventional means otherwise. Mining, oil and gas, chemical processing, power generation, and dredging are a few examples of industries working in slurry environments. The transport of ores, coal, tailings, and processed materials is what slurries are mostly used for in mining (Chung *et al.*, 2021). At coal-based power plants, coal ash and other waste products are transferred through a slurry pipeline (Reddy *et al.*, 2023). Likewise, in oil and gas, slurries like drilling muds are essential for drilling and thriving completion (Zhou *et al.*, 2022). Dredging operations entail removing sediments from water bodies, producing slurry, thus contributing to slurry environments in construction and environmental engineering (Soetan *et al.*, 2023). These systems, while being effective and practical, stand among the harshest environments subjected to those few particles suspended in liquid that are highly abrasive and corrosive. Aggressive particulates in a slurry hit the surfaces of equipment, causing accelerated wearing and degradation of material. Material failure in slurry systems occurs through erosion, corrosion, and erosion–corrosion (Singh *et al.*, 2024a). The deterioration of material and degradation of pipes, pumps, and valves is a matter of great concern since this drastically reduces the working life span of machinery and consequently increases maintenance cost and forced downtimes (Rasse *et al.*, 2021).

Material degradation in slurry environments, resulting from erosion, corrosion, and the synergistic effect of erosion–corrosion, poses a critical problem for industries that rely on slurry transport systems. It is important to understand the mechanisms by which these forms of wear occur so that the economic consequences of material failure can be minimized. The maintenance, repair, and replacement of equipment resulting from slurry-induced wear can be significantly costly in places where slurry systems form the backbone of the operation, such as power generation and mining (Chung *et al.*, 2021).

In particular, the problem of slurry erosion is of high concern, as the process involves the impingement of solid particles suspended within the slurry on the surface of any given material. Erosion due to slurry impact is controlled by parametric factors such as particle size, particle velocity, slurry concentration, and material properties of suspended solids, as well as the target equipment (Rao & Mulky, 2023). Erosion–corrosion appears to chemically aggravate the degradation process, especially in the presence of aggressive chemical agents or fluids acting on the eroded

DOI: 10.1201/9781003561156-3

surfaces (Varga *et al.*, 2021). An integrated action of erosion and corrosion warrants a more detailed study to address the problem more effectively via the choices of suitable materials, coatings, and maintenance strategies, leading to minimized wear, higher efficiency, and longer service lives of major components (Zadeh, 2023).

The safety hazards that arise from material failures must be researched thoroughly, with effective mitigation measures put in place, particularly in critical systems such as pipelines and pumps. Equipment-related issues can result in accidents that cause environmental degradation and operational stoppage, highlighting the importance when it comes to controlling wear processes in slurry systems (Singh *et al.*, 2024b).

This chapter aims to broadly discuss the degradation mechanisms in slurry conditions, with the main focus being on slurry erosion, erosion–abrasion, and erosion–corrosion. The detailed mechanisms that govern these wear processes are examined, looking at degradation in terms of the interactions between solid particles and materials (Javaheri *et al.*, 2018). The chapter investigates the main factors that affect slurry erosion, including size and velocity of particles, concentration of slurry, and characteristics of particles and target materials.

Additionally, the chapter examines relevant measurement methodologies employed in slurry wear evaluation: the slurry pot tester, jet erosion tests, and CFD simulations (Yan & Li, 2023). Such techniques reveal the prediction possibilities and quantification of wear; they allow industries to better execute strategies for minimizing damage due to slurries and maximize equipment performance and reliability.

This chapter provides some insights into consideration of mechanisms and factors that can reduce material degradation in slurry systems and hence in slurry transport and processing with a view to cutting down maintenance cost, reducing downtime, and increasing equipment life.

2.2 SLURRY EROSION MECHANISM

Slurry erosion is the wear or damage caused on a material surface due to mechanical interaction from solid particles suspended in a fluid medium. This kind of erosion occurs when slurries, solid particles suspended in a liquid, flow through pipelines, pumps, and other fittings. The particles keep hitting the surface as the particles abrade the material. Slurry erosion becomes a serious problem in mining, oil, gas, power generation, and chemical processing where these kinds of slurries are commonly used.

Figure 2.1 presents the basic functioning of the slurry erosion process, whereby solid particles impact the substrate at a certain angle (α) determined by the flow direction. These interactions

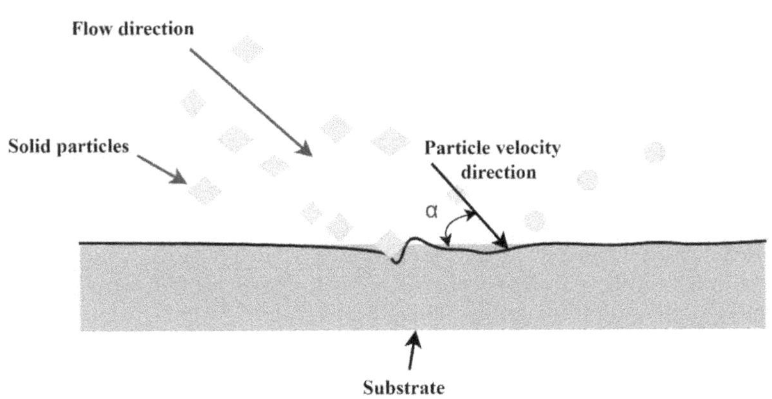

FIGURE 2.1 Overview of the slurry erosion process.

cause destruction of the materials at the surface. The operations of slurry erosion are numerous and consist of various physical processes causing damage to materials. Direct impact, cutting/plowing, indentation/deformation, and fatigue are some basic mechanisms.

2.2.1 Impaction/Direct Impact

This is the simplest erosion mechanism in which erosion is caused when particles suspended in a slurry collide with the surface of a target (Flores *et al.*, 2012). The particles transfer their kinetic energy to the surface and thus cause localized removal of material (Wang *et al.*, 2021). Erosion intensity depends upon particle velocity, impact angle, and hardness of the particles. To explain briefly, high velocities and perpendicular incidence cause the most severe erosion (Xie *et al.*, 2015).

2.2.2 Cutting/Plowing

In this mechanism, particles roll and slide freely on the surface, while they create grooves in it during erosion. Plowing takes place when particles are softer than surface materials and glide across them, displacing the materials without their direct fracture (Tarodiya & Gandhi, 2017). On these grounds, we are convinced that the observed phenomenon is surface deformation instead of material removal by impact.

2.2.3 Indentation/Deformation

Particles from the slurry can cause local plastic deformation on the surface of the target, thereby displacing and removing material. If such an impact is not severe enough to cause fracture, the impact may result in indentation (Messa & Malavasi, 2018). The degree of deformation depends upon the hardness of the particles and the hardness of the surface material along with the velocity of the particles.

2.2.4 Fatigue

Furthermore, erosion may occur due to the repeated impacts exerted by slurry particles upon a surface. These repeated impacts impose cyclic stresses on the material, resulting in the initiation and slow propagation of cracking over time. When the cracks increase in size, the material is weakened, making it much more susceptible to erosion. Erosion of this nature poses great problems in areas where the slurry particles are constantly deposited over the surface at intermediate velocities (Alam & Farhat, 2018). In addition, if left unchecked, this can lead to catastrophic failure of the material in high-stress industrial-use cases.

2.2.5 Factors Influencing Slurry Erosion

There are a significant number of aspects related to particle properties, which are responsible for slurry erosion, followed by slurry and material properties, and finally flow properties. These are some of the factors that determine the extent of the wear acting on equipment and material.

2.2.5.1 Particle Characteristics

Erosion depends much on the physical properties of solid particles that have been suspended in a slurry. The following are important to take into consideration.

2.2.5.1.1 Size

Depending on their size, particles have energies and thus the ability to damage a surface. The larger the particles, the more severe the erosion due to a large amount of kinetic energy generated while colliding on the target surface (Solanki *et al.*, 2023).

2.2.5.1.2 Shape

The shape of particles is another causative factor for erosion. Because of their sharp edges, angular particles create greater abrasive wear on surfaces than spherical particles, as these pointed edges tend to concentrate the force on minute areas of the target surface (Zhao *et al.*, 2022). Also, as the irregularity of particle shape increases, abrasive action intensifies, thus deepening grooves and quickening material removal.

2.2.5.1.3 Hardness and Density

Hard particles like sand or silica tend to erode more because their penetration abilities into the target surface are very high. In the same manner, dense particles would also produce great amounts of kinetic energy and a large impact force on colliding (Nassef *et al.*, 2022).

2.2.5.1.4 Concentration

The concentration of solid particles present in the slurry is among the factors accountable for the erosion rate. That is, the higher the particle concentration, the more frequent the particle impact on the surface, thereby inducing wear (Sekar & Bennett, 2023).

2.2.5.2 Slurry Properties

The essential properties of the slurry are considered when establishing the erosion rate:

2.2.5.2.1 Viscosity

The flow of slurry is decided by its viscosity. Slurries of high viscosity depress particle velocity, thereby reducing erosion. On the other hand, slurries with high viscosity retain a larger number of particles in direct contact with the surface (Li *et al.*, 2019).

2.2.5.2.2 Density

The slurry density dictates the momentum of particles. As the slurry density increases, the mass carried by the solid particles also increases, which accounts for the amount of surface kinetic energy transferred by the particles during impacts.

Slurry density determines the momentum of particles. Increasing the slurry density results in increased mass carried by solid particles, which accounts for the quantity of surface kinetic energy transferred by particles during their impacts (Chung *et al.*, 2021).

2.2.5.2.3 Flow Velocity

Flow velocity defines the velocity of the slurry moving through the pipe. It greatly influences slurry erosion by increasing the velocity of the particles, thus causing aggressive erosion. The erosion rate of the particles impacting the surface is influenced by the impact angles. Abrasion occurs at a shallow angle, while direct impact occurs at steeper angles, thereby leading to localized material removal (Kumar *et al.*, 2023).

Figure 2.2 shows the relationship between slurry flow velocity and erosion rate (mm/yr), presenting the CFD simulation data as well as flow loop test data. The slower degradation of material greatly increases with increasing slurry velocity. This plot also shows an indication of the settling velocity threshold.

2.2.5.2.4 Temperature

Temperature can change the properties of both the slurry and the material. For example, with higher temperatures, the target material may begin to soften and thereby become vulnerable to erosion. Sometimes temperature variations might change the chemical constitution of the slurry, thereby bringing about corrosion–erosion effects (Sridhar & Balasubramanian, 2022).

2.2.5.3 Material Properties

The susceptibility to wearing in slurry conditions largely depends on the material characteristics of the material being eroded.

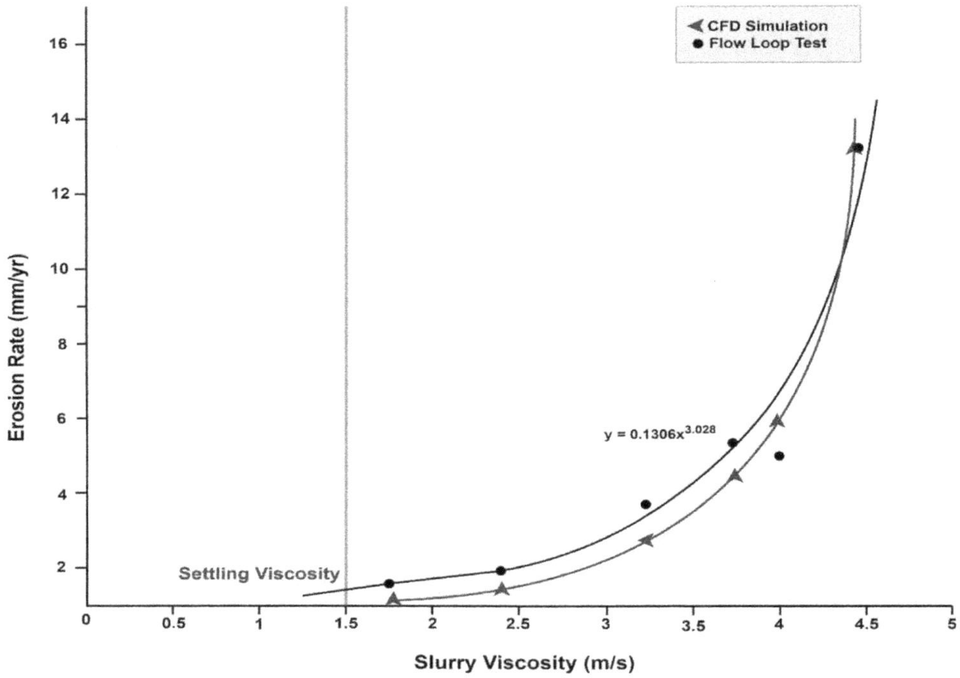

FIGURE 2.2 Effect of slurry flow velocity on erosion rate.
(**Yang** *et al.*, **2018**)

2.2.5.3.1 Hardness

Hardness of the material is considered one of the main criteria to provide resistance against erosion by particles of slurry, meaning the harder the materials are, the less erosion will take its toll (Nie *et al.*, 2025). Yet the hardness of both the material and the particle remain very important in understanding the process of wear.

2.2.5.3.2 Toughness and Ductility

Higher energies are absorbed by materials with increased toughness and ductility. Consequently, erosion happens at a slower rate. More brittle materials will show surface failure and erode with successive impacts (Singh *et al.*, 2024c).

2.2.5.3.3 Microstructure

With respect to erosion resistance, the microstructure of a material is an important parameter to consider. For example, in slurry erosion, materials having coarse microstructure generally face less erosion than those with fine microstructure (Kübarsepp *et al.*, 2021).

2.2.5.3.4 Elastic Modulus

The elastic modulus of a material makes it possible for the material to deform under elastic impact. Materials with a lower elastic modulus tend to have more plastic deformation, which may cause their erosion rate to rise (Rasse *et al.*, 2021).

2.2.5.4 Flow Conditions

The flow condition of slurry inside equipment also affects erosion.

2.2.5.4.1 *Impact Angle*

The types of wear vary with the impact angle between the particles and the target surface. At a low angle of incidence, abrasion is more prevalent on the surface of materials; at a high impact angle, there is an impact and cutting action (Xu *et al.*, 2020a).

2.2.5.4.2 *Flow Regime*

The flow pattern determines the particle impacts on a surface. The particle impact frequency under laminar flow would be low since chaotic particle movement obtains under turbulent flow (Singh *et al.*, 2024d).

2.2.5.4.3 *Presence of Turbulence*

If turbulence exists in the slurry flow, it will vigorously agitate the particles, increasingly impressing their collisions upon the surface with enhanced frequency and energy level. This results in lacerated equipment application in turbulent slurry flows (Sinha *et al.*, 2017).

2.2.5.4.4 *Flow Geometry*

The flow geometry of a slurry transportation system consisting of bends, elbows, and valves affects erosion rates. Places of intense turbulence are bends and are characterized by higher rates of wear due to increased frequency of particle collisions (Yang *et al.*, 2018).

Table 2.1 presents an overview of critical parameters influencing slurry erosion, depicting the synergistic effects of particle characteristics, slurry properties, material properties, and flow conditions

TABLE 2.1
Factors Influencing Slurry Erosion and Their Effects

Factor	Characteristic/ Property	Effect on Erosion Rate
Particle Properties	Size	Larger particles generally cause more severe erosion due to higher kinetic energy.
	Shape	Angular particles generate more abrasive wear than spherical ones due to concentrated force.
	Hardness and Density	More complex and denser particles cause greater penetration and higher impact force.
Slurry Properties	Concentration	Higher particle concentration leads to more frequent impacts and increased wear.
	Viscosity	High viscosity can depress particle velocity but may increase particle–surface contact.
	Density	Higher slurry density increases particle momentum and transferred kinetic energy upon impact.
	Flow Velocity	Increased flow velocity leads to higher particle velocity and more aggressive erosion.
	Temperature	High temperatures can soften materials, increasing susceptibility, and alter slurry chemistry, promoting erosion–corrosion activity.
Material Properties	Hardness	Harder materials generally exhibit better resistance to material loss from erosion.
	Toughness and Ductility	Materials with higher toughness and ductility absorb impact energy, reducing cracking and erosion rates.
	Microstructure	Coarse microstructures are generally less erodible than fine ones in slurry erosion.
	Elastic Modulus	A lower elastic modulus can lead to increased plastic deformation, resulting in higher erosion rates.
Flow Conditions	Impact Angle	Low angles favor impingement abrasion; higher angles emphasize direct impact and cutting.
	Flow Regime	Turbulent flow increases the frequency and energy of particle impacts compared to laminar flow.
	Flow Geometry	Bends, elbows, and valves can create turbulence, resulting in localized high wear rates.

in contributing to the rate and intensity of degradation of a material. Each parameter is listed with an explanation of how it influences erosion. Understanding these parameters is paramount when it comes to forecasting the behavior of materials in abrasive slurry environments and designing industrial systems that can withstand abrasion. This table serves as a guide for researchers and engineers to assist them in reducing wear and maximizing equipment longevity in industries including mining, oil and gas extraction, and power generation, wherever slurry transport takes place.

2.2.6 Erosion Rate Models and Empirical Correlations

The prediction of erosion rates for slurry systems is an interesting and important matter for optimizing the choice of materials and design of systems. There are various models and empirical correlations to estimate the erosion rates of different materials per slurry flows, considering the physical nature of the particles, their target material, and flow conditions.

2.2.6.1 Finnie's Model for Ductile Materials

As one of the earliest and most frequently applied models, Finnie's model predicts the erosion rate for ductile materials undergoing solid particle impact. Finnie (1974) stated that for ductile materials, the rate of erosion varies approximately as the velocity of particles raised to the power of 2.5. The assumption behind this model is that erosion occurs through the kinetic energy transfer from the particles onto the surface, which removes material by the mechanisms of plastic deformation and micro-cutting. The erosion rate, therefore, can be determined by using the following equation:

$$E = k.V^n \tag{2.1}$$

E stands for erosion rate, V for particle velocity, and k and n are constants with values depending on the material under question. The model is meant for materials that plastically deform upon impact, metals and alloys being notable among them.

2.2.6.2 Bitter's Model for Brittle Materials

Brittle materials such as ceramics and hard alloys tend to be subjected to erosion through different mechanisms than what one might find in ductile ones. The Bitter (1963) model gives the erosion rate of brittle materials depending on particle impact angle and particle velocity. Bitter said that velocity squared is a part of the erosion rate, which depends also upon the impact angle of particles on the surface. The erosion rate of brittle materials increases as the velocity and impact angle increase, considering particles cause fracture and remove material by chipping and propagation of cracks:

$$E = kV^2 f\left(\theta\right) \tag{2.2}$$

E is the erosion rate, V is the velocity of the particles, (θ) is the impact angle, k is a constant parameter depending on the properties of the material, and f is a function of the impact angel. This model provides reliable wear prediction for such materials as ceramics and hardened steels.

2.2.6.3 Empirical Equations for Erosion Rates

Apart from theoretical models, some empirical formulations have also been proposed for slurry erosion rates based on experimental results. For example, erosion may be given as a function of particle size and velocity thus:

$$E = a.\, d^m.\, V^n \tag{2.3}$$

where E is the erosion rate, d is the particle diameter, and V is the velocity. The set of parameters a, m, and n are empirical constants which differ with materials and slurries (Nan *et al.*, 2024). Such parametric relations would be very useful for any industrial application whenever a comprehensive set of experimental data is available.

2.2.7 Test Methods of Slurry Erosion

Several experimental methods have come into existence to compare slurry erosion rates and evaluate materials in a controlled environment. Therefore, some of the common equipment for testing is the slurry pot tester, the Coriolis tester, and the jet erosion tester.

2.2.7.1 Slurry Pot Test

The slurry pot test is a basic, inexpensive, and very effective laboratory simulation of slurry erosion. A sample is immersed in a slurry mixture (solid particles, mostly sand, and liquid, mostly water); the slurry is agitated to simulate the operating environment, and slurry flows around the sample, producing particle impacts. Measuring erosion is done by weighing the specimen before and after the test to obtain material loss. For instance, steel specimens in power plants' pipelines are typically tested in slurry of sand and water to simulate erosion caused by the transportation of coal and ash through pipes (Sinha *et al.*, 2017). The slurry pot test is endorsed to evaluate various materials across different slurry concentrations, particle sizes, and flow conditions.

Figure 2.3 illustrates the experimental setup for a slurry pot test, a standard method used to evaluate slurry erosion. Key components include a D.C. motor, stirrer, transparent lid, baffle plates, and a test specimen immersed in the slurry, all designed to simulate particle impacts in a controlled environment.

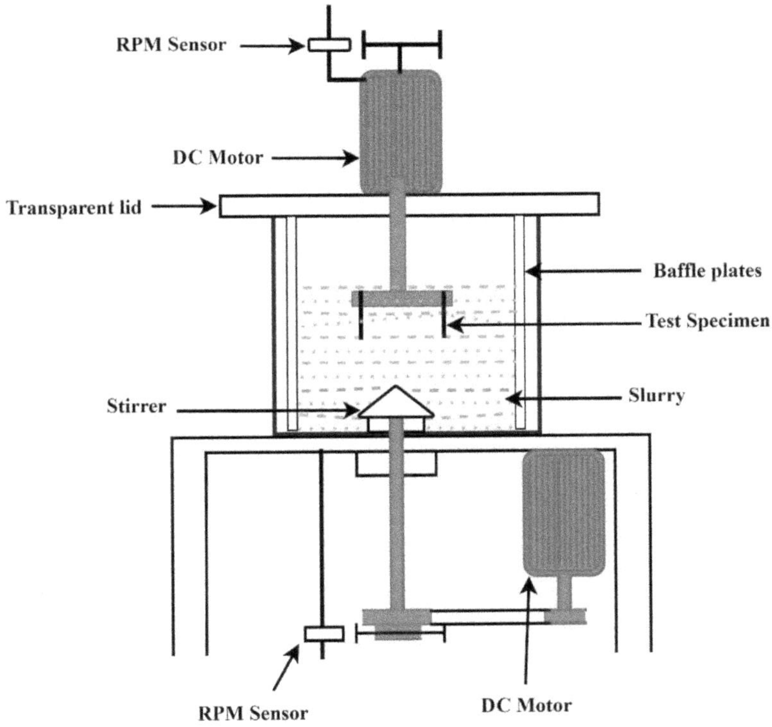

FIGURE 2.3 Testing apparatus for the slurry pot test.

(Created by author)

2.2.7.2 Coriolis Tester

The Coriolis tester operates as a high-velocity, controlled experiment to study slurry erosion, simulating abrasive particle impacts under realistic conditions. In the method, an arm rotates to cause slurry flow, simulating the effect of particles on the target material surface. Velocity, angle of impact, and particle concentration can be controlled to replicate specific industrial conditions. For example, in oil and gas industries, erosion testing is carried out on alloys that are impinged on by fine sand particles at varied impact angles (Kumar *et al.*, 2018). Considering variabilities in operational conditions, Coriolis tests are preferred for evaluating degradation degree in materials.

2.2.7.3 Jet Erosion Test

The jet erosion test uses high-velocity jets of slurry to simulate the dynamic forces acting on materials in turbine, pump, or pipeline systems. In this test, the slurry mixture is pumped at high velocity through a nozzle onto the materials being tested (Figure 2.4). This is in order to simulate the effect of rapidly moving particles of slurry, and erosion rate (material loss) is measured at the conclusion of the test, similar to slurry pot tests but with an emphasis on velocity and impact dynamics. It is usually applied to determine the erosion resistance of materials used in impellers of turbines or pumps subjected to fast-moving slurries (Wang *et al.*, 2025). The jet erosion test is therefore helpful in simulating the high-velocity flow environment in industrial slurry systems.

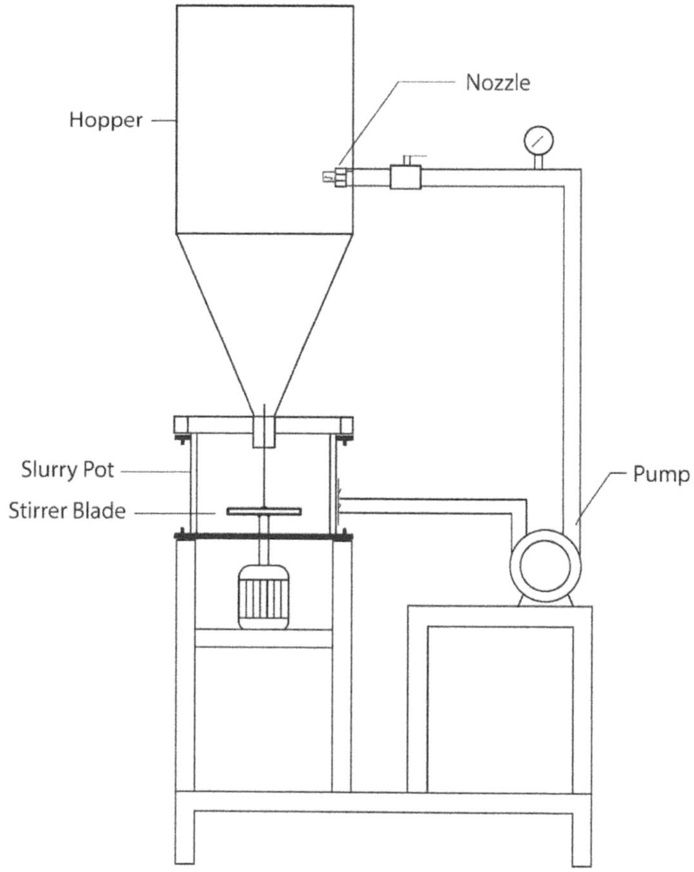

FIGURE 2.4 Testing apparatus for the jet tester.

(Khairnar & Deore, 2019; permissions under CC 4.0)

Table 2.2 presents major experimental methods employed in determining erosion rates and performance of materials in slurry conditions under controlled environments. It contains setup, working principles, and applications of each test with examples such as the slurry pot test, Coriolis tester, and jet erosion test with the closed-loop and open-loop methods. These methods are incumbent upon researchers and engineers for making predictions and quantifications of wear with the accuracy that would allow them to make independent decisions on material choice and system design. By testing these conditions, industrial conditions will be imposed on the subject test, from which useful information can be derived to prevent slurry damage so as to enhance the reliability and life expectancy of crucial equipment used in particular industries.

2.2.8 Test Loops and Their Application

Test loops in laboratory or industrial plant settings simulate long-term slurry flow conditions. These loops simulate the continuous flow of the slurry through the system so that the materials may be tested for longer periods.

2.2.8.1 Closed-Loop Test Systems

Closed-loop test systems simulate the flow of slurry in pipes or any other closed system. In such setups, the slurry is circulated through a loop system, during which the material gets exposed to particle impacts over long durations. This finds utility particularly for testing the wear of pipe fittings or pump casings due to long erosion in mines and power plants (Xu *et al.*, 2020b). The system can be modified for setting flow velocity, particle concentration, and slurry composition within the system.

2.2.8.2 Open-Loop Test Systems

Open-loop systems simulate natural or environmental conditions where the flow of slurry is not recirculated (Solanki *et al.*, 2023), and, typically, it consists of static tests, in which the slurry is allowed to flow over the sample just once upon its introduction. The advantage of open-loop systems is that they can simulate natural variations in slurry flow, temperature, and concentration over time.

TABLE 2.2
Common Test Methods for Slurry Erosion

Test Method	Description	Application/Purpose
Slurry Pot Test	A cost-effective and straightforward laboratory test where a specimen is immersed in a stirred slurry mixture. Material loss is measured by weight.	Evaluating various materials across different slurry concentrations, particle sizes, and flow conditions, such as steel in coal/ash slurries.
Coriolis Tester	A high-velocity, controlled experiment where a rotating arm causes slurry flow, simulating abrasive particle impacts on a target material.	Studying material degradation under specific industrial conditions by controlling velocity, impact angle, and particle concentration, such as alloys against fine sand.
Jet Erosion Test	High-velocity slurry jets are directed at target materials through a nozzle to simulate dynamic forces. Material loss is measured similarly to the slurry pot test.	Testing the erosion resistance of materials used in high-velocity applications like turbine impellers or pumps exposed to fast-moving slurries.
Closed-Loop Systems	Simulate continuous slurry flow through pipes or closed systems, exposing materials to repeated particle impacts over long durations.	Useful for testing wear of pipe fittings or pump casings in mines and power plants due to long-term erosion.
Open-Loop Systems	Simulate natural or environmental conditions where slurry flow is not recirculated, typically involving static tests with a single slurry flow over a sample.	Simulates natural variations in slurry flow, temperature, and concentration over time.

2.2.9 Factors Influencing Erosion Results in Testers

Slurry erosion test results vary widely, dependent upon several parameters such as the properties of the particles, environmental conditions, and the nature of the test material.

2.2.9.1 Particle Shape and Size

Particle size and the shape exert a pronounced impact on wear rate during a slurry test. Large, angular particles wear aggressively owing to kinetic energy and sharp edges, while spherical ones inflict less abrasive damage (Singh *et al.*, 2024a). Another interesting characteristic is the particle size distribution because particle sizes differing from each other interact physically in different ways with a given material; smaller particles may inflict more relentless, low-energy impacts in contrast to large particles rendering less frequent but severe material removal.

2.2.9.2 Test Environment Variables

Test variables like velocity of the slurry, the angle of impact, and the concentration of slurry greatly influence the erosion rate (Rao & Mulky, 2023). An increase in slurry velocity means higher particle energy being translated to severe erosion. Apart from that, the angle of impact also dictates the kind of wear observed—the shallow ones are mostly abrasive wear, whereas at steep angles, most of the damage is due to direct impact.

2.2.9.3 Material Properties

Erosion resistance also depends upon the material properties; these include hardness, fracture toughness, and microstructure. Hard materials tend to offer higher erosion resistance during slurry testing since they are less prone to material removal (Wróbel *et al.*, 2025). Conversely, materials of better toughness or ductility stand against fatigue erosion better because they absorb a level of impact energy without cracking.

2.3 EROSION–ABRASION

The term erosion–abrasion denotes a kind of wear process where the impact of particles and the sliding or rubbing contact of materials lead to the wear of a material. In general, erosion may be described as material removal brought about by direct particle impact on the surface, whereas abrasion comes about when particles slide, scrape, or rub against a surface (Pan *et al.*, 2023). Hence, in erosion–abrasion, the two wear mechanisms work either simultaneously or almost simultaneously to produce material degradation that is much worse than what either wear would produce in isolation. Erosion–abrasion would occur mostly in environments where particle concentration is high and in relative movement between particles and target surfaces. This is true for slurry pipeline systems, dredging operations, mining equipment, and industrial processes, where particle impacts and sliding contacts are generated in a dynamic environment (Javaheri *et al.*, 2018). The synergistic effects of the two wear mechanisms result in far more complicated degradation modes for which prediction and control become difficult. Erosion and erosion–abrasion are two distinct phenomena important in understanding the wear mechanisms responsible for most industrial applications. While pure erosion refers only to the damage caused by particle impact, erosion–abrasion considers the extra damage caused when particles slide and scrape across the surface, thereby increasing the wear rate.

2.3.1 Mechanisms of Combined Erosion and Abrasion

In erosion–abrasion action, erosion and abrasion work together to remove material. The abrasive action of particles trapped between moving surfaces or sliding on the surface increases erosion. Several actions of the joint process include cutting, plowing, and micro-fatigue.

2.3.1.1 Cutting

A huge effect of cutting is caused by sharp-edged particles interacting with the target material. Cutting in an erosion–abrasion environment is boosted due to the simultaneous existence of impact and sliding. Particles in motion repeatedly collide with a given surface, causing material removal through localized deformation. Continuing to slide over the surface, the particles with their sharp edges remove material, thereby accelerating the cutting action compared to pure erosion (Medvedovski *et al.*, 2021).

2.3.1.2 Plowing

Plowing occurs when particles slide on the surface, only pushing away the material without really removing it off the surface. Erosion–abrasion happens when impact and plowing wear act against each other. The repeated sliding across the surface creates grooves or furrows, particularly where the material is softer or harder than the particle (Brownlie *et al.*, 2021). In such wear, metals are the most significant, as deformation by plowing eventually leads to displacement of materials.

2.3.1.3 Micro-Fatigue

Micro-fatigue arises due to repetitive impacts and sliding actions, generating localized stresses on the material surfaces. This cyclic loading helps in initiating and propagating cracks inside the material. The particle impacts, as well as the abrasive sliding, work faster in the micro-fatigue process under erosion–abrasion. With repeated stress on the surface, the microcracks forming would join together, coalescing into larger cracks, which further escalates the wear rate (Banka & Rai, 2024). The combined effect of these three mechanisms—cutting, plowing, and micro-fatigue—creates a highly efficient process for material removal in erosion–abrasion situations. The impact of particles combined with the abrasive rubbing of the surface concerned creates a complex wear pattern that is far more destructive than erosion or abrasion alone.

2.3.2 Factors Influencing Erosion–Abrasion

Some influencing factors of erosion–abrasion are a combination of those that act on pure slurry erosion, with some additional parameters related to the abrasive contact. These include particle characteristics, slurry properties, material properties, and abrasive contact parameters such as normal load, contact pressure, and sliding speed.

2.3.2.1 Particle Characteristics

Similar to slurry erosion, erosion–abrasion rates will be affected by particle-dependent characteristics. Big and angular particles with sharp edges cause much more wear than small spherical ones (Xi *et al.*, 2023). In hard particle impacts, higher hardness and density values increase the erosion rate. More complex and dense materials would have a more aggressive erosion on material surfaces by impact or abrasion than soft particles (Calderon-Hernandez *et al.*, 2020). More particle concentration in the slurry or any other system will also increase the erosion–abrasion rates due to more frequent impacts or sliding contacts of particles.

2.3.2.2 Slurry Properties

The properties of the slurry, or of the fluid medium, have an effect on erosion–abrasion operations. To illustrate, viscosity in a fluid may govern the relative motions of particles and therefore may have an influence on either the impact or sliding action of the particles on the surface. Whereas highly viscous fluids slow down the moving particles and usually reduce the rate of erosion, they can conversely increase the rate of particles coming into contact with the surfaces (Xiaogang, 2024). The impact energy transmitted to the material surface depends on the density of the slurry, the flow velocity of the slurry, and also the size distribution of the particles (Banka & Rai, 2024).

2.3.2.3 Abrasive Contact Parameters

The wear rate in erosion–abrasion depends on the pH of the slurry and the contact conditions of the abrasive particles and surface. These include normal load, that is, the particle force acting on the surface of the material, and contact pressure, force acting on the surface per unit area. The greater the normal load and the contact pressure, the greater the wear due to higher interaction intensity of particles with the surface (Wang *et al.*, 2022). The abrasion will be more aggressive, that is, with greater material removal rates, at higher sliding speeds (Javaheri *et al.*, 2018).

Table 2.3 provides a very detailed description about what factors significantly affect erosion–abrasion while also referring to the important abrasive conditions of contact on certain conditions. It groups the factors into four categories: particle characteristics, slurry properties, material properties, and specific abrasive contact parameters, such as normal load, contact pressure, and sliding speed. The table shows how each factor affects the rate of material degradation with the support of relevant recent literature. This table is important for understanding the complicated interactions that govern wear in combined erosion and abrasion environment so that suitable materials might be developed along with processes optimized for each application in industry.

2.3.3 Synergistic Effects in Erosion–Abrasion

The erosion–abrasion process has a synergistic effect with degradation, wherein the sum of degradation resulting from erosion and abrasion is greater than the separate degradation caused by erosion and abrasion. In other words, the combination of particle impingement with sliding contact enhances the wear rate to an extent that is more than what would have developed in the case where the two processes acted separately.

The complementary nature of impact and abrasion forces results in more efficient material removal. Intermittent material removal takes place by erosion through impact on surface sites, while

TABLE 2.3
Factors Influencing Erosion–Abrasion

Factor	Characteristic/ Property	Effect on Erosion–Abrasion Rate
Particle Characteristics	Size and Shape	Larger and more angular particles cause significantly more wear due to increased kinetic energy and sharp edges.
	Hardness and Density	More complex and denser particles result in more aggressive material removal through impact and abrasion.
	Concentration	Higher particle concentrations result in more frequent impacts and sliding contacts, increasing wear.
Slurry Properties	Viscosity	Can control the relative motion of particles; high viscosity might reduce erosion but increase contact.
	Density and Flow Velocity	Affect the impact energy transferred to the material surface.
Material Properties	Hardness	Harder materials generally resist erosion and abrasion better.
	Toughness and Ductility	High toughness and ductility help absorb impact energy and resist fracturing under combined conditions.
	Microstructure	Fine-grained materials tend to be more resistant to abrasive wear.
Abrasive Contact Parameters	Normal Load and Contact Pressure	Higher load and pressure lead to heavier wear due to increased intensity of particle interaction.
	Sliding Speed	Higher sliding speed results in more aggressive abrasion and higher material removal rates.

at the same time, particles are sliding over the surface through abrasion, which continuously contributes to the formation of grooves and furrows on the surface. This continuous action of cutting, plowing, and micro-fatigue consequently speeds up the degradation of materials (Kumar *et al.*, 2023).

The presence of an erosion–abrasion environment can very well turn the wear rate unpredictable by not only adding the erosion rate to the abrasion rate. Both mechanisms acting together may tend to form positive feedback that in the end accelerates material damage (Zhao *et al.*, 2022). For example, the initiation of surface cracks by erosion is further propagated by particle abrasion, thereby making material loss greater than when either erosion or abrasion predominates alone.

A synergy between erosion and abrasion is evident in the mining industry, where high particle impacts and abrasive sliding cause significant damage to equipment. For instance, the combined action of erosion and abrasion causes severe wear of pipeline materials in slurry transportation systems or dredging operations, resulting in more frequent maintenance and replacements (Javaheri *et al.*, 2018).

2.4 EROSION–CORROSION

Erosion–corrosion describes the rapid degradation of a material because of the combined effects of mechanical erosion and electrochemical corrosion. Pure erosion involves mechanical removal of material as a result of impacts or friction caused by particles, and pure corrosion pertains to chemical deterioration of a material as it reacts with the surrounding environment (typically in the presence of corrosive substances like oxygen or acids). Erosion–corrosion occurs when both processes act together, often synergistically, leading to more severe material degradation than either process individually (Wang *et al.*, 2022).

In erosion–corrosion, the erosion forces disrupt the material's surface, removing the protective oxide or passive films and exposing fresh, reactive metal to the corrosive environment, thereby rendering the material susceptible to electrochemical attack. It is generally regarded as an acceleration phenomenon, worse than the simple additive effect of erosion and corrosion (Zhao *et al.*, 2022). This makes it one of the most formidable problems in industries where mechanical wear and corrosive environments coexist; examples include marine, chemical processing, and power generation industries (Khan *et al.*, 2024).

2.4.1 Mechanisms of Combined Erosion and Corrosion

Several mechanisms come into play when the damage from erosion and corrosion operates in unison. They are mainly (1) mechanical removal of protective layers, (2) mass transport enhancement, (3) local corrosion, and (4) mechanical erosion, causing susceptibility to corrosion.

2.4.1.1 Mechanical Removal of Passive Films

One particular mechanism in erosion–corrosion is the mechanical removal of passive oxide or protective films. In many metals, including stainless steel and aluminum, protective films are formed on their surfaces, allowing them to resist corrosion to some extent. However, erosion, particularly through particle impact or abrasion, continually removes these protective films, exposing fresh, reactive metal to the surrounding corrosive medium (Li *et al.*, 2025). This process accelerates corrosion because the metal surface is now more susceptible to electrochemical reactions with corrosive agents, such as oxygen, chloride ions, and other aggressive species.

2.4.1.2 Enhanced Mass Transport

Erosion increases mass transport at the material surface. The mechanical forces generated by particle impacts or friction acting in the fluid create turbulence, thereby enhancing the transport of corrosive species (such as oxygen, acids, or salts) to the material surface. This turbulent flow further assists in washing out corrosion products, such as metal ions or hydroxides, which may act as protective barriers against corrosion on that surface. The increased flow and removal of corrosion products mean that the material is continually exposed to a fresh supply of corrosive agents, speeding up the degradation process (Xu *et al.*, 2021).

2.4.1.3 Localized Corrosion

The erosion damage, in the form of pits, crevices, or grooves, may provide the best initiation sites for localized corrosion. Local removal of material by erosion could lead to microscopically rough regions that may interfere with the uniformity of the passive layer or even increase the local electrochemical activity. Consequently, it localizes such events that occur on a small-scale, leading to material-specific failure.

2.4.1.4 Mechanical Damage Enhancing Corrosion

Erosion may also result in mechanical damage to a material's surface, causing the formation of dislocations, microcracks, and other types of defects, thereby increasing the material's surface energy and making it more reactive and prone to corrosion. The mechanical forces imparted during erosion establish a stress field that acts to alter the material's microstructure, facilitating the initiation and propagation of corrosion at these defective sites (Xie *et al.*, 2015). This synergy between mechanical damage and corrosion favors further degradation of the material, particularly in aggressive environments where erosion and corrosion combine.

2.4.2 Factors Influencing Erosion–Corrosion

The rate and severity of erosion–corrosion depend on several factors, including all those related to slurry erosion, such as particle size, shape, and concentration, as well as the specifics of the corrosive environment and the electrochemical properties of the material.

2.4.2.1 Factors from Slurry Erosion

The same factors that influence slurry erosion—particle characteristics, slurry properties, material properties, and flow conditions—also play a vital role in erosion–corrosion. For example, the size and shape of particles will influence the impact energy and the abrasive action that removes protective oxide films. Slurry properties, such as fluid viscosity, flow velocity, and temperature, will affect the interaction between particles and the material surface, influencing the rate of localized corrosion (Kumar *et al.*, 2023).

2.4.2.2 Corrosive Environment Properties

The corrosive environment's properties hugely dictate the erosion–corrosion rate. It involves factors such as the following.

2.4.2.2.1 pH

If the medium is acidic or alkaline, the electrochemical action on a material varies accordingly. Generally, a lower pH (acidic) environment accelerates corrosion of metals like steel or aluminum (Medvedovski *et al.*, 2021).

2.4.2.2.2 Dissolved Oxygen

The amount of dissolved oxygen in the surrounding fluid affects the oxidation rate and the formation of corrosion products on the material surface. Thus, the higher the oxygen concentration, the faster the corrosion (Wang *et al.*, 2021).

2.4.2.2.3 Presence of Aggressive Ions

Ions such as chloride, sulfide, and nitrate can accelerate the corrosion process. Chloride ions can disrupt the passive oxide film, leading to pitting and crevice corrosion (Banka & Rai, 2024).

2.4.2.2.4 *Temperature and Conductivity*

High temperatures increase both the corrosion rate and ion mobility in solution, thereby accelerating erosion–corrosion. The same applies to solution conductivity, which affects electrochemical reactions (Chen *et al.*, 2023).

2.4.2.3 Electrochemical Aspects

Another crucial factor in erosion–corrosion is the electrochemical nature of the materials involved. Each material possesses a characteristic electrochemical potential, which determines its corrosion resistance. The mechanical removal of passive films in erosion–corrosion exposes fresh material, which likely has no electrochemical potential and initiates localized corrosion. The polarization behavior of the material under consideration generally reflects how easily it can develop a protective oxide layer and influences its susceptibility to corrosion in the presence of an erosive environment (Zhao *et al.*, 2022).

2.4.3 Synergistic and Antagonistic Effects in Erosion–Corrosion

Erosion corrosion can be broadly characterized by synergistic effects, where the combined degradation rate is significantly higher than the sum of the individual erosion and corrosion rates.

2.4.3.1 Synergism between Erosion and Corrosion

In most cases, erosion and corrosion cause accelerated degradation of metals. The mechanical removal of protective layers due to erosion exposes fresh metal to corrosive environments, and the increased mass transport facilitates the removal of corrosion products, further enhancing the electrochemical attack on the material surface. The synergistic effect can cause swift failure of materials, especially in environments where mechanical wear and corrosive conditions coexist, such as the marine environment and chemical processing systems (Xu *et al.*, 2021).

Some examples of this effect include the rapid deterioration of seawater pipelines since erosion–corrosion damage occurs due to the presence of abrasive sand particles and aggressive saline conditions. Similarly, heat exchangers and turbines are industrial systems that experience erosion–corrosion primarily due to the combined effects of high-velocity fluid flow and corrosive chemical agents (Khan *et al.*, 2024).

2.4.3.2 Antagonistic Effect

In rare cases, when erosion and corrosion conflict with each other, then one may impede the other. Erosion could, for instance, produce a thick layer of corrosion products that protect the surface and thus lower corrosion and slow down overall degradation. On the other hand, erosion may also remove corrosive species for some materials and hence provide a short-lived decrease in corrosion rates (Wang *et al.*, 2021). That being said, antagonistic effects tend to be less common and less significant than synergistic effects in erosion–corrosion.

2.4.4 Electrochemical Measurement Techniques in Erosion–Corrosion

Several electrochemical methods are often used to study and measure erosion–corrosion processes. These techniques give a fundamental notion of the erosion mechanism involved and help to evaluate the ability of a material to resist erosion.

2.4.4.1 Potentiodynamic Polarization

Potentiodynamic polarization is one of the most-used electrochemical techniques for corrosion investigations. Potentiodynamic methods involve sweeping the applied potential and recording the current, from which corrosion potentials, corrosion current densities, Tafel slopes, and so on may be resolved, providing a general insight as to whether the compound is anodically or cathodically being polarized in the resisting environment (Zhang *et al.*, 2021).

2.4.4.2 Electrochemical Impedance Spectroscopy

Electrochemical impedance spectroscopy (EIS) is a powerful tool that can perform analysis of material behavior in real time. By determining the impedance of the material applied by an alternating current (AC) signal, EIS can inform one about the electrochemical properties of the materials, particularly their resistive and capacitive properties and the formation of any protective layers. EIS is primarily useful in erosion–corrosion studies and in following the measurable corrosion resistance of a material while it is mechanically worn away (Xie *et al.*, 2015).

2.4.4.3 Electrochemical Noise Analysis

The working principle of electrochemical noise analysis (ENA) is to detect and record the fluctuations of electrochemical current and potential during corrosion attack. These fluctuations themselves give direct information on the mechanisms of corrosion, mainly for the initiation of pitting or other forms of localized corrosion, which are primarily important for erosion–corrosion cases (Bai *et al.*, 2024).

2.4.4.4 Galvanic Current Measurements

Galvanic-current measurements require two different materials to form a galvanic couple. A measurement of the current between the members of this galvanic couple is made to understand the electrochemical interference occurring between them, especially in mixed-material systems common in erosion–corrosion applications (Kumar *et al.*, 2023).

2.5 COMPARATIVE ANALYSIS OF TEST METHODS

Multiple test setups are used to study the erosive effects of slurry and its interaction with wear mechanisms. The most common techniques applied are slurry pots, Coriolis, and jet erosion testing. Each has its advantages and disadvantages, and it thus becomes important to choose the one most suitable to the problem or the given industrial application.

2.5.1 Slurry Pot Test

This particular test is simple and is quite widely applied due to ease of implementation. In this test, a specimen is kept immersed in a slurry mixture, and the slurry mixture is stirred to simulate dynamic flow conditions. One of its main advantages is the simulation of erosion at the level of particle size and concentration. It also enables easy observation and calculation of material loss with time (Khan *et al.*, 2024). Yet the slurry pot test does have some limitations. It usually does not consider the velocities of turbulence associated with real slurry transport schemes such as pipelines and industrial equipment (Lin *et al.*, 2023). Also, it cannot properly simulate complex flow geometries such as bends or elbows, which are very common in industrial systems.

2.5.2 Coriolis Tester

The Coriolis test apparatus offers a high degree of control and testing sophistication for slurry erosion under heavy flow conditions. The arm revolves, simulating slurry flow and allowing the operator to set the velocity of impact, concentration of particles, and impact angles. The real strength of the Coriolis tester lies in its ability to truly emulate dynamic flow scenarios under high-velocity particle impacts (Zhao *et al.*, 2022). The Coriolis tester has its disadvantages. First, the complexity and cost may preclude many potential users from giving this method a try, unlike the simpler slurry pot test. Second, it may not be able to replicate all possible flow geometries or environmental conditions that may exist in the actual case—for example, a complex slurry system or a highly turbulent environment (Zhang *et al.*, 2021).

2.5.3 Jet Erosion Test

The jet erosion test involves the disruption of a high-speed jet of slurry upon any given material surface simulating high-speed impacts. This operation is mostly employed for analyzing materials under dynamic slurry conditions, for example, pump impellers or turbine blades (Xie *et al.*, 2015). The velocity and angle of impact in this test can be very precisely controlled when determining a material's property for various flow regimes. But because the jet erosion test involves long-term wear or environments where multiple mechanisms interact, such as erosion–abrasion and erosion–corrosion, an obvious limitation is set upon it. The jet erosion test mainly looks at particle impacts, which may fail to take into account other mechanisms in real slurry systems, such as abrasive sliding (Sharma & Gandhi, 2020).

2.5.4 Correlations of Real-World Data and Lab Tests

Although laboratory tests are beneficial in studying slurry erosion, the results obtained in controlled environment settings often do not reflect the cases in real life; real-world slurry systems are complicated and include variables such as particle velocity, flow dynamics, temperature, and material properties (Li *et al.*, 2025). Thus, errors are inherently built into the extrapolation of lab data for field performance. Other major determining factors affecting how much foresight can be given through the slurry pot test, the Coriolis tester, and the jet erosion test will be the degree to which they can replicate reality. While doing so, the presence of multi-mechanism wear (erosion, abrasion, and corrosion), complex flow geometries, and environmental conditions should provide the backdrop against which lab results should be read. Calibration of these lab test results against field data is thus necessary to improve prediction scores for these particular test results (Kumar *et al.*, 2023).

2.6 INTERACTIONS AND SYNERGISTIC EFFECTS AMONG SLURRY EROSION, EROSION–ABRASION, AND EROSION–CORROSION

2.6.1 Complex Interplay in Real-World Scenarios

In real-life slurry conditions, materials undergo wear by more than one method simultaneously. It may be due to erosion by slurry, erosion–abrasion, and erosion–corrosion, making it very difficult to formulate the overall wear rate.

2.6.1.1 Slurry Erosion, Erosion–Abrasion, and Erosion–Corrosion Occurring Simultaneously

The slurry materials in many industrial systems undergo erosion, abrasion, and erosion–corrosion simultaneously. For instance, in mining slurry pipelines, the solid particles in the slurry cause erosion by impacting pipe surfaces, while the solid particles in the slurry move by sliding action over pipe surfaces, resulting in erosion–abrasion. Where corrosive agents such as acids and salts are present, erosion–corrosion contributes to degradation by removing protective oxide films, exposing fresh metal surfaces to corrosion (Javaheri *et al.*, 2018).

When more than one mechanism acts simultaneously, it creates much more complicated patterns of wear and accelerated material degradation than if each had to work alone. For the interrelated case, erosion–corrosion is accelerated in turbulent slurry flow conditions. Particles abrade corrosion products, continuously exposing fresh metal for the corrosive agent to act upon for further deterioration (Zhao *et al.*, 2022).

2.6.2 Quantifying Synergism

The synergistic effect occasioned by erosion, abrasion, and corrosion is challenging to quantify because it is impossible to isolate each contribution by different mechanisms. The erosion–abrasion–corrosion environment does not result in a wear rate which is a summation of the erosion, abrasion, and corrosion; instead, the rate is far greater, because of a kind of synergistic interaction between mechanisms (Li *et al.*, 2025).

2.6.2.1 Challenges in Separating and Quantifying Contributions

Separating and quantifying the individual contributions of erosion and abrasion or erosion and corrosion in multi-mechanism environments offer a challenge since these processes usually occur together and influence each other. For example, erosion may expose the fresh material to corrosion, whereas corrosion would cripple the material to make it susceptible to abrasion. Each of these processes creates a feedback loop that prevents measuring and quantifying the contribution of each mechanism independently (Xie *et al.*, 2015).

2.6.2.2 Methods or Models Used to Estimate Synergistic Effects

In estimating the synergistic effects of concurrent wear mechanisms, various researchers rely on multi-mechanism degradation models. Generally, these models use experimental data and methods to establish and analyze occurrences in interacting complex systems, which may either be an empirical correlation or numerical simulation incorporating the relative contribution of each mechanism as a function of operating conditions (Islam *et al.*, 2024).

2.6.3 Case Examples of Multi-Mechanism Degradation

In typical multi-mechanisms degrading situations as in mining industries, power generation, and chemical processing, slurry transport systems often experience a decline in reliability when erosion, abrasion, and corrosion occur simultaneously.

2.6.3.1 Pump Impellers

In mining slurry systems, pump impellers experiences erosion due to the relative velocity of particle wakes and abrasion caused by the sliding motion of particles. In some cases, where corrosive substances (such as acids in mining slurries) are present, erosion–corrosion will accelerate material loss. Multi-mechanism degradation places impellers into rapid material loss, which reduces working life and increases frequency of maintenance (Wang *et al.*, 2022).

2.6.3.2 Pipelines

Considering the combined effects of extremely high particle velocities and corrosive agents such as chlorine or sulfur compounds, slurry pipelines used in mining or power generation systems experience erosion. This erosion, along with abrasion and corrosion, can ultimately lead to pipeline failure. This combination of wearing mechanisms leads to costly repairs and replacements of pipelines (Zhao *et al.*, 2022).

2.6.3.3 Valves in Mining Slurries

Erosion by impingement and abrasion caused by sliding particles are common mechanisms through which valves wear in a mixed slurry transport system. Moreover, the manufacturers of these corrosive slurries further promote erosion–corrosion to these valves. The cumulative effect of these mechanisms results in the intense degradation of valve seats and other components, leading to leakage and system failure (Li *et al.*, 2019).

2.7 MATERIAL SUSCEPTIBILITY AND FAILURE MODES

2.7.1 Response to Different Material Classes

In slurry environments, the material performs differently with regard to hardness, corrosion resistance, and microstructure. Different material classes exhibit dissimilar responses to the complex wearing mechanisms of slurry erosion, erosion–abrasion, and erosion–corrosion.

2.7.1.1 Metals and Alloys

The strength and durability offered by metals and alloys, which include carbon steels, stainless steels, cast irons, nickel alloys, and aluminum alloys, have warranted their use in slurry transport systems and mining equipment. The wear performance of such materials depends on their hardness and corrosion resistance. Carbon steels, for example, have high toughness and strength but may be susceptible to corrosion in either acidic or alkaline slurry environments, thereby degrading quickly (Kang *et al.*, 2024). Stainless steels provide excellent corrosion resistance and, therefore, may be used in areas that undergo both corrosion and erosion; however, they may lack wear resistance compared to complex alloys like nickel alloys, which are highly resistant to both erosion and corrosion in aggressive service environments (Zhang *et al.*, 2021).

Cast iron represents a classic material chosen to resist slurry abrasion, wears-resistant, and it is easy to maintain. Although, it is prone to brittle fracture in high-impact situations as compared to steel. Carbides present in the microstructure, grain size, and phase distribution in these materials arrest wear. Nickel alloys, for instance, Inconel, are considered to have corrosion resistance to high temperatures, in addition to mechanical wear resistance, and hence are suitable for very high-demand applications (He *et al.*, 2024).

Aluminum alloys are, in general, considered the ideal choice for light engineering products and applications with less abrasive slurry environments, mainly because of their lightness and good corrosion resistance; however, their hardness is lower than that of steels, which works against them in resisting wear and erosion, especially in highly abrasive or erosive slurry environments (Islam *et al.*, 2024).

2.7.1.2 Ceramics

Alumina, silicon carbide, and zirconia are ceramic materials used for wear resistance and corrosion resistance, thus making them suitable where highly erosive conditions exist. These materials are hard, making them very resistant against abrasive wear and erosion in slurry systems where particle velocity is very high. Their inherent brittleness has thus restricted their applications. Ceramics crack and later fracture in the presence of impact or dynamic loading, thus restricting their applications for systems where either of those mechanisms or cyclic loading is common (Li *et al.*, 2019). Examples of where silicon carbide was utilized include pumps and valves requiring high wear resistance, but it becomes impractical in high-impact slurry environments because of its brittleness.

2.7.1.3 Polymers and Composites

In slurry systems, the use of polymers and composites is increasingly talked about, including elastomers, polyurethanes, and fiber-reinforced composites, due mainly to their versatility with respect to resilience, flexibility, and resistance to a certain kind of wear. Elastomers offer abrasion resistance and flex to absorb the dynamic impacts resulting from the solid particles in the slurry, especially when sealing applications are required or wherever flexibility is required (Wang *et al.*, 2022). Polyurethanes are mainly used in slurry pumps and in pipes, where maximum impact resistance with relative abrasion resistance is desired. However, they suffer from an increased rate of oxidative degradation and chemical attack, especially in high-temperature environments.

Fiber-reinforced composites strike an almost perfect balance between strength, stiffness, and corrosion resistance, areas where metallic materials might be too heavy or susceptible to corrosion. But their drawback is in situations requiring very high-impact loads, where the composite matrix will fracture or degrade, especially when exposed to very high concentrations of particles (Messa & Malavasi, 2018).

2.7.2 Common Failure Modes

Slurry media cause numerous types of erosions, abrasions, and corrosive failures, sometimes acting in unison. It becomes necessary, therefore, for the one concerned with correct material selection and system design to understand these modes of failures.

2.7.2.1 Wall Thinning

Wall thinning takes place due to the gradual removal of the surface material of a component by erosive wear or corrosion. Within the slurry setup, concerns are raised with wall thinning most often within the pipe or valve in which the incessant movement of abrasive particles in slurry causes the gradual disintegration of the materials. Destruction begins by the removal of the surface layer in erosion, and corrosion action aggravates it by rendering the material more susceptible to degradation. With the decrease of wall thickness, the component becomes weak in the sense of structure and may be faced with failure regarding leakage or rupture (Chen *et al.*, 2023).

2.7.2.2 Pitting

Pitting represents a localized form of corrosion, typically due to an impact or abrasion that destroys the protective oxide layer of a material. Every such action removes the oxide layer and exposes fresh metal to further electrochemical degradation, where tiny deep pits form. This is commonly encountered in stainless steels and nickel alloys submitted to corrosive slurry environments, where chloride or sulfate ions foster localized corrosion (He *et al.*, 2024). Over time, these pits become large enough to reduce the material further, causing its total destruction.

2.7.2.3 Grooving

Grooving creates narrow, elongated depressions or grooves on a work surface as the abrasive particles slide against it continuously. Grooving is a typical erosion–abrasion phenomenon and is particularly pronounced in materials subjected to extreme particle concentrations in slurry. The chances of localized corrosion can, however, be increased in the presence of grooves, as these can trap corrosive agents and cause deterioration in the material in those regions with rapid intensity (Wang *et al.*, 2022). Grooving turns out to be worse in systems with components that work in rotation, where abrasive particles continuously have sliding contact.

2.7.2.4 Localized Material Loss

Localized material loss can occur due to the synergistic effect of erosion and corrosion, resulting in pits, grooves, or other forms of surface deformation. This type of wear can significantly reduce the operational capacity of critical components, such as pipes, pumps, and valves, in slurry-transporting systems. Erosion–corrosion preferentially creates localized weaker points, aspects that have posed specific difficulties in predicting the life of equipment and have so far led to unexpected failure (Zhang *et al.*, 2021).

2.7.2.5 Component Fracture

Component fracture is the worst type of failure in slurry systems. Such failure usually occurs in materials that are inherently brittle, such as ceramics or cast irons, and is often the result of combined erosion, abrasion, and corrosion. In such cases, the mechanical impacts from the slurry particles initiate cracks or fractures, which are then further grown by the corrosive environment, leading to rapid material failure. For equipment working under cyclic loading or dynamical stressing, like turbine blades or impellers, this is the primary cause of sudden catastrophic failure of the system (Li *et al.*, 2022).

2.8 FUTURE DIRECTIONS AND INNOVATIONS IN TESTING

2.8.1 The Role of Simulation and Computational Modeling in Refining Experimental Methods

Slurry testing methods are likely to become extinct or at least face extinction in favor of simulation and computational modeling, mostly CFD. CFD simulation can model particle flow and interaction with materials in slurry systems with great accuracy, allowing a better understanding of the wear mechanism under various working conditions. When combined with experimentation, CFD models can simulate several slurry conditions, particle velocities, and impact angles to provide a more realistic prediction of wear rates (Messa & Malavasi, 2018).

2.8.2 Development of Multifunctional Testers

Another interesting direction for development is to create multifunctional testers that combine erosion, abrasion, and corrosion tests on a single platform. This equipment considers the simultaneous existence of various wear-interaction phenomena, thus opening the door for a serious holistic investigation of material degradation (Qin *et al.*, 2024).

Integrating all these functions in a single system will enable industries to generate more comprehensive data on material performance, which in turn will yield informed material choices and improved designs for slurry systems.

2.9 CONCLUSION

From slurry erosion to erosion–abrasion to erosion–corrosion, these are different yet closely intertwined mechanisms that considerably degrade materials in industrial systems. Nevertheless, briefly put, slurry erosion is the impact of particles, erosion–abrasion is particles sliding over surfaces, and erosion–corrosion is the concomitant wear of surfaces with electrochemical corrosion. A mix can occur with varying degrees, depending on particle and slurry characteristics, material properties, and environmental conditions. Setting forth these mechanisms is important for better prediction of the degradation of materials and for selection of materials in slurry-handling systems.

To come up with good selection criteria for materials and their protection, degradation processes should be taken as a whole. An understanding of the usually complicated interrelations between these different factors would help further modeling work, hopefully using CFD, with a view to more accurately predict multi-wear mechanisms. Improved materials, in-situ monitoring methods, and novel coatings will also be needed to provide the best protection for slurry-exposed parts.

REFERENCES

Alam, T., & Farhat, Z. N. (2018). Slurry erosion surface damage under normal impact for Pipeline steels. *Engineering Failure Analysis*, *90*, 116–128. https://doi.org/10.1016/j.engfailanal.2018.03.019.

Bai, J., Ma, X., Wang, S., Niu, J., Ma, R., Du, A., . . . & Fan, Y. (2024) Corrosion and wear behaviour of slaconcentrationsce f Cr, Ni, and Mo under different chlo concentration. *séria (Rio de Janeiro)*, *29*(2), e20240095. https://doi.org/10.1590/1517-7076-RMAT-2024-0095.

Banka, J., & Rai, A. K. (2024). Erosion and flow visualization in centrifugal slurry pumps: A comprehensive review of recent developments and future outlook. *Particulate Science and Technology*, *42*(3), 427–459. https://doi.org/10.1080/02726351.2023.2259336.

Bitter, J. G. A. (1963). A study of erosion phenomena part I. *Wear*, *6*(1), 5–21. https://doi.org/10.1016/0043-1648(63)90003-6

Brownlie, F., Hodgkiess, T., Pearson, A., & Galloway, A. M. (2021). A study on the erosion-corrosion behaviour of engineering materials used in the geothermal industry. *Wear*, *477*, 203821. https://doi.org/10.1016/j.wear.2021.203821.

Calderon-Hernandez, J. W., Sinatora, A., de Melo, H. G., Chaves, A. P., Mano, E. S., Leal Filho, L. S., . . . & Pinto, T. C. S. (2020). Hydraulic convey of iron ore slurry: Pipeline wear and ore particle degradation in function of pumping time. *Wear*, *450*, 203272. https://doi.org/10.1016/j.wear.2020.203272.

Chen, Q., Zhou, H., Wang, Y., Wang, D., Zhang, Q., & Liu, Y. (2023). Erosion wear at the bend of pipe during tailings slurry transportation: Numerical study considering inlet velocity, particle size, and bend angle. *International Journal of Minerals, Metallurgy and Materials*, *30*(8), 1608–1620. https://doi.org/10.1007/s12613-023-2672-z.

Chung, R. J., Jiang, J., Pang, C., Yu, B., Eadie, R., & Li, D. Y. (2021). Erosion–corrosion behaviour of steels used in slurry pipelines. *Wear*, *477*, 203771. https://doi.org/10.1016/j.wear.2021.203771.

Finnie, I. (1974). *Studies of the Mechanical Behavior of Materials*. University of Glasgow (United Kingdom).

Flores, J. F., Neville, A., Kapur, N., & Gnanavelu, A. (2012). Corrosion and erosion–corrosion processes of metal-matrix composites in slurry conditions. *Journal of Materials Engineering and Performance*, *21*, 395–405. https://doi.org/10.1007/s11665-011-9926-z.

He, Z., Wu, Z., Gao, Z., Qin, Z., Yuan, J., Gao, Y., . . . & Hu, W. (2024). Erosion–corrosion behaviour of Ni-Al-Cu coating on nickel-aluminium bronze alloy in 3.5 wt% NaCl solution. *Corrosion Science*, *238*, 112380. https://doi.org/10.1016/j.corsci.2024.112380.

Islam, M. A., Jiang, J. J., & Xie, Y. (2024). Optimizing material selection: A study of erosion-corrosion performance in homogeneous and carbide-containing materials. *Wear*, *536*, 205181. https://doi.org/10.1016/j.wear.2023.205181.

Javaheri, V., Porter, D., & Kuokkala, V. T. (2018). Slurry erosion of steel-review of tests, mechanisms and materials. *Wear*, *408*, 248–273. https://doi.org/10.1016/j.wear.2018.05.010.

Kang, C., Li, M., Teng, S., Liu, H., Chen, Z., & Li, C. (2024). Erosive wear caused by large solid particles carried by a flowing liquid: A comprehensive review. *Processes*, *12*(6), 1150. https://doi.org/10.3390/pr12061150.

Khairnar, K., & Deore, E. R. (2019). Modification of jet erosion test rig to erosion test due to solid-liquid mixture. *International Journal of Engineering Research & Technology*, *8*(12), 424–427.

Khan, R., Mourad, A. H. I., Wieczorowski, M., Damjanović, D., Pao, W., Elsheikh, A., & Seikh, A. H. (2024). Erosion–corrosion failure analysis of the elbow pipe of the steam distribution manifold. *Engineering Failure Analysis*, *160*, 108177. https://doi.org/10.1016/j.engfailanal.2024.108177.

Kübarsepp, J., Juhani, K., & Tarraste, M. (2021). Abrasion and erosion resistance of cermets: A review. *Materials*, *15*(1), 69. https://doi.org/10.3390/ma15010069.

Kumar Yadav, Y., Kumar Singh, A., & Siddhartha. (2023). Insights to improve the tribo-performance of materials used under slurry erosion applications: A review. *Proceedings of the Institution of Mechanical Engineers, Part L: Journal of Materials: Design and Applications*, *237*(1), 3–32. https://doi.org/10.1177/14644207221108565.

Kumar, R., Singh, R., & Kumar, S. (2018). Erosion and hot corrosion phenomena in thermal power plant and their preventive methods: A study. *Asian Review of Mechanical Engineering*, *7*(1), 38–45. https://doi.org/10.51983/arme-2018.7.1.2436.

Li, B., Liu, X. J., Zhu, H. W., Guan, H. P., & Guo, R. T. (2025). Recent progress of round-the-world clock photocatalytic system in environmental and energy applications: A review. *ChemPhysChem*, *26*(9), e202401144. https://doi.org/10.1002/cphc.202401144.

Li, P., Zhao, Y., & Wang, L. (2019). Research on the erosion–corrosion rate of 304 stainless steel in acidic slurry via the experimental design method. *Materials*, *12*(14), 2330. https://doi.org/10.3390/ma12142330.

Lin, S., Li, D., Zhou, Q., Chu, M., & Sun, H. (2023). Copper-nickel reaction behaviour of copper-nickel alloy pipe during service in a marine environment. *Engineering Failure Analysis*, *153*, 107628. https://doi.org/10.1016/j.engfailanal.2023.107628.

Medvedovski, E., Leal Mendoza, G., & Vargas, G. (2021). Influence of boronizing on steel performance under erosion–abrasion-corrosion conditions simulating downhole oil production. *Corrosion and Materials Degradation*, *2*(2), 293–324. https://doi.org/10.3390/cmd2020016.

Messa, G. V., & Malavasi, S. (2018). A CFD-based method for slurry erosion Prediction. *Wear*, *398*, 127–145. https://doi.org/10.1016/j.wear.2017.11.025.

Nan, H., Han, W., Li, R., Dong, Y., Shen, X., Song, H., . . . & Guo, X. (2024). Erosion Characteristics and mechanisms of different particle sizes under the synergy effect of cavitation and particle erosion. *Physics of Fluids*, *36*(12). https://doi.org/10.1063/5.0244598.

Nassef, A., Keller, M., Hassani, S., Shirazi, S., & Roberts, K. (2022). A review of erosion-corrosion models for the oil and gas industry applications. *Recent Developments in Analytical Techniques for Corrosion Research*, 205–233. https://doi.org/10.1007/978-3-030-89101-5_10.

Nie, S., Shi, L., He, H., Ji, H., Yin, F., & Ma, Z. (2025). Fractal and multifractal characterization of wear surface dynamics for tribopairs in a bi-directional seawater pump. *Measurement*, 117763. https://doi.org/10.1016/j.measurement.2025.117763.

Pan, K., Deng, J., Qu, G., Wei, K., & Liang, Y. (2023). Synergistic assisted hydrogen production of tailings slurry and coal slurry: Process and mechanism investigation. *Journal of Cleaner Production*, *422*, 138582. https://doi.org/10.1016/j.jclepro.2023.138582.

Qin, Z., Yuan, Y., Chen, Z., Li, Y., & Xia, Y. (2024). Combined preparation and application of geopolymer pavement materials from coal slurry-slag powder-fly ash mining solid waste: A case study. *Construction and Building Materials*, *441*, 137510. https://doi.org/10.1016/j.conbuildmat.2024.137510.

Rao, P., & Mulky, L. (2023). Erosion-corrosion of materials in industrial equipment: A review. *ChemElectroChem*, *10*(16), e202300152. https://doi.org/10.1002/celc.202300152.

Rasse, C., Mary, N., Abe, H., Watanabe, Y., & Normand, B. (2021). Role of the jet angle, particle size and particle concentration in the degradation behaviour of carbon steel under slow slurry erosion–corrosion conditions. *Metals*, *11*(8), 1152. https://doi.org/10.3390/met11081152.

Reddy, N. V. K., Pothal, J. K., Barik, R., & Senapati, P. K. (2023). Pipeline slurry transportation system: An overview. *Journal of Pipeline Systems Engineering and Practice*, *14*(3), 03123001. https://doi.org/10.1061/JPSEA2.PSENG-1391.

Sekar, A., & Bennet, A. R. (2023). A comprehensive review of synergistic and individual effects of erosion–corrosion in ferrous piping materials. *Corrosion Reviews*, *41*(4), 399–416. https://doi.org/10.1515/corrrev-2022-0063.

Sharma, S., & Gandhi, B. K. (2020). Erosion wear behaviour of martensitic stainless steel under the hydro-abrasive condition of hydropower plants. *Journal of Materials Engineering and Performance*, *29*, 7544–7554. https://doi.org/10.1007/s11665-020-05238-2.

Singh, J., Kannaram, S., Singh, J. P., Gill, H. S., & Joshi, A. (2024a). Review on slurry erosion in slurry transportation systems. In *AIP Conference Proceedings* (Vol. 2986, No. 1). AIP Publishing. https://doi.org/10.1063/5.0192573.

Singh, J., Kumar, S., & Gill, H. S. (2024b). Review of testing facilities assisting in development numerical models for erosion calculation in centrifugal slurry pumps. *International Journal on Interactive Design and Manufacturing (IJIDeM)*, *18*(8), 5301–5327. https://doi.org/10.1007/s12008-023-01282-z.

Singh, J., Vasudev, H., & Chohan, J. S. (2024c). Review on computational fluid dynamics-based analysis of surface erosion in hydraulic machinery. *International Journal on Interactive Design and Manufacturing (IJIDeM)*, *18*(8), 5353–5380. https://doi.org/10.1007/s12008-023-01336-2.

Singh, S., Khan, N. A., Ramadan, R., Shehata, N., Kapoor, D., Dhanjal, D. S., . . . & Ramamurthy, P. C. (2024d). Environmental fate, toxicological impact, and advanced treatment approaches: Atrazine degradation with an emphasis on circular economy strategies. *Desalination and Water Treatment*, 100201. https://doi.org/10.1016/j.dwt.2024.100201.

Sinha, S. L., Dewangan, S. K., & Sharma, A. (2017). A review on particulate slurry erosive wear of industrial materials: In the context of pipeline transportation of mineral slurry. *Particulate Science and Technology*, *35*(1), 103–118. https://doi.org/10.1080/02726351.2015.1131792.

Soetan, O., Nie, J., Viteritto, M., & Feng, H. (2023). Evaluation of sediment dredging in remediating toxic metal contamination—a systematic review. *Environmental Science and Pollution Research*, *30*(27), 69837–69856. https://doi.org/10.1007/s11356-023-27489-x.

Solanki, P., Jain, B., Hu, X., & Sancheti, G. (2023, September). A review of beneficial use and Management of dredged material. *Waste*, *1*(3), 815–840. MDPI. https://doi.org/10.3390/waste1030048.

Sridhar, K., & Balasubramanian, V. (2022). Erosion–corrosion-resistant coatings for seawater piping components—a review. *A Treatise on Corrosion Science, Engineering and Technology*, 591–610. https://doi.org/10.1007/978-981-16-9302-1_30.
Tarodiya, R., & Gandhi, B. K. (2017). Hydraulic performance and erosive wear of centrifugal slurry pumps-a review. *Powder Technology*, *305*, 27–38. https://doi.org/10.1016/j.powtec.2016.09.048.
Varga, M., Rojacz, H., Widder, L., & Antonov, M. (2021). High temperature erosion–corrosion of wear protection materials. *Journal of Bio-and Tribo-Corrosion*, *7*(3), 8. https://doi.org/10.1007/s40735-021-00504-9.
Wang, H. K., Deng, L., Zhang, Y., Yu, Y., & Gao, J. (2025). A comprehensive review on erosion–corrosion characteristics of steel subjected to tensile stress. *Journal of Iron and Steel Research International*, 1–22. https://doi.org/10.1007/s42243-024-01415-y.
Wang, Q., Huang, Q., Sun, X., Zhang, J., Karimi, S., & Shirazi, S. A. (2021). Experimental and numerical evaluation of the effect of particle size on slurry erosion prediction. *Journal of Energy Resources Technology*, *143*(7), 073101. https://doi.org/10.1115/1.4049417.
Wang, X., Wang, H., Wu, A., & Kang, G. (2022). Wear law of Q345 steel under the abrasion—corrosion synergistic effect of cemented paste backfill. *Construction and Building Materials*, *332*, 127283. https://doi.org/10.1016/j.conbuildmat.2022.127283.
Wróbel, J., Pietrusiak, D., Rozmus, R., Roicki, R., Zarzycki, B., & Stefanek, P. (2025). Failure analysis and guidelines for further exploitation of centrifugal slurry pumps used for copper flotation waste transport: A case study. *Engineering Failure Analysis*, *174*, 109479. https://doi.org/10.1016/j.engfailanal.2025.109479.
Xi, X., Tao, Y., Jiang, S., & Yin, C. (2023). Study on the formation mechanism and mechanical properties of composite foam slurry material for mine plugging. *Energy*, *281*, 128295. https://doi.org/10.1016/j.energy.2023.128295.
Xiaogang, L. (2024). Erosion-corrosion. In *The ECPH Encyclopedia of Mining and Metallurgy* (pp. 615–616). Singapore: Springer Nature Singapore. https://doi.org/10.1007/978-981-99-2086-0_1334.
Xie, Y., Jiang, J. J., Tufa, K. Y., & Yick, S. (2015). Wear resistance of the materials used for slurry: Transport. *Wear*, *332*, 1104–1110. https://doi.org/10.1016/j.wear.2015.01.005.
Xu, Y., Liu, L., Xu, C., Wang, X., Tan, M. Y., & Huang, Y. (2020a). Electrochemical characteristics of the dynamic progression of erosion–corrosion under different flow conditions and their effects on corrosion rate calculation. *Journal of Solid State Electrochemistry*, *24*, 2511–2524. https://doi.org/10.1007/s10008-020-04795-9.
Xu, Y., Liu, L., Zhou, Q., Wang, X., Tan, M. Y., & Huang, Y. (2020b). An overview of major experimental methods and apparatus for measuring and investigating erosion–corrosion of ferrous-based steels. *Metals*, *10*(2), 180. https://doi.org/10.3390/met10020180.
Xu, Y., Zhang, Q., Zhou, Q., Gao, S., Wang, B., Wang, X., & Huang, Y. (2021). Flow accelerated corrosion and erosion—corrosion behaviour of marine carbon steel in natural seawater. *NPJ Materials Degradation*, *5*(1), 56. https://doi.org/10.1038/s41529-021-00205-1.
Yan, J., & Li, F. (2023). Effects of sediment dredging on freshwater systems: A comprehensive review. *Environmental Science and Pollution Research*, *30*(57), 119612–119626. https://doi.org/10.1007/s11356-023-30851-8.
Yang, G., Tsai, K., Serate, D., & Wu, M. (2018, April 15–19). Wear modeling of dense slurry flow in oil sands coarse tailings (CT) pipelines. In *Proceedings of CORROSION 2018* (pp. 1–14). Phoenix, AZ: AMPP. https://doi.org/10.5006/C2018-10637.
Zadeh, S. A. (2023). Erosion–corrosion. In *Introduction to Corrosion-Basics and Advances*. IntechOpen. https://doi.org/10.5772/intechopen.109106.
Zhang, S., Hu, K., Liu, X., & Yang, Y. (2021). Corrosion-erosion mechanism and research the prospect of bare materials and protective coatings for a power generation boiler. *Acta Metall Sin*, *58*(3), 272–294. https://doi.org/10.11900/0412.1961.2021.00464.
Zhao, J., Song, C., & Li, G. (2022). Fast-charging strategies for lithium-ion batteries: Advances and perspectives. *ChemPlusChem*, *87*(7), e202200155. https://doi.org/10.1002/cplu.202200155.
Zhou, H., Ji, Q., Liu, W., Ma, H., Lei, Y., & Zhu, K. (2022). Experimental study on erosion-corrosion behaviour of liquid-solid swirling flow in a pipeline. *Materials & Design*, *214*, 110376. https://doi.org/10.1016/j.matdes.2021.110376.

3 Head Losses Associated with Slurry Flows

Jashanpreet Singh, Jatinder Pal Singh, Virinder Kumar, and Miroslaw Szala

3.1 INTRODUCTION

Coal-based power plants serve as the predominant source of electricity generation in India. Automotive/locomotives and conduits are the two methods by which coal is transported in thermal power facilities. According to Woodcock and Mason (1987), pipelines are widely recognised as an effective means of conveying bulk solid materials across short and long distances with the help of carrier fluids. Depending on the specific industrial application, solid materials are conveyed using various carrier fluids such as water, oxygen, or oil (Li et al. 2002). The widespread use of pipelines for transporting solid materials is attributed to their year-round availability, economic efficacy, reliability, and low maintenance requirements. When a pipeline system is operated optimally, it minimises energy consumption and has minimal environmental impact (Singh et al. 2016a). In various industrial applications, pipelines convey solid particulates through carrier fluids. Typical applications include transporting mineral ores in the mining sector, disposing of mineral residues in processing facilities, and handling fly and bottom ash in thermal power plants (Williams 1983).

A significant aspect of solid material handling in thermal power plants involves disposing of solid waste from coal combustion. Using coal in thermal power plants amounts to high quantities of solid wastes, primarily as fly and bottom ashes (Singh et al. 2016b). The proper disposal of the large amount of fly ash and bottom ash yielded by the coal-fired power plants has become a main environmental issue. While fly ash has found increasing applications in sectors such as cement manufacturing, road embankment construction, and high-volume composite materials, the utilisation of bottom ash remains minimal. Currently, bottom ash is primarily confined to landfilling practices. Despite the stabilisation in coal ash generation following coal combustion, both bottom ash and fly ash are generally transported to designated ash dykes through pumping and pipeline systems situated far from the power station (Kumar et al. 2014). This results in substantial energy usage for operating slurry pumps, necessitating the development of a cost-effective slurry disposal system. Two-phase solid–liquid flows often exhibit complex behaviour, primarily arising from the interactions between solid particles, which form internal structures and result in intricate flow behaviour.

This chapter provides an extensive literature review that explores previous research works. The studies reviewed cover a wide range of topics such as the physico-chemical and characteristics, leaching patterns, and flow behaviour of various slurries. Also, the experimental studies of pressure drop, the influence of additives on pressure reduction, and computational studies concerning the slurry flow are considered. These studies are explored in depth in the subsequent section.

3.2 RHEOLOGICAL PROPERTIES OF DIFFERENT MINERAL AND WASTE SLURRIES

Rheology is the study of the flow properties and how fluids respond to external forces. Rheological properties of the solid–liquid slurries are critical for accurately designing and evaluating the performance of the slurry pipelines. Several flow characteristics also affect the slurry rheology, which

DOI: 10.1201/9781003561156-4

depends on parameters like solid concentration, porosity, fluid viscosity, specific gravity, shear rate, and water absorption capacity. Several researchers (Zhu et al. 2013; Panda and Swain 2014; Pani et al. 2015) have explored the rheology of multi-particulate slurries at varying solid concentrations under conditions with and without the addition of chemical additives. Variations in shear rate significantly influence the slurry's microstructure, leading to progressive dispersion of solid particles within the liquid medium. As shear rate increases, complete dispersion of the solid phase tends to occur. Figure 3.1 schematically represents the effect of shear rate on particle dispersion.

The rheological properties of various slurries have been examined by several researchers (Mooney 1951; Roscoe 1952; Williams 1953; Einstein 1956; Rutgers 1962; Landel et al. 1965; Thomas 1965; Krieger 1968; Schrick et al. 1972; Wilson 1982). Numerous studies have examined the rheological properties of coal–water slurry suspensions (Shao et al. 2012; Ganeshan and Panda 2017). The flow behaviour of slurries is governed by factors such as particle size and shape, solid concentration, shear rate, yield stress, and the physicochemical characteristics of the liquid phase (Moreland 1963; Horsley 1982; Sive and Lazarus 1987).

Bournonville and Nzihou (2002) investigated the influence of solid content, yield stress, and related parameters on the rheological properties of fly ash produced from solid municipal trash incinerators. During the experiments, they varied solid concentration from 35–55% (by weight). Rheological measurements were performed using a parallel plate RS150 rheometer (shown in Figure 3.2). The fly ash–water slurry exhibited shear-thinning and thixotropic characteristics throughout the testing.

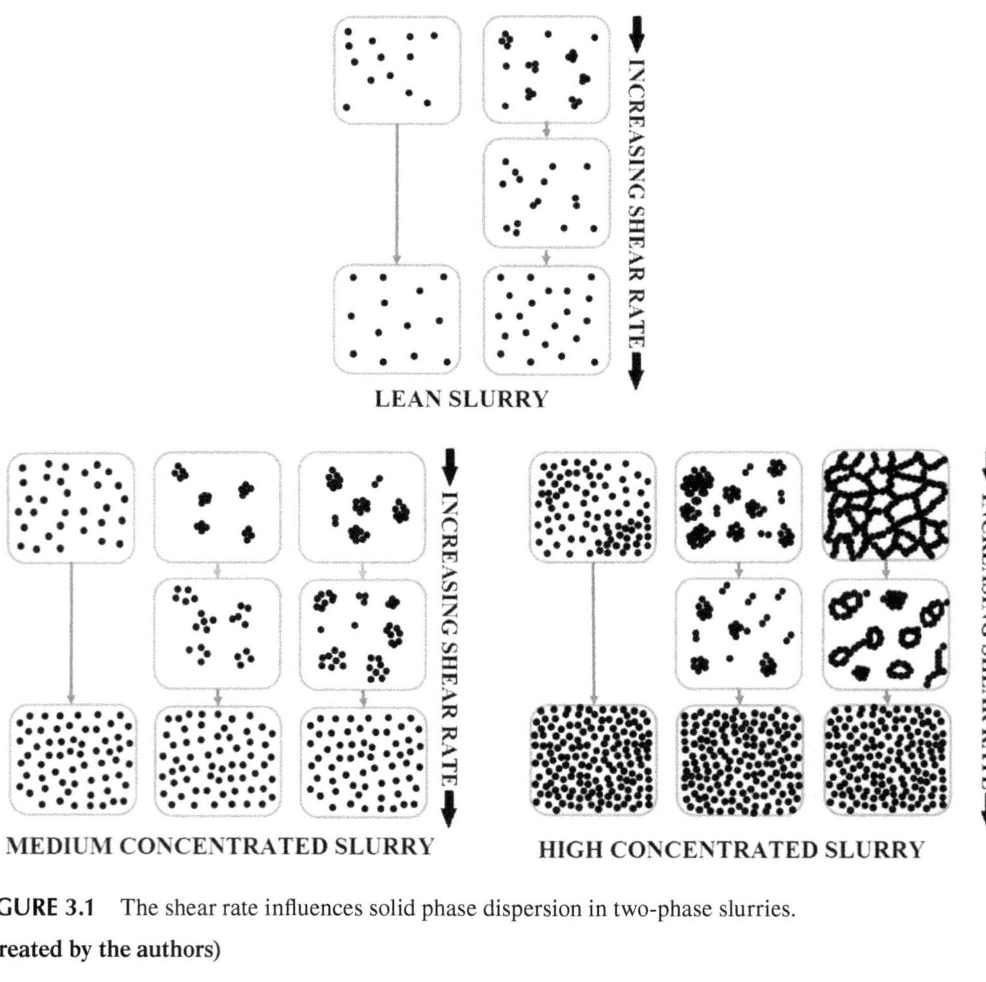

FIGURE 3.1 The shear rate influences solid phase dispersion in two-phase slurries.

(Created by the authors)

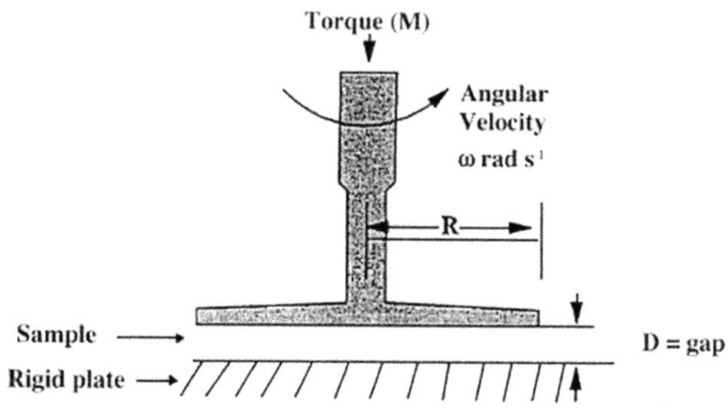

FIGURE 3.2 Schematic geometry of plate–plate flow rheometer.

(Bournonville and Nzihou 2002)

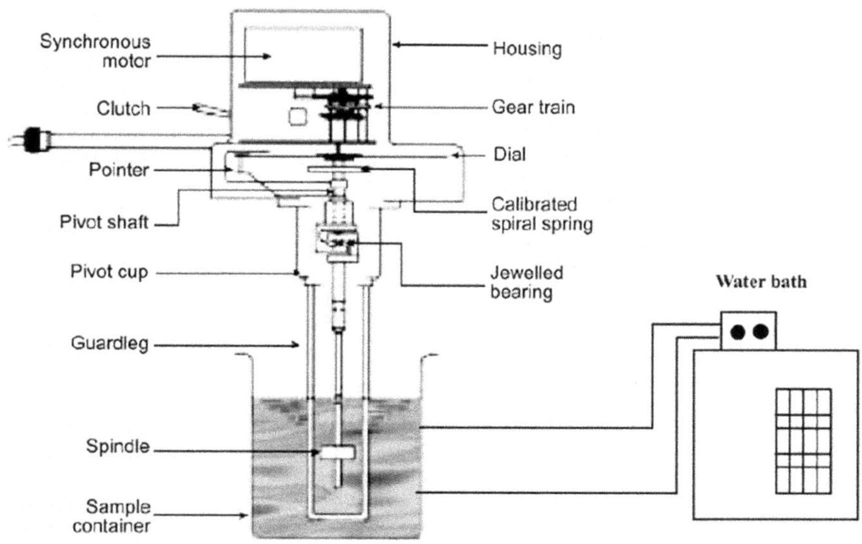

FIGURE 3.3 Schematic of the viscometer setup.

(Yoon et al. 2009)

Kwak et al. (2005) have considered the effect of concentrations relating to the rheological behavior and depositional processes of tailings and kaolinite suspension by carrying out a special flume experiment to simulate the conditions of pasted deposition. The findings indicated that residual stress based on a conventional cyclic loading–unloading stress time curve was ineffective in giving a reliable estimate of yield stress. Instead, a more accurate and repeatable estimate of the measurements was obtained at the transition point of the rheological behavior between elastic and viscoelastic regions along with the corresponding maximum stress value. Moreover, both were found to be more precise and independent of the vane geometry used in the rheometer. Verma et al. (2006) studied the influence of particle-size distribution in the rheological behavior of coal ash suspensions, with the observation that fly ash suspensions show non-Newtonian rheological properties when the solid content in this suspension is more than 40% by weight. Continuing with such findings, Yoon et al. (2009) focused on how chemical additives influenced the rheology of coal–water slurries. Solid concentration varied between 60% and 70%, and 20°C to 50°C was used as the temperature for the experiments. Figure 3.3 is a schematic diagram of the viscometer used. Four additives, including methyl methacrylate, a compound formed out of polypropylene glycol, polypropylene, and sodium naphthalene, were examined in a concentration range of 0% to 0.8% by weight. Among these, sodium naphthalene was the best compound to enhance the flow behaviour of coal–water slurries with high solid contents.

Senapati et al. (2013) conducted an experiment to study the high solid concentration (62–65 gm/L in weight) rheology of fly ash-water slurries and determined that their flow behaviour and nature were that of a pseudo-plastic rheology. Panda and Swain (2014) conducted an in-depth study of the rheological properties of coal–water slurries with the help of low-grade Indian coal, putting a quantitative relationship between solid concentration and apparent viscosity. Their values indicated that viscosity rises exponentially as the solid content is more than 50% by weight. In a later investigation, Lee et al. (2017) used a viscometer test to describe the rheological characteristics of tailings slurries mixed with fly ash. They formulated predictive modelling of rheology behaviour and checked it against the root mean square error (RMSE). They found that increasing the amount of fly ash induced higher shear stress, and increased temperature resulted in reduced shear stress values. Kumar et al. (2017a) evaluated how various additives altered bottom ash rheology findings, finding that sodium sulphate was more effective than Henko detergent. Notably, a 0.4% load of sodium sulphate significantly increased the flow properties of bottom ash.

There have been widespread studies focused on applying chemical additives in the rheological modification of slurries. Mosa et al. (2008) evaluated the influence of sulphonic acid, sodium tri-polyphosphate, and sodium carbonate on the rheological behavior of coal–water slurries, concluding that sulphonic acid is the best agent to alter the coal water slurry rheology by enhancing the flow property. Senapati et al. (2008) studied the rheological characteristics of coal fly ash and bottom ash slurries. Results showed that the power law model gave the best fit when compared to the viscosity of the bottom ash slurry, but the fly ash slurry exhibited pseudo-plastic flow behaviour. Chandel et al. (2009) examined the rheological behavior of fly ash-water slurries in the presence of chemical additives. Their study showed that adding sodium hexametaphosphate notably decreased the apparent viscosity of the slurry. Zhu et al. (2013) assessed how starch-based dispersants influence the physical and chemical characteristics of coal–water slurries. They found that a small amount of additives may significantly lower the apparent viscosity of the mixture. Styrene was the most effective in lowering viscosity among the three tested dispersants, styrene, naphthalene sulfonate, and natural sodium lignosulfonate. On the other hand, its addition resulted in the coal–water slurry having an enhanced tendency to sediment in static conditions.

Pani et al. (2015) completed a research study regarding the rheological characteristics of Jindal Power, Rajgarh-generated fly ash through fly ash–water slurries ranging from 50% to 60% weight concentration. Wetting agents, Ghari detergent powder and sodium silicate, were added to the slurry to evaluate their impact on the flow characteristics that followed their addition. The findings confirmed that the slurry had Bingham plastic behaviour, and apparent viscosity reduced after adding the two additives. Sodium silicate was the best agent in enhancing flowability. Slatter and Seddon (2015) studied the rheological flow of the coarse and fine varieties of tailings. In these experiments, the slurries were injected with two different varieties of flocculants. The experiment showed that the rheology of sample B (fine slurry) was also unchanged regarding both flocculants.

Very limited literature exists that can examine the hydraulic behaviour of bottom ash, fly ash, and coal–water slurry at high solid concentration. Since the transmission of high-concentration slurries has incurred increasing costs, there is a need to develop good methods of reducing the associated transport expenses. Moreover, the design of slurry-conveying pipes relies on various parameters, including the solid and liquid phases densities, the volumetric flow rate, and the overall flow behaviour. It is therefore imperative to have a detailed understanding of the rheological properties of these slurries when it comes to the successful design and the smooth operation of coal–slurry disposal systems.

3.3 EXPERIMENTAL STUDY ON HEAD LOSS IN SLURRIES

During the past decades, many experiments have been conducted to understand the head-loss behaviour of solid–liquid suspension under transportation in pipelines. Table 3.1 summarises the various slurry pipelines throughout the world. Various studies have assessed the head-loss characteristics of different slurry types using pilot-scale pipeline test loops (Mosa et al. 2008; Seitshiro et al. 2012; Ravelet et al. 2013). A conventional hydraulic conveyance system includes a pipeline loop; a large pump; and ancillary equipment, including flow meters, pressure transducers, and data-acquisition devices (Chandel et al. 2010). The experimental outcome of several researchers is summarised in the subsequent overview.

Some researchers have studied how the loss of head affects the flow of multi-particulate slurries at an initial stage of the research of slurry transportation (Sive and Lazarus 1987, Chandel et al. 2010; Sahoo et al. 2010). Bantin and Streat (1970) conducted an experiment examining dense-phase slurry in vertical and horizontal lines, showing that high-velocity transport is possible with minimal head loss within a limited pressure range. Mukhtar et al. (1995) conducted an experiment on head loss when an iron ore and zinc tailing leachate entered the 90° pipe bend. They observed that bend loss coefficients for slurry flows were lower than those for water, with bend losses increasing as flow

TABLE 3.1
Various Slurry Pipelines throughout the World

Solid Material	Location	Throughput (Million Metric Tons per Year)	Length (Miles)	Starting Year
Coal	Ohio, US state	1.2	105	1957
	Russia	4.0	7	1966
	France	1.5	6	1954
	Arizona	5.0	273	1970
	Japan	0.3	16	1965
Iron concentrate	Tasmania	2.3	53	1967
	Mexico (Pena Colorada)	1.6	30	1974
	Mexico (Las Truchas)	1.4	17	1975
	Argentina	1.9	20	1977
	Brazil	12.0	247	1978
	India	7.5	42	1980
	Mexico (La Perla)	4.5	237	1983
Limestone	Trinidad	0.5	6	1959
	England	1.5	57	1964
	California	1.4	17	1971
	Australia	2.0	15	1981
Phosphate concentrate	Brazil	2.0	70	1979
	Utah	1.8	94	1986
	India	0.1	7	1983
Iron–sand–copper concentrate	New Zealand	1.0	4	1971
	New Guinea	1.0	17	1972
	Indonesia	0.3	69	1973
	Arizona	0.4	11	1974
	New Mexico	0.7	7	1982
	Turkey	0.9	40	1973
	Utah	0.7	17	1987

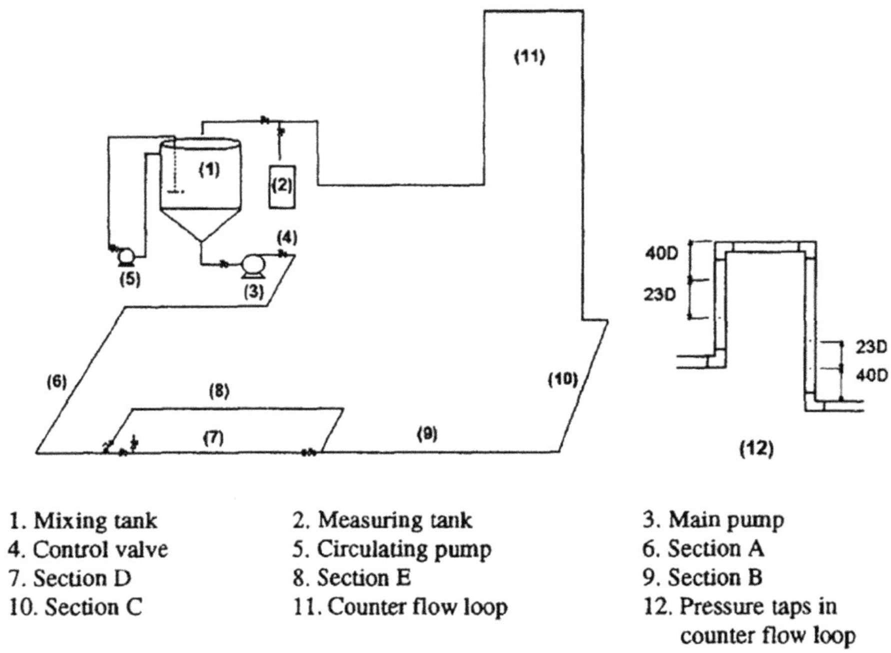

FIGURE 3.4 Pilot-scale test loop configuration used by Fangary et al. (1997).

velocity rose. Fangary et al. (1997) investigated how fine particles influence the flow behaviour of polydisperse phosphate slurries using a pilot-scale pipeline loop (as shown in Figure 3.4). The study utilised pipes with diameters of 1, 1.5, and 2 inches. The results showed that slurries containing finer particles experienced greater head losses than those with coarser particles, particularly at higher flow velocities.

The head loss is also reduced, even at greater solid concentrations, due to mixing coarser solid particles in fine particulate slurry. Mishra et al. (1998) analysed head loss behaviour in standard and modified pipe bends. Their results showed that while modified bends caused slightly higher head losses than conventional designs, those with an area ratio of 2 led to a minimal increase in head loss. These modifications substantially reduced material wear, highlighting a balance between energy efficiency and equipment longevity.

Kumar et al. (2000) investigated the head loss of coal ash–water slurry in a horizontal pipeline in a pilot-scale test loop installed at the National Capital Power Plant, Dadri, India. The experimental setup featured a 155-mm-diameter pipeline with a total length of 160 meters. Their findings demonstrated that it effectively transported fly ash and bottom ash at high solid concentrations up to 50 wt%. Adding bottom ash to fly ash slurry notably reduced head loss at elevated concentration. Additionally, high-concentration slurry flow helped minimise uneven erosion of the pipeline material. Li et al. (2002) have also shown stabilisation of fly ash–water slurry by adding several additives. In order to determine deposition stability, four agents were used: S-130 and S-194 (both rhamsan polysaccharides), vanzan, and carboxymethyl cellulose. S-194 showed the best stabilising activity at 0.2% weight level. The enhanced stability was attributed to forming an extensive network structure facilitated by the long molecular chains of S-194, which reduced particle settling. The study also proposed a slurry transport system integrated with an additive injection mechanism for improved performance.

Lu and Zhang (2002) investigated frictional losses in horizontal pipelines during the coal–water slurry flow with solid concentrations varying between 22.1% and 33.5% by weight. Their study proposed a generalized Reynolds number applicable to non-Newtonian slurry flows. This generalized Reynolds number was subsequently used to determine the friction loss factor, providing a more accurate characterization of flow resistance in such complex fluid systems. Kumar et al. (2003) critically examined the head loss in fly ash–water slurries flowing along the horizontal pipes on a pilot scale. The experiment was performed in a 160-meter test loop (Figure 3.5), which covered solid concentrations between 10% and 60% by weight. The results demonstrated a direct relationship between solid concentration, flow velocity, and head loss. Furthermore, incorporating bottom ash particles (coarse) into the fly ash slurry significantly reduced the specific energy needed to transport the slurry. Most preferred energy potentials were attained when the bottom ash to fly ash ratio held any value between 4:1 and 3:2.

Kaushal et al. (2005) investigated how a slurry containing glass particles affects head loss in horizontal pipelines. The study utilised glass beads with mean sizes of 440 and 125 μm. Experiments were conducted using a 22-meter-long test setup (see Figure 3.6), focusing on conditions with a 50% solids concentration and a flow speed of 5 m/s. Findings showed that, in general, higher slurry concentrations resulted in reduced head loss, although this pattern did not persist at the 5 m/s velocity. Additionally, the slurry containing 440-μm beads exhibited flow characteristics consistent with a sliding bed regime.

Verma et al. (2006) analysed the behaviour of the head loss of a horizontal pipeline of 50-mm diameter and 90° bend shape made to transport a fly ash and water combination. In the study, solid concentrations were between 50% and 65% and flow velocities between 1.0 and 3.5 m/s. The results indicated systematic growth of relative head loss in both high flow velocity and solid concentrations, especially in the area near the bend. In a separate study, Chen et al. (2009) analysed how particle concentration, fluid temperature, and particle size influenced the slip velocity in coal–water slurry systems. The experiments were conducted with solid concentrations ranging from 65.3% to 68.2% (by weight). The investigation employed three conduits with 25-, 32-, and 50-mm diameters. Figure 3.7 illustrates the pilot setup schematic diagram. They discovered that the slip velocity is reduced as the solid concentration, particle size, and slurry temperature are increased. Vlasak and Chara (2009) studied the flow dynamics and head loss characteristics of various slurry mixtures. Their findings indicated that the slurries began to exhibit Newtonian fluid characteristics at higher solid concentrations.

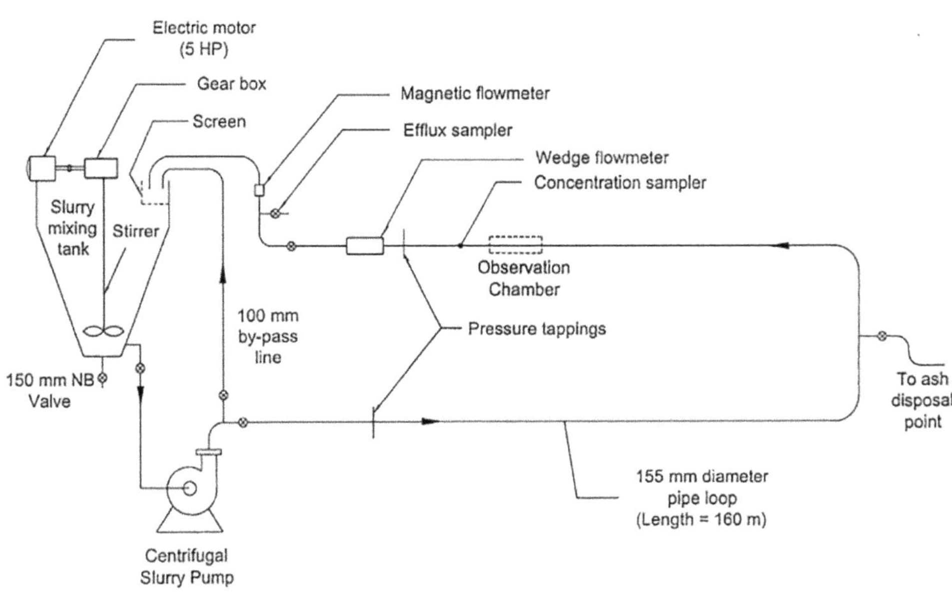

FIGURE 3.5 Pilot-scale test loop configuration used by Kumar et al. (2003).

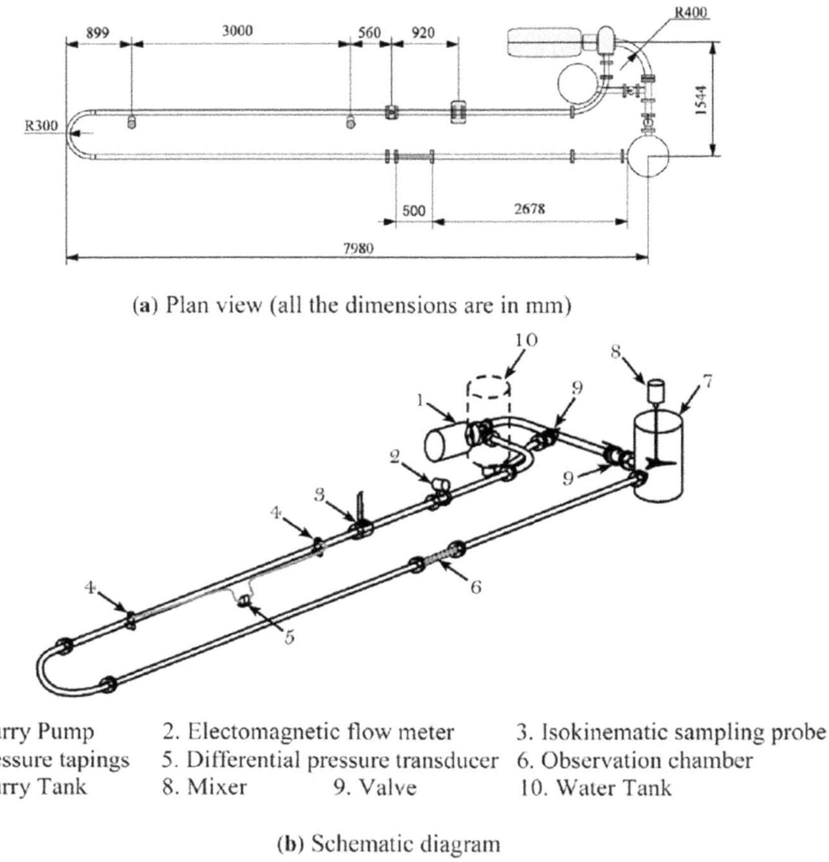

FIGURE 3.6 Pilot-scale test loop configuration used by Kaushal et al. (2005).

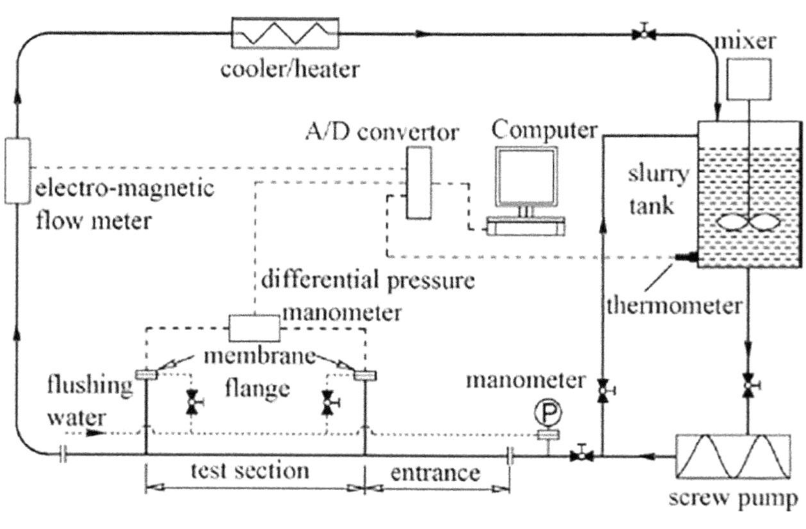

FIGURE 3.7 Pilot-scale test loop configuration used by Chen et al. (2009).

Chandel et al. (2009) examined the influence of additives in modifying the rheological characteristics and the calibre-based head losses to a fly ash–water slurry moving through a straight pipe 40 mm in diameter under comparatively high regimes of solid concentration. Their results revealed that Darby and Melson's (1981) model showed strong agreement with the experimental observations.

Chandel et al. (2010) carried out a study to clarify the head-loss performance of coal–ash slurries with large solids moving in horizontal pipes. The slurry was a coal–ash particle suspension in water; arranged at 50% to up to 70% (by weight) solid concentration; the experiment consisted of utilising a pilot-scale test loop, 50 meters in length, with an internal diameter of 42 mm (as shown in Figure 3.8). They stated that the concentration of solids and the flow velocity were positively correlated with the rate of cranium loss. Additionally, the effective conveyance of slurry was achieved by combining fly and bottom ash in a 4:1 ratio.

A reduction in specific energy consumption was observed up to a solid concentration of 65%. Vlasak and Chara (2011) analysed the flow characteristics and head loss in three types of slurries: sand, stony dust, and a blend of both, across solid concentrations ranging from 6% to 51%. The experiments were performed using a laboratory-scale prototype loop, as shown in Figure 3.9. Their findings indicated that slurries containing coarser particles exhibit higher head losses than finer ones. Moreover, a coarser mix, which included finer sand particles, reduced hydraulic loss in transport by as much as 50%.

Slatter (2011) introduced an innovative modelling technique to characterise fluid suspension, focusing on clay-based slurries at solids up to 10% by volume. The procedure highlighted the relevance of the dynamic similarity of the fine particulate suspensions. Eltoukhy (2013) tested the head loss in a mix of sand and water that passed through a straight pipe with a 50.8-mm diameter. The findings showed that as the concentrations of solids rose and the irregularities on the pipes rose, the head loss rose. The study also decided that the slurries with finer sand particles had a higher head loss than those with coarse particles. Hashemi et al. (2014) examined the process of modifying the behaviour of sand–water slurries because of changes in solid concentration and flow velocity. The experiments were carried out in a pilot-scale test facility using solid concentrations between 20% and 35% by weight, with a flow velocity of 2.5 m/s. The study concluded that particle–wall interactions at higher solid concentrations led to increased turbulence fluctuations in the region near the pipe wall.

Wu et al. (2015) carried out experimental and computational analyses investigating information on the head loss character of cement-fly ash slurry during its flow through a horizontal pipeline. The pilot-scale loop (120 mm in diameter) where the testing was carried out was 35 meters long. Their results showed that when the solid concentration was within 78% and 79.5% by volume, the slurries

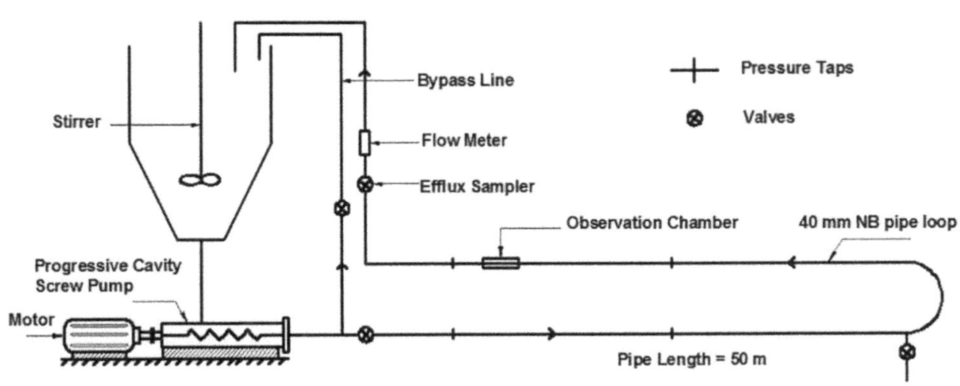

FIGURE 3.8 Pilot-scale test loop configuration used by Chandel et al. (2010).

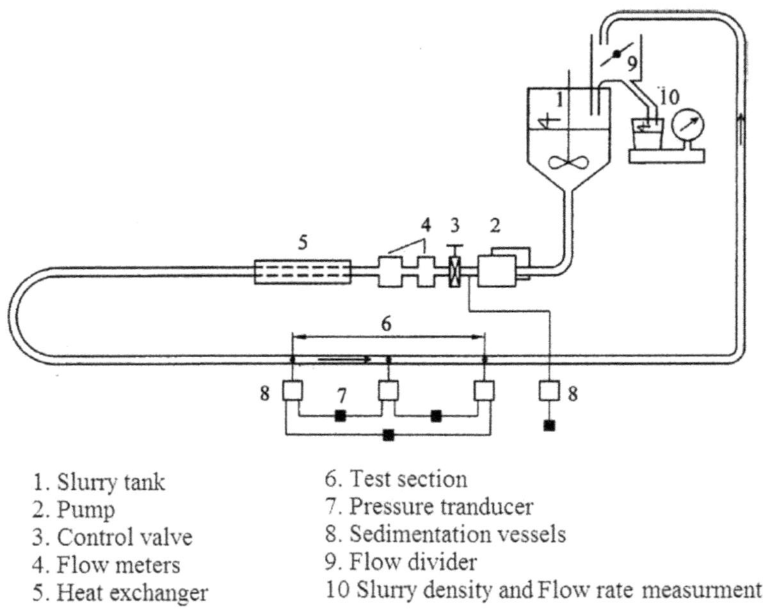

FIGURE 3.9 Pilot-scale test loop configuration used by Vlasak and Chara (2011).

were of the character of Bingham plastic. There was also a markedly high value of head loss as the volume of slurry increased. Senapati and Mishra (2015) conducted a further experimental study to investigate the rheological characteristics and head-loss behaviour of a fly ash–water slurry by including bottom ash. Their findings indicated that adding coarser bottom ash particles to fine fly ash slurry reduced apparent viscosity and the head loss generated as it flowed through the pipeline.

Aleksandrov et al. (2017) conducted a methodical study of the performance profile of polyurethane-coated pipelines compared to traditional metal pipes (steel). Putting more focus on the operational costs of the two systems, they used definite energy computations to measure performance. The findings revealed that the pipelines lined with polyurethane have the best effectiveness over conventional steel pipes. The flow behaviour and the head loss of fly ash water slurry to which bottom ash was added were investigated by Kumar et al. (2017b). Their research exhibited that bottom ash is more effective as an additive to reduce head loss than sodium sulphate and detergent, which only showed tangible advantages.

The transportation of heavy solid materials via pipeline has become a more managed and popular technology, as outlined in recent literature. Recently, this technique has achieved prominence, particularly in treating high-solid content slurries, since it is efficient in operation and less hazardous to the environment (Jacobs 2014; Bain and Bonnington 1970). Hydraulic transportation of viscous products has many advantages. The pipeline systems are covered and reduce the dust released, taking up less area and requiring fewer people to operate them effectively. Nevertheless, I-PES systems require high-performing devices such as superior control modules, flow meters, pressure gauges, and compressors to streamline their efficiency and effectiveness (Schaan et al. 2000; Wu et al. 2015; Qi et al. 2018). Solid particulate transport in pipelines has many advantages compared to traditional modes, which are mainly attributed to its ability to reduce pollution to the environment. Furthermore, pipelines may be located in complex geographical areas, which lowers the traffic, requires minimum maintenance, and operates throughout the year Charles and Redberger 1962; Singh et al. 2018a).

A review of experimental studies on pressure drop shows that head loss during slurry suspension flow through pipelines is a complex phenomenon. Additionally, current research offers limited insight into head loss behaviour within pipe bends. As a result, there should be an additional study on the head loss behaviour of the coal–water slurry flow, especially on curved or non-linear conduit sections.

3.4 EVALUATION OF HEAD LOSS THROUGH COMPUTATIONAL METHODS

Estimating head loss in slurry pipelines differs significantly from single-phase flow due to the presence of solid particles, which can substantially affect the overall flow behaviour and dynamics. Numerous researchers have explored this through pilot-scale setups and experimental studies to understand head loss behaviour in slurry transport systems (Biswas et al. 2000; Gandhi et al. 2001; Verma et al. 2006; Chandel et al. 2009). The gradual advancement of computational solutions such as computational fluid dynamics (CFD) has reduced the cost and effort of studying complex flow behaviours to a great extent. The CFD simulations are respectively recognised as a cost-effective and efficient research tool in solving complex fluid flow problems (Wang et al. 2013). Simulations of head loss in conduits may provide several valid estimates of fluid behaviour, which can be used in a piping system design and optimisation procedure. An intensive literature search was conducted looking into the aspects of head loss of diverse types of slurry. Various authors have concentrated on examining the loss of head in straight pipeline length and at the pipe bends, especially when the pipeline carries different solids (Mishra et al. 1998; Verma et al. 2006; Chandel et al. 2010).

Different authors (Ling et al. 2003; Kaushal et al. 2012; Ma et al. 2015) have examined numerical simulations of the behaviour of head loss in horizontal pipelines that convey solid–liquid mixtures. Ling et al. (2003) used the RNG model k-mode of the turbulence to analyse the flow dynamics of a silica/sand–water slurry in a horizontal conduit. A three-dimensional algebraic slip multiphase model promoted a fully developed flow regime. The governing equations were computed using the control volume, combined with a block-structured mesh, which was used to discretise the computational area (as illustrated in Figure 3.10). Their simulation outcomes were validated against experimental observations, demonstrating strong agreement. Furthermore, the post-processing capabilities of CFD provided more profound insights into the behaviour of the two-phase flow system.

Lahiri and Ghanta (2008) performed a CFD simulation to study the behaviour of zinc tailings mixed with water. Their numerical findings were validated using experimental results provided by Kaushal and Tomita (2002). A two-dimensional CFD procedure for a horizontal pipeline was estimated and tested by integrating the Euler–Euler multiphase formulation with the standard k-epsilon turbulence model. The outcomes demonstrated good correlation with the experimental data. Their study also indicated that the settling velocity and drag coefficient are influenced by local concentration variations, particularly across the vertical cross-section of the pipe.

Chen et al. (2009) investigated the head-loss behaviour in the coal–water slurry that flowed in horizontal pipes. When they integrated the Eulerian multiphase turbulence model with the RNG k-e approach, their numerical simulations correlated well with existing experimental data. The

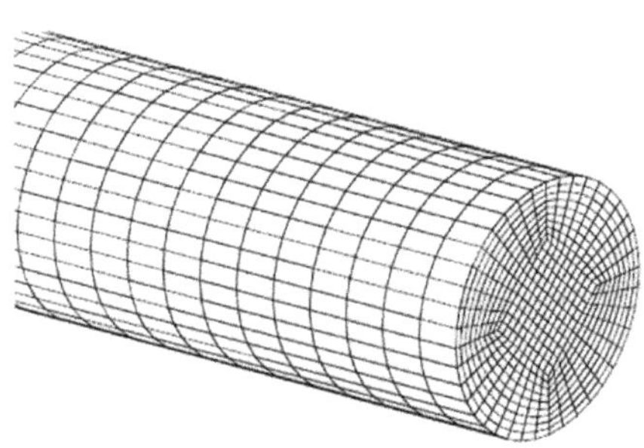

FIGURE 3.10 Unstructured horizontal pipe meshing by Ling et al. (2003).

study aimed more at the effect of particle size and flow velocity on the head loss. It was observed that at higher solid concentrations, particles exhibited a more uniform spatial distribution. Additionally, fine particles showed more consistent vertical dispersion than larger ones. Their findings highlighted that, beyond gravity, particle–particle interactions significantly affected the distribution of the concentration in solids and the profiles of axial velocities. Similarly, Ekambara et al. (2009) utilised CFD to evaluate the flow dynamics of a solid–liquid slurry in a horizontal pipeline. Their simulations covered many parameters, including solid volume concentrations between 8% and 45%, particle sizes varying from 90 to 500 µm, and pipeline diameters spanning from 50 to 500 mm. The CFD results were validated against experimental data compiled from multiple sources (as presented in Table 3.2), showing excellent agreement. They concluded that solid loading, flow velocity, and particle size influenced the solid volume fraction distribution within the pipeline. Eesa and Barigou (2009) also conducted a numerical investigation into solid–liquid slurry flow using ANSYS CFX 10.0. A numerical simulation was carried out to examine the hydraulic conditions of a 45-mm line carrying a dense solid–liquid suspension (41 vol. % solids content) whose particles had diameters between 5 and 10 mm. The Eulerian/Eulerian modelling framework was used to model the highly non-linear particle/fluid interactions.

The numerical outcomes closely matched the simulation data, revealing that the highest concentration of solid particles occurred near the bottom of the pipeline. Interestingly, the maximum particle velocity was not located along the pipe's centreline but appeared in its upper region. Lahiri and Ghanta (2010) proposed a generalised CFD-based flow model for solid–liquid transport, which they validated using experimental data from Kaushal et al. (2005), involving a glass particle suspension. The simulations employed the Euler–Euler multiphase framework alongside the standard k–ε turbulence model. A simulations were performed using FLUENT 6.2, and the authors concluded that the developed model effectively captured the flow behaviour under the tested conditions.

TABLE 3.2
Experimental Data Sets Used for Confirmation of Numerical Results

						Measuring Equipment		
Investigators	**Pipe Diameter (mm)**	**Particle Size (µm)**	**Solid Concentration (% by volume)**	**Specific Gravity of Solids**	**Flow Velocity (m/s)**	**Head Loss**	**Solid Concentration**	**Velocity**
Roco and Shook (1983)	50.7 51.5 263 495	165 480 520 1300	6–35	2.65	1.5–4.5		γ-ray absorption	Electromagnetic flow meter
Schaan et al. (2000)	50 150	85 90 100	15–45	2.65	0.8–5.0	Pressure transducers	γ-ray absorption	Electromagnetic flow meter
Gillies and Shook (2000)	105 264 495	420 420	26–47	2.65	1.8–5.8	Pressure transducers	γ-ray absorption	Electromagnetic flow meter
Gillies et al. (2004)	103	90 270	10–45	2.65	2.0–8.0	Pressure transducers	γ-ray absorption	Electromagnetic flow meter
Kaushal and Tomita (2002)	54.9	125 440	5–50	2.47	1.0–5.0	Pressure transducers	γ-ray absorption	Electromagnetic flow meter

Kaushal et al. (2012) conducted an experimental study devoted to the flow behaviour of a suspension of glass beads in water through a horizontal pipeline, with the numerical analysis executed by using the k-omega shear-strain transport (SST) turbulence model. They also compared experimental observations with the results of their efforts to justify their simulation results. The solid concentrations varied between 0% and 50%, whereas the superficial velocity of fluids remained at 5 m/s throughout the experiment. According to their findings, the Eulerian approach matched the experimental data more closely than the mixture model. Figure 3.11 is a three-dimensional plot of the solid concentration at 40% volume fraction at various flow velocities. The authors also concluded that the SST k-omega model is the most suitable for simulating flows with high particle concentrations. In a study by Kumar et al. (2012), a CFD simulation was conducted to examine two-phase slurry flow in a horizontal pipeline. Their model involved a solid concentration of 30% and an average particle diameter of 440 μm. A non-uniform, unstructured mesh structure was used to establish the computational domain and study the flow dynamics; a fully coupled RNG k–e turbulence model and the algebraic slip model were used. The simulation results demonstrated a high level of consistency with the experimental data. Messa and Malavasi (2013) characterised the

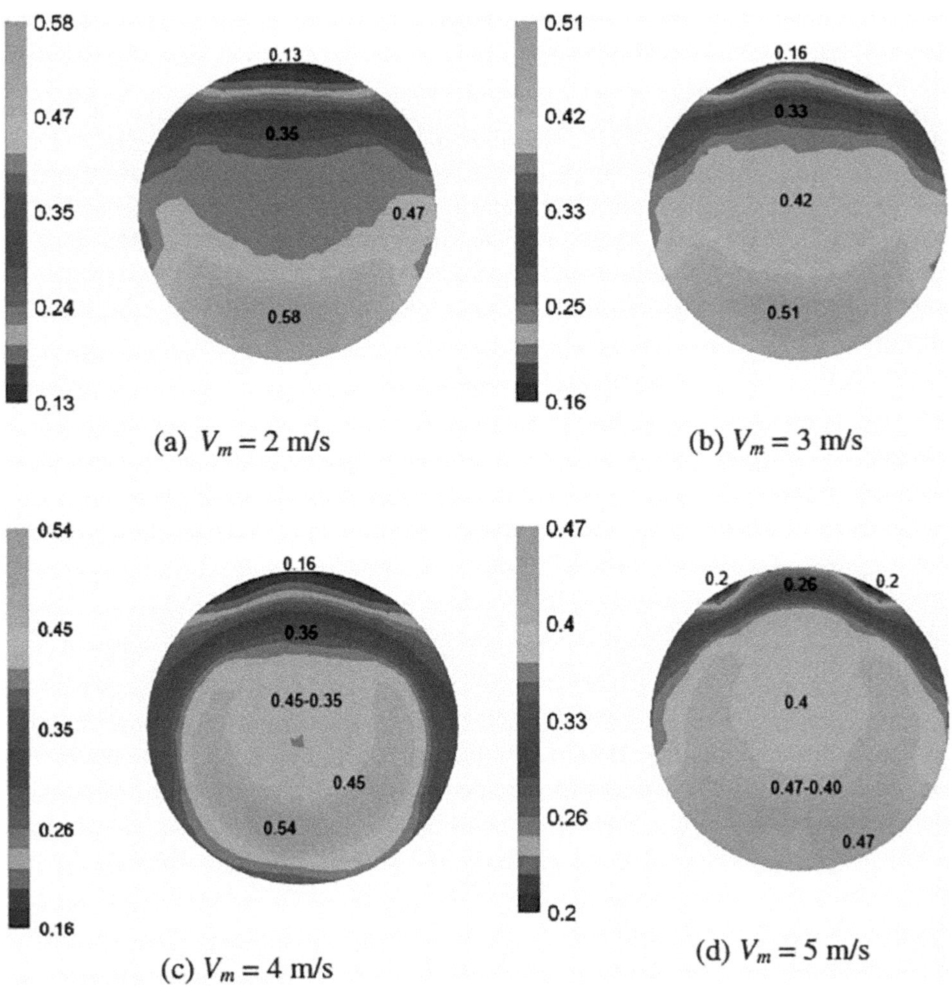

FIGURE 3.11 Concentration contours at 40% volume fraction.
(Kaushal et al. 2012)

flow of solid–liquid slurry in an upward-sloping opening through numerical simulations with the subsequent variation of the volume percentage between 5% and 20%. Their results showed that the observations made during the experiment were closely related to the ones computed with the help of the Eulerian multiphase system. Kim and Han (2013) also conducted CFD simulations to study the flow of a sand–water mixture through a horizontal pipeline. They utilised the Eulerian granular multiphase framework combined with the k–ε turbulence model and employed a non-uniform hexahedral mesh for domain discretisation. The sand had a density of 2650 kg/m³, with an average particle diameter of 0.54 mm. Their discussion showed that head losses in the conduit were greater at high particle concentrations and low Reynolds numbers. Similarly, Nabil et al. (2013) conducted a report on numerical simulation studies on sand–water slurry pipeline flow using a pilot-scale experimental validation. The authors developed a model using an Eulerian multiphase approach coupled with the k–ε turbulence method to examine the particles of diameters 0.2, 0.7, and 1.4 mm across the velocity scale of 0.5 to 5.0 m/s and solid volume concentrations of 5% to 30%. The simulation and the experimentally observed data agreed demonstrably, although there was specific agreement in the medium and coarse particles in the lower concentration values (5–10%). Pinto et al. (2014) conducted a steady-state CFD study incorporating population balance models alongside a two-phase Eulerian framework with a k–ε turbulence model. Their research focused on modelling large-scale tailing thickeners with a diameter and height of 120 m. The study concluded that increased turbulence intensity leads to both a reduction in the torque range and fragmentation of particle aggregates.

Gopaliya and Kaushal (2015) presented an experimental study to investigate the flow behaviour of sand–water slurries through the horizontal pipeline with specific attention to the effects of particle size. Solid matter concentrations between 15 and 45% by volume and four diameter ranges (0.18, 0.29, 0.55, and 2.4 mm) were studied. It was a GAMBIT-designed pipeline with an inner diameter of 53.2 mm and a length of 2.7 m. The computational mesh comprised hexahedral elements, with simulations conducted until a convergence threshold of 10^{-3} was achieved. They reported that the head loss increased with the concentration of solids and flow velocity, regardless of the particle size.

Ofei and Ismail (2016) studied the influence of particle size distribution and solids concentration on the velocity distributions, concentration profiles and frictional gradients within a horizontal two-phase slurry flow. A comparison between various CFD methods demonstrated that the Eulerian model combined with a k–ε model of turbulence was the most adequate approximation of two-phase systems. They also reported in their research the role of alterations in flow parameters of the process of agglomeration of particles and possible wear on the pipeline infrastructure. Rawat et al. (2016) studied how coal ash affects head loss during transport through a straight pipeline. They validated several turbulence models by simulation using ANSYS FLUENT and found that the SST k-omega model delivered the closest result to what was seen experimentally. Their findings showed increased discharge velocity led to a corresponding rise in pressure (or head) loss. To analyse the transport behaviour of a backfill slurry in a pipeline system, Chen et al. (2017) created a three-dimensional CFD model. They also demonstrated by simulations that the three-phase foam slurry backfill (TFSB) could ensure a stable flow state during the conveyance system. Che et al. (2017) adopted an experimental approach in the laboratory and at pilot scale to study the nature of deposition of iron ore residue paste. They came up with the conclusion that dry-surface spreading requires a slurry concentration greater than 70%. Besides, the material deposited at tailings storage facilities (TSFs) exhibited uniform flow behaviour with low segregation, yielding a 2.0–2.5% repose angle. Mishra et al. (2017) applied CFD to simulate the fluid transport of coal and copper-ore slurries through a straight pipe. The range of particle size of the coal was 70.5–275.7 μm, and the range of particle sizes of copper ore was 40.1–278.5 μm. The slurry volume fraction was maintained at a span between 0.08 and 0.19. Their findings showed maximum head loss was experienced when coarse copper ore or fine coal particles were introduced into the mixture.

The use of CFD simulations has been universally recognised as a cost-effective and time-saving technique for solving most complicated fluid movement situations, as seen by Gandhi et al. (2001). There can be numerical modelling of head loss in piping systems that can be useful in knowing the behaviour of fluids and thus helps in the design and optimisation of the pipelines. The studies have aimed to model head loss in slurry pipelines and cutting corners where solid–liquid flow poses special complications. Hossain et al. (2011) investigated how the solid particles are deposited in horizontal conduits under turbulent flow conditions. They used FLUENT 6.2 and researched through a three-dimensional helical shape with four 90° turns. The results proved that the deposition tendency was significantly affected by the size of particles and flow velocity; the deposition of larger particles was greater than that of smaller. The maximum particle accumulation was 60, where the downstream segment of the bend was, whereas in the upstream part, deposition mainly occurred at the bottom of the pipe bend.

Numerical simulation was adopted by Ma and Zhang (2012) to study the behaviour of head loss in tetrabutylammonium bromide slurry in pipes of various element formats such as straight pipes, bends, abrupt contractions and expansions. The concentration of slurry was either 10 or 30% weight basis. Their findings indicated that loss coefficients of bends, expansions, and contractions decreased with increase in the flow velocity. On the contrary, the change in mass fraction caused the increase in the values of loss coefficients,

CFD has been used by various researchers (Ling et al. 2003; Pinto et al. 2014) to simulate and forecast the head loss in both the horizontal and the vertical pipelines that carry different types of slurries such as sand and water, coal and water, fly ash and water and mineral ore and water mixture. Using the literature, it is common to indicate that the Eulerian multiphase model is the most suitable one to be used to model dense slurry flows (Kaushal et al. 2012). This model successfully portrays the interactions between solid-to-solid particles and fluid-to-solid and wall-to-solid impacts. Its application mainly entails the physical characteristics of the solid phase and the carrier fluid. Further, past studies highlight that among the various turbulence modelling strategies, the RNG k–ε and SST k–ω models have proved the most reliable for simulating the flow behaviour of high-concentration slurries (Rawat et al. 2016).

3.5 CONCLUSION

This chapter has offered an in-depth insight into the head loss behaviour of solid–liquid slurry flow, focusing specifically on coal, fly ash, and bottom ash suspensions through pipeline systems, with special attention to 90° pipe bends. The literature survey revealed significant gaps, particularly in predicting head loss in bends, the influence of additives, and the general lack of universal models applicable across different slurry types and geometries. Experimental studies have consistently shown that head loss in slurry pipelines rises with increasing solid concentration and flow velocity. However, the strategic blending of fly ash with bottom ash and the use of specific additives demonstrated potential in mitigating these losses, enhancing slurry stability, and reducing energy consumption. Notably, the slurry behaviour in pipe bends was highly sensitive to geometrical factors, especially the ratio of bend radius to pipe diameter (r/D). CFD simulations were carried out using ANSYS Fluent to address the constraints associated with physical experimentation. In conjunction with the SST k–ω turbulence model, the Eulerian multiphase model was validated against experimental observations and demonstrated strong predictive capability in capturing flow behaviour and optimising pipe bend configurations. The simulation outcomes corroborated experimental trends and provided a more detailed understanding of flow dynamics and particle distribution within the slurry. This study enhances the current understanding of the complex dynamics associated with slurry flow and establishes a robust experimental–numerical framework for designing efficient slurry transport systems. The findings offer practical insights for optimising pipeline geometry, minimising head loss, and enhancing the overall operational efficiency of slurry handling in industrial settings.

REFERENCES

Aleksandrov, V.I., Vasilyeva, M.A. and Pomeranets, I.B., 2017, October. Estimation of efficiency of hydrotransport pipelines polyurethane coating application in comparison with steel pipelines. In *IOP Conference Series: Earth and Environmental Science* (Vol. 87, No. 2, p. 022001).

Bain, A.G. and Bonnington, S.T., 1970. *The Hydraulic Transport of Solids by Pipeline* (No. BOOK). Pergamon Press.

Bantin, R.A. and Streat, M., 1970. Dense-phase flow of solid-water mixtures in pipelines. *Proceedings of Hydro transport, 1*, pp. 1–24.

Biswas, A., Gandhi, B.K., Singh, S.N. and Seshadri, V., 2000. Characteristics of coal ash and their role in hydraulic design of ash disposal pipelines. *Indian Journal of Engineering and Material Science, 7*, pp. 1–7.

Bournonville, B. and Nzihou, A., 2002. Rheology of non-Newtonian suspensions of fly ash: effect of concentration, yield stress and hydrodynamic interactions. *Powder Technology, 128*(2–3), pp. 148–158.

Chandel, S., Seshadri, V. and Singh, S.N., 2009. Effect of additive on pressure drop and rheological characteristics of fly ash slurry at high concentration. *Particulate Science and Technology, 27*(3), pp. 271–284.

Chandel, S., Singh, S.N. and Seshadri, V., 2010. Transportation of high concentration coal ash slurries through pipelines. *International Archive of Applied Science and Technology, 1*(1), pp. 1–9.

Charles, M.E. and Redberger, P.J., 1962. The reduction of pressure gradients in oil pipelines by the addition of water: Numerical analysis of stratified flow. *The Canadian Journal of Chemical Engineering, 40*(2), pp. 70–75.

Che, Y., Feng, M. and Xue, T., 2017, June. Pilot plant study of dry deposition of paste tailings in Baogang West Mine. In *Proceedings of the 20th International Seminar on Paste and Thickened Tailings* (pp. 260–266). University of Science and Technology, Beijing.

Chen, L., Duan, Y., Pu, W. and Zhao, C., 2009. CFD simulation of coal-water slurry flowing in horizontal pipelines. *Korean Journal of Chemical Engineering, 26*(4), pp. 1144–1154.

Chen, X., Zhou, J., Chen, Q., Shi, X. and Gou, Y., 2017. CFD simulation of pipeline transport properties of mine tailings three-phase foam slurry backfill. *Minerals, 7*(8), p. 149.

Darby, R. and Melson, J., 1981. How to predict the friction factor for flow of Bingham plastics. *Chemical Engineering, 88*(26), pp. 59–61.

Eesa, M. and Barigou, M., 2009. CFD investigation of the pipe transport of coarse solids in laminar power law fluids. *Chemical Engineering Science, 64*(2), pp. 322–333.

Einstein, A., 1956. *Investigation on the Theory of the Brownian Movement*. New York: Dover Publications.

Ekambara, K., Sanders, R.S., Nandakumar, K. and Masliyah, J.H., 2009. Hydrodynamic simulation of horizontal slurry pipeline flow using ANSYS-CFX. *Industrial & Engineering Chemistry Research, 48*(17), pp. 8159–8171.

Eltoukhy, M.A.R., 2013. Hydraulic gradient of sand-water mixture flow for different sands and pipes. *Life Science Journal, 10*(4), pp. 3490–3495.

Fangary, Y.S., Ghani, A.A., El Haggar, S.M. and Williams, R.A., 1997. The effect of fine particles on slurry transport processes. *Minerals Engineering, 10*(4), pp. 427–439.

Gandhi, B.K., Singh, S.N. and Seshadri, V., 2001. Performance characteristics of centrifugal slurry pumps. *Journal of Fluids Engineering, 123*(2), pp. 271–280.

Ganeshan, G. and Panda, D., 2017. Effect of oil viscosity on rheology of coal-oil mixtures. *International Journal of Coal Preparation and Utilization, 37*(5), pp. 227–236.

Gillies, R.G. and Shook, C.A., 2000. Modelling high concentration settling slurry flows. *The Canadian Journal of Chemical Engineering, 78*(4), pp. 709–716.

Gillies, R.G., Shook, C.A. and Xu, J., 2004. Modelling heterogeneous slurry flows at high velocities. *The Canadian Journal of Chemical Engineering, 82*(5), pp. 1060–1065.

Gopaliya, M.K. and Kaushal, D.R., 2015. Analysis of effect of grain size on various parameters of slurry flow through pipeline using CFD. *Particulate Science and Technology, 33*(4), pp. 369–384.

Hashemi, S.A., Sadighian, A., Shah, S.I.A. and Sanders, R.S., 2014. Solid velocity and concentration fluctuations in highly concentrated liquid-solid (slurry) pipe flows. *International Journal of Multiphase Flow, 66*, pp. 46–61.

Horsley, R.R., 1982. Viscometer and pipe loop tests on gold slime slurry at very high concentrations by weight, with and without additives. In *Proceedings of Hydrotransport* (Vol. 8).

Hossain, A., Naser, J. and Imteaz, M.A., 2011. CFD investigation of particle deposition in a horizontal looped turbulent pipe flow. *Environmental Modeling & Assessment, 16*(4), pp. 359–367.

Jacobs, B.E., 2014. *Design of Slurry Transport Systems*. CRC Press.

Kaushal, D.R., Sato, K., Toyota, T., Funatsu, K. and Tomita, Y., 2005. Effect of particle size distribution on pressure drop and concentration profile in pipeline flow of highly concentrated slurry. *International Journal of Multiphase Flow*, *31*, pp. 809–823.

Kaushal, D.R., Thinglas, T., Tomita, Y., Kuchii, S. and Tsukamoto, H., 2012. CFD modeling for pipeline flow of fine particles at high concentration. *International Journal of Multiphase Flow*, *43*, pp. 85–100.

Kaushal, D.R. and Tomita, Y., 2002. Solids concentration profiles and pressure drop in pipeline flow of multi-sized particulate slurries. *International Journal of Multiphase Flow*, *28*(10), pp. 1697–1717.

Kim, C. and Han, C., 2013. Numerical simulation of hydraulic transport of sand-water mixtures in pipelines. *Open Journal of Fluid Dynamics*, *3*(4), p. 266.

Krieger, I.M., 1968. Shear rate in the Couetta viscometer. *Transaction Society of Rheology*, *12*, pp. 5–13.

Kumar, K., Kumar, S., Gupta, M., Garg, H.C. and Singh, G., 2017b. Measurement of flow characteristics for multiparticulate bottom ash-water suspension with additives. *Journal of Residuals Science & Technology*, *14*, pp. 11–17.

Kumar, K., Kumar, S., Singh, G., Singh, J.P. and Singh, J., 2017a. Erosion wear investigation of HVOF sprayed WC-10Co4Cr coating on slurry pipeline materials. *Coatings*, *7*(4), pp. 54/1–11.

Kumar, N., Kaushal, D.R. and Kumar, A., 2012. Computational study of the two phase flow in horizontal slurry pipeline. *International Journal of Mechanical and Production Engineering*, *1*(1), pp. 2315–4489.

Kumar, S., Gandhi, B.K. and Mohapatra, S.K., 2014. Performance characteristics of centrifugal slurry pump with multi-sized particulate bottom and fly ash mixtures. *Particulate Science and Technology*, *32*(5), pp. 466–476.

Kumar, U., Gandhi, B.K., Singh, S.N., Seshadri, V. and Gupta, V.K., 2000, November. Performance characteristics of demonstration unit for transportation of solid-liquid flow at high concentration setup at NCPP, Dadri. In *Proceedings International Seminar on Material handling Systems Indian Institute of Plant Engineers (Delhi-Haryana Chapter), November 16–18, New Delhi* (pp. 126–136).

Kumar, U., Mishra, R., Singh, S.N. and Seshadri, V., 2003. Effect of particle gradation on flow characteristics of ash disposal pipelines. *Powder Technology*, *132*(1), pp. 39–51.

Kwak, M., James, D.F. and Klein, K.A., 2005. Flow behaviour of tailings paste for surface disposal. *International Journal of Mineral Processing*, *77*(3), pp. 139–153.

Lahiri, S.K. and Ghanta, K.C., 2008. Prediction of pressure drop of slurry flow in pipeline by hybrid support vector regression and genetic algorithm model. *Chinese Journal of Chemical Engineering*, *16*(6), pp. 841–848.

Lahiri, S.K. and Ghanta, K.C., 2010. Regime identification of slurry transport in pipelines: A novel modelling approach using ANN & differential evolution. *Chemical Industry and Chemical Engineering Quarterly*, *16*(4), pp. 329–343.

Landel, R.F., Moser, B.G. and Baumann, A.J., 1965. Rheology of concentrated suspension: effect of surfactant, In *4th International Congress on Rheology* (pp. 663–673).

Lee, J.K., Ko, J. and Kim, Y.S., 2017. Rheology of fly ash mixed tailings slurries and applicability of prediction models. *Minerals*, *7*(9), pp. 165–176.

Li, L., Usui, H. and Suzuki, H., 2002. Study of pipeline transportation of dense fly ash-water slurry. *Coal Preparation*, *22*(2), pp. 65–80.

Ling, J., Skudarnov, P.V., Lin, C.X. and Ebadian, M.A., 2003. Numerical investigations of liquid-solid slurry flows in a fully developed turbulent flow region. *International Journal of Heat and Fluid Flow*, *24*(3), pp. 389–398.

Lu, P. and Zhang, M., 2002. Resistance properties of coal-water paste flowing in pipes. *Fuel*, *81*(7), pp. 877–881.

Ma, L., Huang, C., Xie, Y., Jiang, J., Tufa, K.Y., Hui, R. and Liu, Z.S., 2015. Modeling of erodent particle trajectories in slurry flow. *Wear*, *334*, pp. 49–55.

Ma, Z.W. and Zhang, P., 2012. Pressure drops and loss coefficients of a phase change material slurry in pipe fittings. *International Journal of Refrigeration*, *35*(4), pp. 992–1002.

Messa, G.V. and Malavasi, S., 2013. Numerical investigation of solid-liquid slurry flow through an upward-facing step. *Journal of Hydrology and Hydromechanics*, *61*(2), pp. 126–133.

Mishra, R., Ghanta, K.C., Mullick, A.N. and Sinha, S.L., 2017. Numerical prediction of flow behavior and erosion prediction of coal water and copper ore water slurries. *Journal of Advanced Research in Dynamical and Control System*, *4*, pp. 2368–2388.

Mishra, R., Singh, S.N. and Seshadri, V., 1998. Pressure drop across conventional and diverging-converging pipe bends in the flow of multi-sized particulate slurries. *Indian Journal of Engineering and Materials sciences*, *5*, pp. 9–14.

Mooney, M., 1951. The viscosity of a concentrated suspension of spherical particles. *Journal of Colloid Science*, *6*(2), pp. 162–170.

Moreland, C., 1963. Viscosity of suspensions of coal in mineral oil. *The Canadian Journal of Chemical Engineering*, *41*(1), pp. 24–28.

Mosa, E.S., Saleh, A.H.M., Taha, T.A. and El-Molla, A.M., 2008. Effect of chemical additives on flow characteristics of coal slurries. *Physicochemical Problems of Mineral Processing*, *42*, pp. 107–118.

Mukhtar, A., Singh, S.N. and Seshadri, V., 1995. Pressure drop in a long radius 90° horizontal bend for the flow of multisized heterogeneous slurries. *International Journal of Multiphase Flow*, *21*(2), pp. 329–334.

Nabil, T., El-Sawaf, I. and El-Nahhas, K., 2013, November. Computational fluid dynamics simulation of the solid-liquid slurry flow in a pipeline. In *Proceedings of 17th International Water Technologies Conference IWTC17* (Vol. 57).

Ofei, T.N. and Ismail, A.Y., 2016. Eulerian-eulerian simulation of particle-liquid slurry flow in horizontal pipe. *Journal of Petroleum Engineering*, p. 5743471/1–10.

Panda, S. and Swain, R.K., 2014. Some aspects of the rheological behaviour of coal water slurries. *International Journal of Thermal and Environment Engineering*, *8*(2), pp. 103–107.

Pani, G.K., Rath, P., Barik, R. and Senapati, P.K., 2015. The effect of selective additives on the rheological behavior of power plant ash slurry. *Particulate Science and Technology*, *33*(4), pp. 418–422.

Pinto, M., Kemp, I., Bermingham, S., Hartwig, T. and Bisten, A., 2014. Development of an axisymmetric population balance model for spray drying and validation against experimental data and CFD simulations. *Chemical Engineering Research and Design*, *92*(4), pp. 619–634.

Qi, C., Chen, Q., Fourie, A., Zhao, J. and Zhang, Q., 2018. Pressure drop in pipe flow of cemented paste backfill: Experimental and modeling study. *Powder Technology*, *333*, pp. 9–18.

Ravelet, F., Bakir, F., Khelladi, S. and Rey, R., 2013. Experimental study of hydraulic transport of large particles in horizontal pipes. *Experimental Thermal and Fluid Science*, *45*, pp. 187–197.

Rawat, A., Singh, S.N. and Seshadri, V., 2016. Computational methodology for determination of head loss in both laminar and turbulent regimes for the flow of high concentration coal ash slurries through pipeline. *Particulate Science and Technology*, *34*(3), pp. 289–300.

Roco, M.C. and Shook, C.A., 1983. Modeling of slurry flow: The effect of particle size. *The Canadian Journal of Chemical Engineering*, *61*(4), pp. 494–503.

Roscoe, R., 1952. The viscosity of suspensions of rigid spheres. *British Journal of Applied Physics*, *3*(8), p. 267.

Rutgers, I.R., 1962. Relative viscosity of suspensions of rigid spheres in Newtonian liquids. *Rheologica Acta*, *2*(3), pp. 202–210.

Sahoo, B.K., De, S., Carsky, M. and Meikap, B.C., 2010. Enhancement of rheological behavior of Indian high ash coal-water suspension by using microwave pretreatment. *Industrial & Engineering Chemistry Research*, *49*(6), pp. 3015–3021.

Schaan, J., Sumner, R.J., Gillies, R.G. and Shook, C.A., 2000. The effect of particle shape on pipeline friction for Newtonian slurries of fine particles. *The Canadian Journal of Chemical Engineering*, *78*(4), pp. 717–725.

Schrick, W., Smith, L.G., Hass, D.B. and Husband, W.H.W., 1972. *Experimental Studies on the Hydraulic Transport of Coal*. Engineering Division of Saskatchewan Research Council, Canada.

Seitshiro, I., Fujii, S., Yokoyama, N., Sato, I. and Sato, H., 2012. The multi-sized slurry flows in horizontal pipes: Innovated models and verification. *International Journal of the Society of Materials Engineering for Resources*, *19*(1–2), pp. 24–31.

Senapati, P.K., Das, D., Nayak, A. and Mishra, P.K., 2008. Studies on preparation of coal water slurry using a natural additive. *Energy Sources, Part A*, *30*(19), pp. 1788–1796.

Senapati, P.K. and Mishra, B.K., 2015. Bulk hydraulic disposal of highly concentrated fly ash and bottom ash-water slurries. *Particulate Science and Technology*, *33*(2), pp. 124–131.

Senapati, P.K., Mishra, B.K. and Parida, A., 2013. Analysis of friction mechanism and homogeneity of suspended load for high concentration fly ash & bottom ash mixture slurry using rheological and pipeline experimental data. *Powder Technology*, *250*, pp. 154–163.

Shao, S., Chen, X., Liu, H. and Wang, F., 2012. Preparation of coal slurry with alcohol fermentation wastewater. *Energy Sources, Part A: Recovery, Utilisation, and Environmental Effects*, *34*(10), pp. 919–928.

Singh, G., Kumar, S., Singh, M.K. and Mohapatra, S.K., 2016a. Environmental impact assessment of ash disposal system of a thermal power plant. *International Journal of Hydrogen Energy*, *41*(35), pp. 15887–15891.

Singh, J., Kumar, S., Singh, J.P., Kumar, P. and Mohapatra, S.K., 2018a. CFD modeling of erosion wear in pipe bend for the flow of bottom ash suspension. *Particulate Science and Technology*, pp. 1–11.

Singh, M.K., Ratha, D., Kumar, S. and Kumar, D., 2016b. Influence of particle-size distribution and temperature on rheological behavior of coal slurry. *International Journal of Coal Preparation and Utilization*, *36*(1), pp. 44–54.

Slatter, P.T., 2011. The engineering hydrodynamics of viscoplastic suspensions. *Particulate Science and Technology*, *29*(2), pp. 139–150.

Slatter, P.T. and Seddon, K.D., 2015, April. Evaluation of the rheology of pipe head flocculated tailings. In *Proceedings of the 18th International Seminar on Paste and Thickened Tailings* (pp. 131–137). Australian Centre for Geomechanics.

Thomas, D.G., 1965. Transport characteristics of suspensions: VIII a note on the viscosity of Newtonian suspensions of uniform spherical particles. *Journal of Colloid Science*, *20*, p. 265.

Verma, A.K., Singh, S.N. and Seshadri, V., 2006. Pressure drop for the flow of high concentration solid-liquid mixture across 90° horizontal conventional circular pipe bend. *Indian Journal of Engineering and Materials Sciences*, *13*, pp. 477–483.

Vlasak, P. and Chara, Z., 2009. Conveying of solid particles in Newtonian and non-Newtonian carriers. *Particulate Science and Technology*, *27*(5), pp. 428–443.

Vlasak, P. and Chara, Z., 2011. Effect of particle size distribution and concentration on flow behavior of dense slurries. *Particulate Science and Technology*, *29*(1), pp. 53–65.

Wang, J., Zhang, T. and Wang, S., 2013. Heterogeneous ice slurry flow and concentration distribution in horizontal pipes. *International Journal of Heat and Fluid Flow*, *44*, pp. 425–434.

Williams, O.A., 1983. *Pneumatic and Hydraulic Conveying of Solids*. M. Dekker.

Williams, P.S., 1953. Flow of concentrated suspensions. *Journal of Applied Chemistry*, *3*, pp. 120–127.

Wilson, K.C., 1982. A dense phase option for coarse coal pipelining. *Journal of Pipelines*, *5*, pp. 251–257.

Woodcock, C.R. and Mason, J.S., 1987. Hydraulic conveying. In *Bulk Solids Handling* (pp. 478–493). Dordrecht: Springer.

Wu, D., Yang, B. and Liu, Y., 2015. Pressure drop in loop pipe flow of fresh cemented coal gangue-fly ash slurry: Experiment and simulation. *Advanced Powder Technology*, *26*(3), pp. 920–927.

Yoon, S.J., Choi, Y.C. and Lee, J.G., 2009. The effect of additive chemicals on the viscosity of coal-petroleum coke-water slurry fuel for a gasification process. *Korean Journal of Chemical Engineering*, *26*(5), pp. 1259–1264.

Zhu, J., Zhang, G., Li, J. and Zhao, F., 2013. Synthesis, adsorption and dispersion of a dispersant based on starch for coal-water slurry. *Colloids and Surfaces A: Physicochemical and Engineering Aspects*, *422*, pp. 165–171.

4 Numerical and Computational Analysis of Slurry Flows in Mineral Beneficiation and Processing

Sukhwinder Singh Sekhon, Virinder Kumar, Parminder Singh, and Ashif Equbal

4.1 INTRODUCTION

In the mining industry, the gravitational concentration of minerals has been accepted as an economical and sustainable separation process [1]. Among the different gravitation-based separation techniques, spiral separators have become significant equipment for separating minerals, especially for particle sizes in the 45-μm to 3-mm range, where there are differences between the densities of valuable mineral particles and the gangue medium [2]. Since the inception of spiral separators/concentrators in 1943, they have been used for the preconcentration of coal or other essential minerals based on densities and the size of particles [3].

An adequate particle-laden slurry mixture flows through spiral concentrators through the helical trough. During the process, a thin film flow develops, and particle separation occurs in the radial direction under various forces such as gravitational, centrifugal, Bagnold, and friction [4]. In earlier stages of their application in mineral processing or beneficiation, the development of mineral-specific spiral separators was based on a trial-and-error method without much consideration given to understanding flow dynamics [5]. Further, the focus shifted towards understanding slurry flow behaviour and particle segregation mechanisms in the spiral separator to enhance the design and separation efficiency on the industrial scale.

Recently, interest in understanding complex slurry flow patterns has developed among researchers to fill the gap of the lack of knowledge of flow dynamics and particle behaviours in spiral separators. Many researchers have attempted to develop different numerical and physical models to explain the dynamics of slurry flow and predict its performance for various controlling parameters. This chapter will discuss comprehensive information on various numerical and physical models and model-solving computational tools such as CFD, DEM, and hybrid CFD-DEM to provide a thorough understanding of slurry flow patterns and particle separation behaviours. Before, diving into the model realm, first, there is a need to understand the geometrical design parameters and application of the spiral concentrator.

4.2 DESCRIPTION OF DESIGN PARAMETERS AND SEPARATION MECHANISMS OF SPIRAL CONCENTRATORS

4.2.1 Design Parameter

A spiral concentrator consists of an open-channel trough wrapped in a helix around a central pillar, as shown in Figures 4.1 and 4.2 [7]. The design parameters of the spiral concentrator comprise the

DOI: 10.1201/9781003561156-5

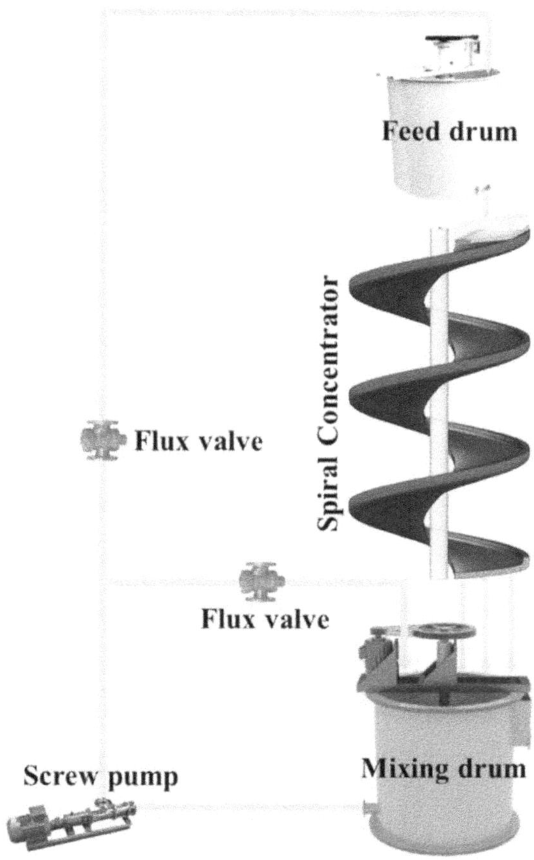

FIGURE 4.1 Working schematic of a spiral concentrator [6].

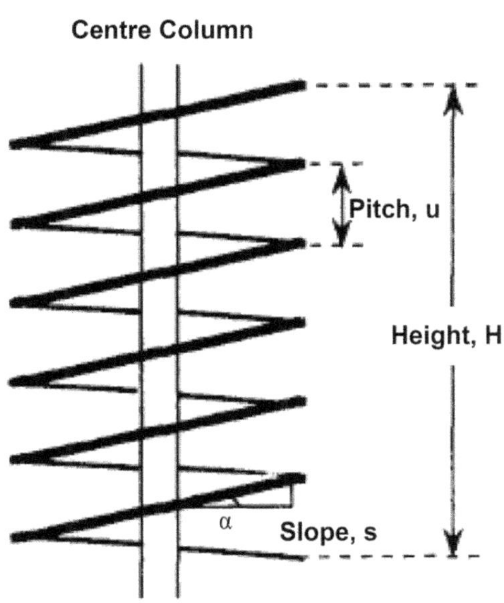

FIGURE 4.2 Schematic of a spiral concentrator [2].

spiral pitch (u), spiral trough shape included at the slope (α), length of the trough, and inner (r_i) and outer radii (r_o) of the spiral, as shown in Figure 4.3.

4.2.2 Working Principles for Separation

During the operation of the spiral concentrator, a slurry mixture of mineral particles with adequate density is fed in through a slurry pump. A fluid film is formed on the trough, with primary flow in a downward direction under the influence of gravitational force and secondary flow in the radial direction and the difference in frictional forces in the upper and lower layers of the fluid film. Therefore, two zones separated by transition zone are developed, where higher specific gravity particles settle in the inner zone, and lower specific gravity particles settle in the transition and outer recovery zone, as shown in Figure 4.4 [1].

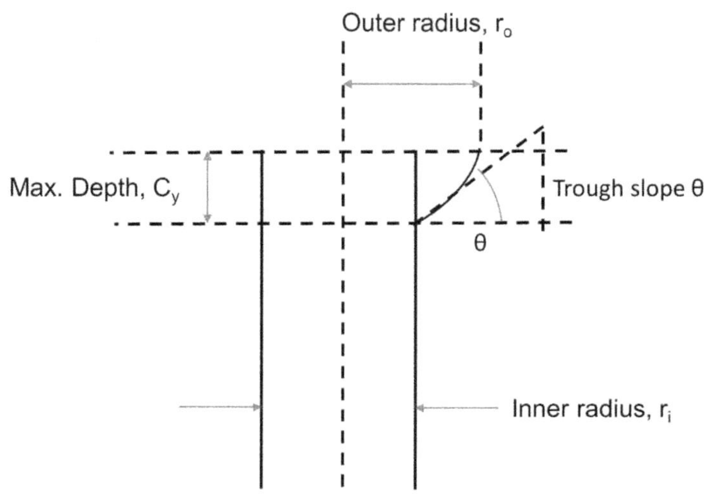

FIGURE 4.3 Cross-sectional view of a trough [2].

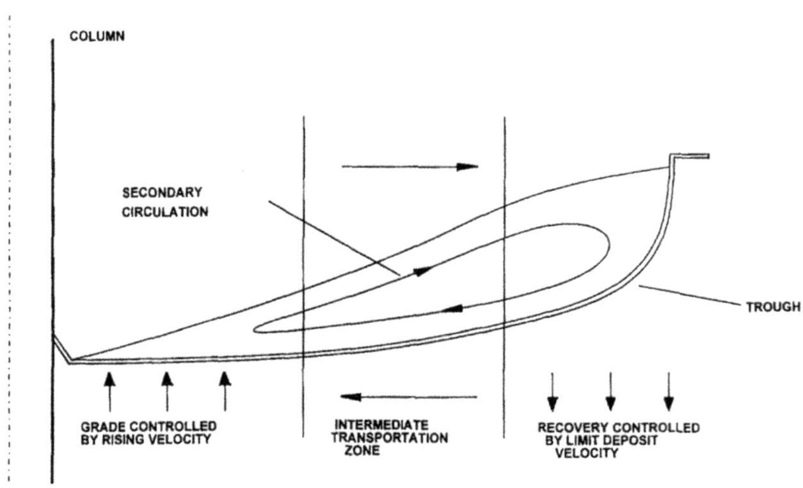

FIGURE 4.4 Secondary flow and separation mechanism [1].

Particles under the dominance of gravitational force settle down and drift towards the inner region of the trough, whereas particles under the influence of centrifugal force drift towards the upper area of the fluids and the outer region [8].

4.2.3 Application of Spiral Concentrators

Spiral concentrators are practical and economical equipment used in mineral processing industries to separate fine and ultra-fine particles. Their application was first documented in the beneficiation of clean coal from gangue media; the performance of spiral concentrators varied with size, density of slurry, and flow characteristics in coal fines and bituminous coal separation [9–11]. Further, modifications with multi-stage and water wash to optimize the flow were found to be successful in enhancing the yield and recovery of target ultra-fine coals [12,13]. Spiral concentrators were further found to be an effective tool in the processing of sand, mica, and silica. Through selective density-wise separation with controlled feed conditions, effective separation of mica-feldspar and silica gangue was reported at an industrial scale [14,15]. Similarly, spiral concentrators aid in pre-concentration as well as the final beneficiation of iron ores (low-grade goethite and hematite ores) and slag of the mineral industry [16–18]. The utility of spiral concentrators was further explored for chromite ores, where through design consideration of spiral geometry and flow dynamics, successful recovery of fine and ultra-fine chromite particles was achieved [19,20]. Thus, a gravity-based spiral concentrator was reported as adequate preconcentration equipment in recovering rare earth element (REE) minerals from complex, unconventional REE-bearing mineral deposits such as beach placer garnet and monazite-rich sands [21–23].

4.3 NUMERICAL MODELLING AND COMPUTATIONAL TECHNIQUES FOR ANALYSIS OF SLURRY FLOW IN SPIRAL CONCENTRATORS

Numerical modelling and computation analysis of slurry flow in spiral concentrators requires a robust understanding of constitutive models. Constitutive models are a foundational, fundamental block for modelling and simulating physical processes by establishing a relationship between conservation laws and the kinematics of the physical system under study. Constitutive models are generally expressed as linear or non-linear ordinary differential equations (ODEs) [24]. Constitutive models are categorized into physical and empirical approaches. Further, physical models are based on the governing laws of physics and are solved through tools like CFD, DEM, and FEM. On the other hand, empirical models are based on experimental data and are solved by tools like RSM, Taguchi, and so on. In this chapter, we will focus only on the physical constitutive models, with CFD as the tool to provide insight into the flow dynamics of the slurry flow and particle separation mechanism.

4.3.1 Governing Principles and Numerical Approaches

The simulation of slurry flow in a spiral involves establishing a multiphase system, which usually includes water, particles, and air. Analytical or numerical models are used to investigate the slurry flow through the volume of fluid (VOF) method to track the interaction between surfaces of air and slurry fluid. These models are governed by the first principles, such as Navier–Stokes equations and Newton's laws, to simulate the fluid flow, particle interaction, and fluid flow turbulence [25]. Refer to equations 4.1 and 4.2 for governing equations; the general conservation of mass and momentum equation is provided:

$$\frac{\partial(\rho u_i)}{\partial x_i} = 0 \tag{4.1}$$

$$\frac{\partial\left(\rho u_i u_j\right)}{\partial x_j} = -\frac{\partial(P)}{\partial x_i} + \rho g_i + \frac{\partial}{\partial x_j}\left[\mu_{eff}\left(\frac{\partial u_i}{\partial x_j} + \frac{\partial u_j}{\partial x_i}\right)\right] \tag{4.2}$$

Here, ρ is density, μ_{eff} is effective viscosity, P is static pressure, $u_{i,j}$ is mean velocity, and g_i is gravitation acceleration.

Modelling of the particle phase is done through various approaches:

Euler–Lagrangian approach: In multiphase flow, the Euler–Lagrangian approach is employed, where fluid is considered a continuum under the Eulerian framework and particles are tracked under the Lagrangian framework. This approach is beneficial in dilute slurry flows where particle–particle interactions are less [26]; refer to equation 4.3.

$$\frac{du_p^i}{dt} = \left(\frac{\rho_f}{\rho_p}\right)u_p^i\frac{\partial u_f^i}{\partial x_i} + \frac{1}{2}\frac{\rho_f}{\rho_p}\frac{d}{dt}\left(u_f^i - u_p^i\right) + \frac{3\mu_f C_D Re_p}{4\rho_p D_p^2}\left(u_f^i - u_p^i\right) + \left(\frac{\rho_p - \rho_f}{\rho_p}\right)g_i \tag{4.3}$$

Euler–Euler approach: When fluid and particle media are considered interpenetrating continua, the Euler–Euler approach is used to obtain fluid–particle interaction. This approach is favourable in the case of dense slurries [27]; refer to equations 4.4 and 4.5.

$$\frac{\partial\left(\alpha_q\rho_q\right)}{\partial t} + \nabla.(\alpha_q\rho_q\vec{v}_q) = 0 \tag{4.4}$$

$$\frac{\partial\left(\alpha_q\rho_q\vec{v}_q\right)}{\partial t} + \nabla.(\alpha_q\rho_q\vec{v}_q\vec{v}_q) = -\alpha_q\nabla p - \nabla p_q + \nabla.\bar{\tau}_q + \alpha_q\rho_q\vec{g} + \vec{F}_d + \vec{F}_b \tag{4.5}$$

Eulerian multi-fluid VOF approach: A further development with an amalgamation of VOF with the Euler–Euler model is also used to simulate the interaction between different phases (water and particle phase) [28].

4.3.2 Turbulence Modelling

Tracking and capturing the effect of turbulence models play a crucial role in CFD analysis. The standard k–ε model is among the most commonly employed turbulence models in the case of fully developed turbulent flow with a high Reynolds number [29]; refer to equation 4.6. The renormalization group (RNG) k–ε model is an improved version used for its accuracy in analysing flow having swirling and rapid strains [30]; refer to equations 4.8–4.12. The shear-stress transport (SST) k–ω model is appropriate to simulate flow turbulence under adverse pressure gradients; refer to equations 4.13 and 4.14. The Reynolds stress model (RSM) is most relevant and accurate in capturing anisotropic turbulence effects and complex flow features [31]; refer to equation 4.15.

$$\frac{\partial(\rho k)}{\partial t} + \frac{\partial(\rho k \mu_i)}{\partial x_i} = \frac{\partial}{\partial x_j}\left[\left(\mu + \frac{\mu_t}{\sigma_k}\right)\frac{\partial k}{\partial x_j}\right] + G_k - \rho\varepsilon \tag{4.6}$$

$$\frac{\partial(\rho\varepsilon)}{\partial t} + \frac{\partial(\rho\varepsilon\mu_i)}{\partial x_i} = \frac{\partial}{\partial x_j}\left[\left(\mu + \frac{\mu_t}{\sigma_\varepsilon}\right)\frac{\partial\varepsilon}{\partial x_j}\right] + C_{1\mu}\frac{\varepsilon}{k}G_k - C_{2\varepsilon}\rho\frac{\varepsilon^2}{k} \tag{4.7}$$

$$\frac{\partial(\rho k)}{\partial t} + \frac{\partial(\rho k \mu_i)}{\partial x_i} = \frac{\partial}{\partial x_j}\left(\alpha_k\mu_{eff}\frac{\partial(k)}{\partial x_j}\right) + G_k - \varepsilon \tag{4.8}$$

$$\frac{\partial(\rho\varepsilon)}{\partial t}+\frac{\partial(\rho\varepsilon\mu_i)}{\partial x_i}=\frac{\partial}{\partial x_j}\left(\alpha_\varepsilon\mu_{eff}\frac{\partial(\varepsilon)}{\partial x_j}\right)+C_{1\varepsilon}\frac{\varepsilon}{k}\mathrm{G_k}-C_{2\varepsilon}\rho\frac{\varepsilon^2}{k}-R_\varepsilon \tag{4.9}$$

$$\mathrm{G_k}=-\overline{\mu_i\,\mu_j}\frac{\partial(\mu_j)}{\partial x_i} \tag{4.10}$$

$$R_\varepsilon=\frac{C_\mu\rho\eta^3\left(1-\frac{\eta}{\eta_o}\right)}{1+\beta\eta^3}\frac{\varepsilon^2}{\kappa} \tag{4.11}$$

$$\eta=\frac{S\kappa}{\varepsilon} \tag{4.12}$$

Here, $\mathrm{G_k}$ is kinetic energy generation, R_ε is for the low Reynolds number, and S is the mean rate of strain tensor.

$$\frac{\partial(\rho k)}{\partial t}+\frac{\partial(\rho k\mu_j)}{\partial x_j}=\frac{\partial}{\partial x_j}\left[(\Gamma_k)\frac{\partial(k)}{\partial x_j}\right]+G_k-Y_k \tag{4.13}$$

$$\frac{\partial(\rho\omega)}{\partial t}+\frac{\partial(\rho\omega\mu_j)}{\partial x_j}=\frac{\partial}{\partial x_j}\left[(\Gamma_\omega)\frac{\partial(\omega)}{\partial x_j}\right]+G_\omega-Y_\omega \tag{4.14}$$

Here, Γ_k and Γ_ω are effective diffusivity, and Y_k and Y_ω are dissipation for k and ω, respectively.

$$\begin{aligned}&\frac{\partial\overline{u_iu_j}}{\partial t}+\frac{\partial U_k\,\overline{u_iu_j}}{\partial x_k}\{Convection\,C_{ij}\}=Tubulent\,diffusion\left(D_{T.ij}\right)\\&+Molecular\,diffusion\left(D_{L.ij}\right)+Stress\,producion\left(P_{ij}\right)\\&+Pressure\,strain\left(\phi_{ij}\right)+Dissipation\left(\varepsilon_{ij}\right)\end{aligned} \tag{4.15}$$

Computational research studies are discussed in the following to provide a holistic understanding of applying numerical modelling and computation analysis procedures.

4.4 ANALYSIS AND PREDICTION OF SLURRY FLOW CHARACTERISTICS, FLUID FLOW DYNAMICS, AND PARTICLE SEGREGATION USING A NUMERICAL MODELLING AND CFD APPROACH

Numerical modelling and CFD simulation results provide a detailed understanding of slurry fluid flow dynamics and particle segregation mechanisms. Also, the effect of spiral geometry, such as slope, pitch, and input variables like slurry density, can be captured to optimize spiral concentrator processes.

4.4.1 Development of Fluid Flow Field

Slurry fluid flow film develops into a steady film flow from its earlier unsteady flow state as it passes down a spiral trough. Slurry fluid flow characteristics such as the development of film flow thickness, which increases from the inner section towards the outer region of the trough; radial flux; and tangential velocities are studied to correlate with the spiral geometry and input conditions to optimize the spiral flow.

Referring to Figures 4.5–4.8, numerical modelling and corresponding CFD-based analysis showed that in the spiral concentrator, water depth increases in the radial direction due to the action of centrifugal force, and the RSM model was found to be more accurate in predictions [32,33].

Also, these models effectively predict turbulence intensity; as seen in Figure 4.9, where turbulence intensity prediction using the RSM turbulence model shows an increase in turbulence intensity of slurry fluid flow along the radial direction and with an increase in flow rate [7]. Therefore, through numerical modelling, CFD tools can help predict slurry flow depth, fluid distribution, turbulence intensity, and velocity distribution along the trough profile.

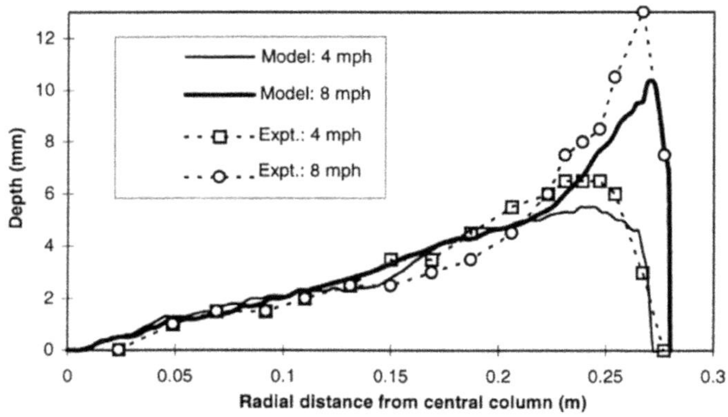

FIGURE 4.5 Comparison of CFD simulated and experimental flow depth [32].

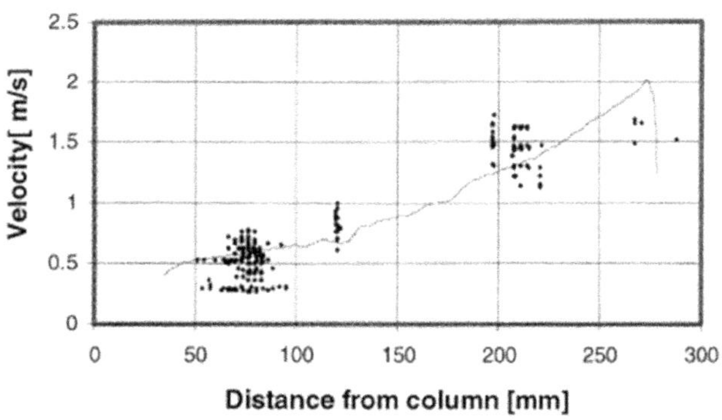

FIGURE 4.6 Prediction of fluid stream velocities along the trough [32].

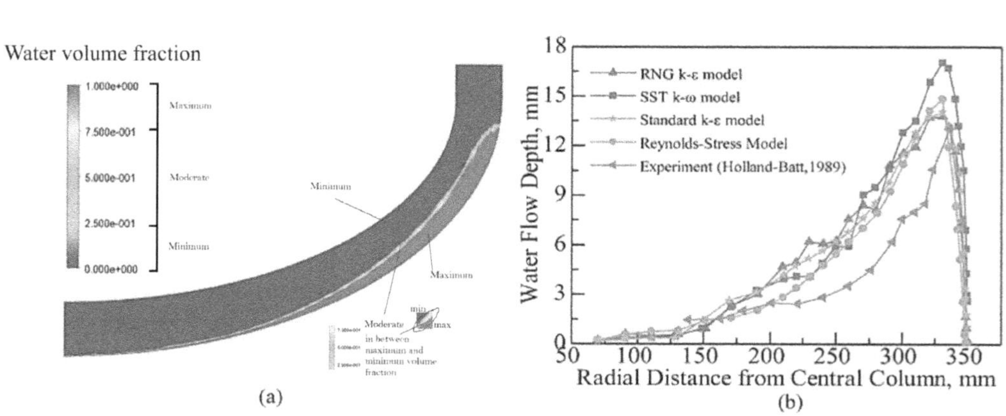

FIGURE 4.7 Simulation of water volume fraction and flow depth for various turbulence models [33].

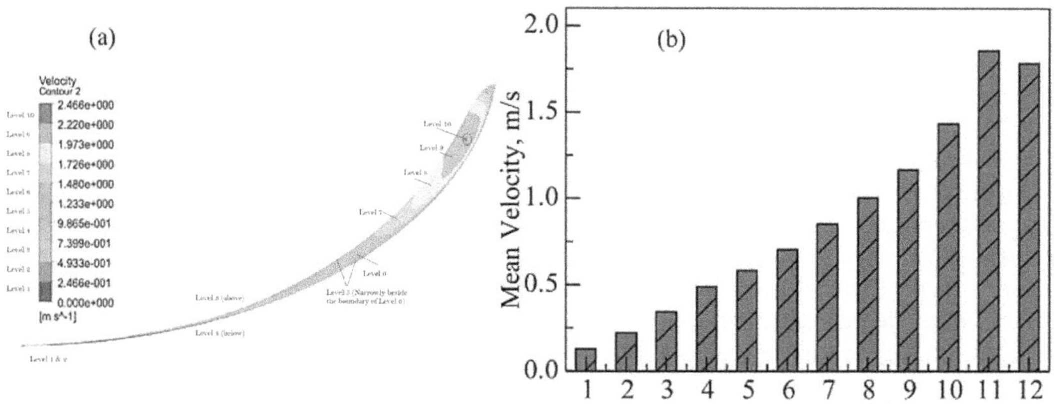

FIGURE 4.8 Simulation of velocity distribution [33].

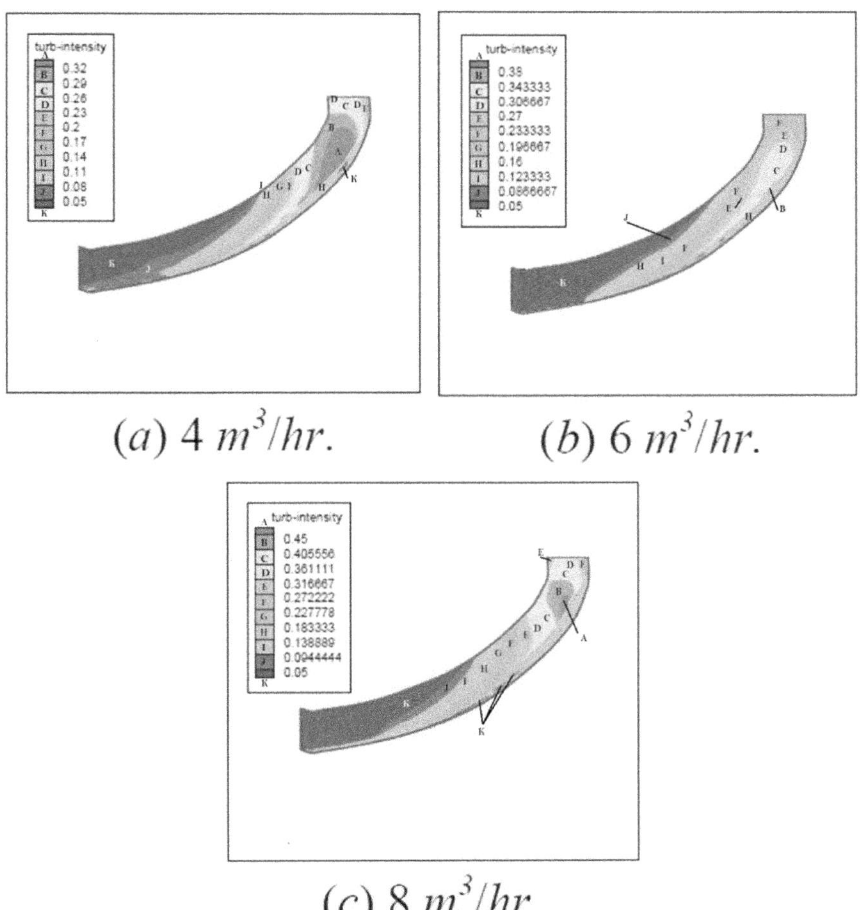

FIGURE 4.9 Prediction of turbulence intensity by RSM modal at different flow rates [7].

4.4.2 Particle Segregation Mechanism

Experimental and simulation studies have shown that relatively coarse particles tend to settle in the inner region of the spiral trough, and fine particles tend to distribute in the middle and outer regions where the secondary flow further segregates the particles based on their specific gravity. The segregation of particles is attributed to the density, particle size, and effect of secondary flow.

Figure 4.10 shows the CFD modelling and prediction-based simulation for fluid flow field and particle segregation analysis, and the figure represents the cumulative frequency percentage of particles (size range 0.075–0.75 mm) along the radial direction of the spiral trough. As seen in Figure 4.10, the segregation mechanism of all the particle sizes has a similar S-shaped distribution pattern, where dense, coarse particles drift to the inner section of the trough and less dense particles are seen settling towards the outer sections of the trough. Also, as seen in the figure, both coal and quartz of around 0.075 mm in size are in suspension and segregated to the outer sections. In contrast, when the size rises above 0.5 mm, the outward movement of coal and quartz particles reduces at both flow rates, with particles settling and moving inwards to the inner sections [34].

4.4.3 Influence of Geometric Parameters

The cross-sectional shape of a spiral trough, that is, parabola, straight, or elliptical, has a significant impact on the flow dynamics and particle movement. Through CFD, the effect of operational parameters like spiral geometry and solid particle concentration can be analysed, and predictions can also be made for the development and optimization of design.

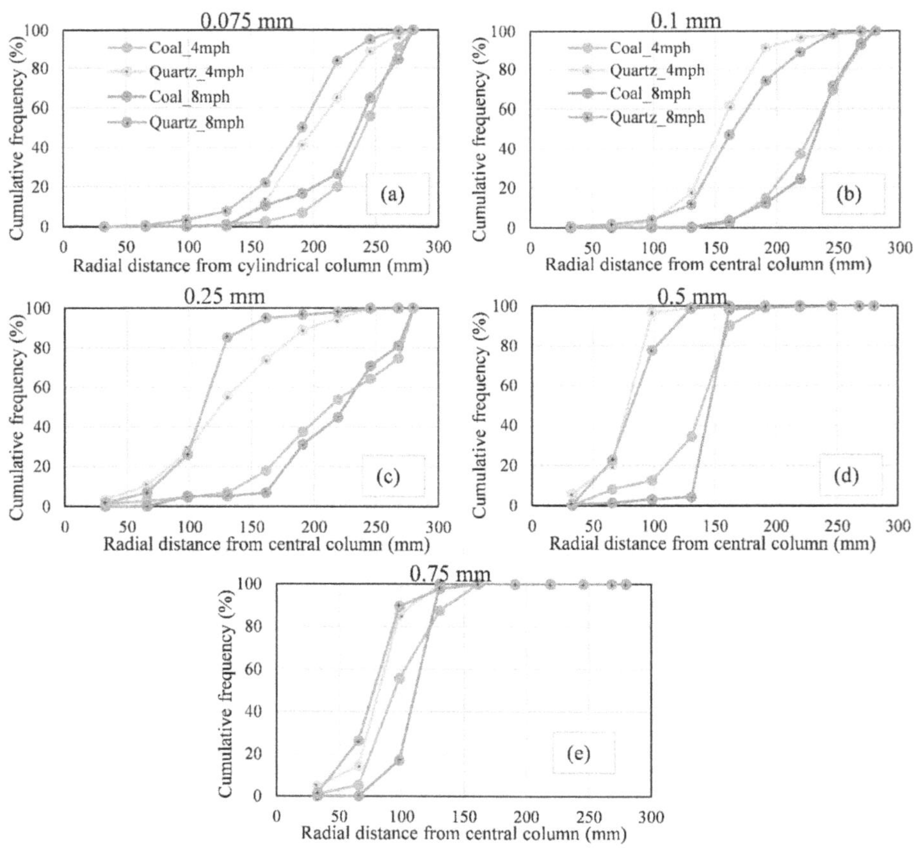

FIGURE 4.10 Particle segregation prediction on sixth turns for coal and quartz non-spherical particles at 4 and 8 m^3/hr: (a) 0.075 mm, (b) 0.1 mm, (c) 0.25 mm, (d) 0.5 mm, (e) 0.75 mm [32].

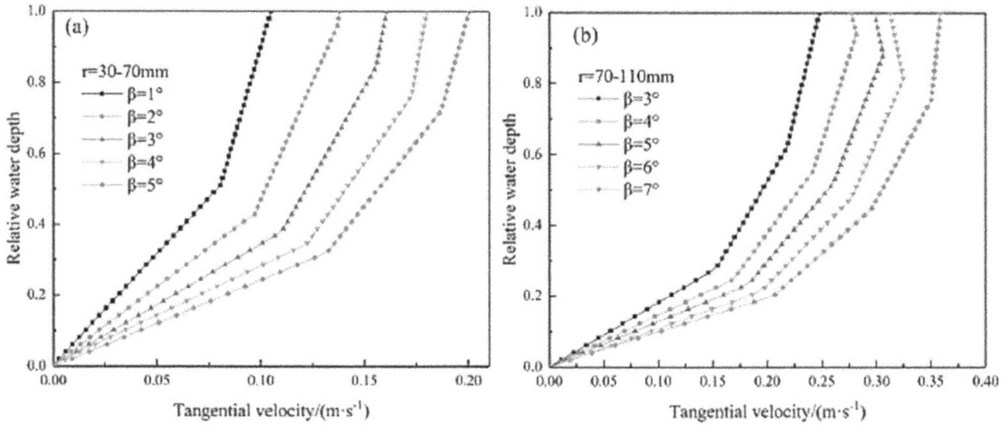

FIGURE 4.11 Fluid flow depth and tangential velocity variation at different trough slopes [35].

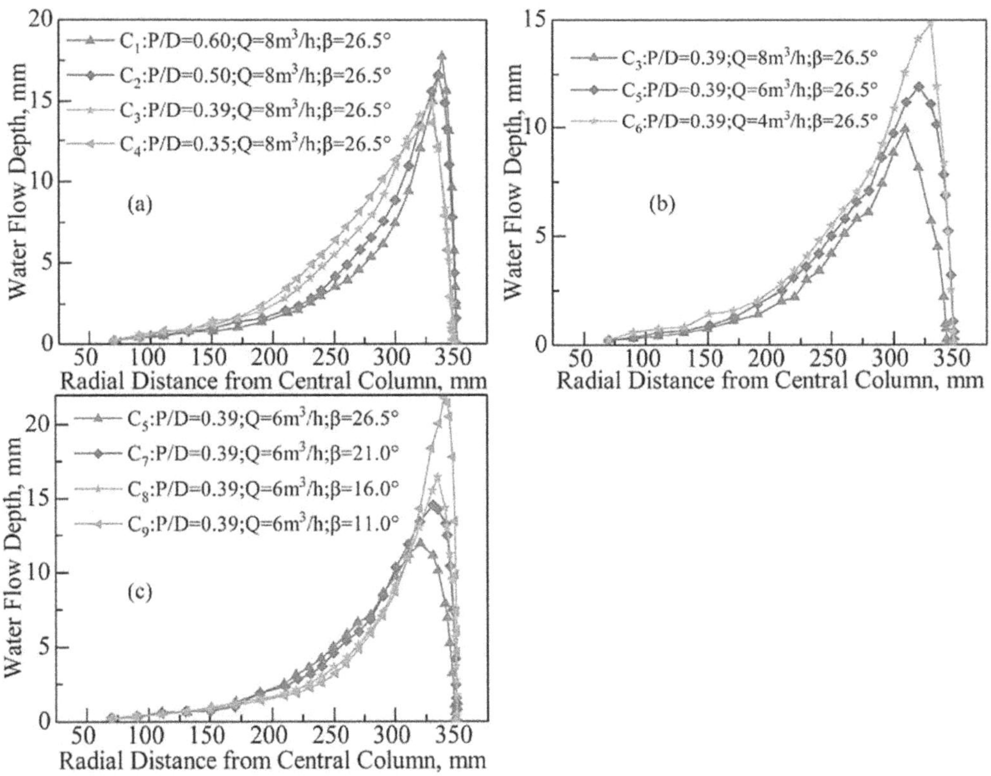

FIGURE 4.12 Impact of operational parameters: (a) pitch, (b) flow rate, and (c) transverse angle on fluid flow depth [33].

As observed in Figure 4.11, with the increased trough slope (β), the numerical simulation showed an increase in the intensity of primary and secondary flow, which increases the slurry fluid flow depth in the outer trough region. The same effect of trough slope is predicted for the tangential velocity [35]. Such results show that an increase in trough slope has a prolonged impact on the development of stable fluid flow. Furthermore, simulation results from another study showed the impact of operational parameters on slurry fluid flow dynamics focusing on depth of flow. As seen in the simulation results (Figure 4.12), in the inner regions of the trough, operational parameters such as pitch, flow rate, and transverse angle have little effect on the water depth, but a significant effect was seen in the water depth for the outer regions of the trough [33].

4.5 CONCLUSION

These numerical modelling and computational studies provide a fundamental understanding of hydrodynamics and particulate phenomena related to slurry in the spiral concentrator. Application of numerical modelling CFD analysis has predictive capabilities for complex slurry flows in spiral concentrators. The system application approach, including first principles such as conservation laws and Navier–Stokes equations coupled with a multiphase phase flow approach encompassing VOF, Eulerian–Lagrangian, and Eulerian–Eulerian methods, enables researchers to capture the complex interaction of slurry fluid flow dynamics, particle characteristics, and operation parameters. Further, these studies compared various turbulence models, from k–ε variants to the more complex Reynolds stress model, providing more refined and accurate predictions for flow fields, turbulence intensities, and particle trajectories. Therefore, these numerical modelling and CFD analyses are cost-effective and robust approaches to studying slurry flows for improved designs and optimized operational parameters that can be achieved for effective mineral beneficiation and processing.

REFERENCES

[1] Holland-Batt AB. Some design considerations for spiral separators. Minerals Engineering. 1995;8(11):1381–1395.
[2] Doheim MA, Abdel Gawad AF, Mahran GMA, et al. Numerical simulation of particulate-flow in spiral separators: Part I. Low solids concentration (0.3% & 3% solids). Applied Mathematical Modelling. 2013;37(1):198–215.
[3] Burt RO. Gravity concentration technology [Internet]. New York, NY: Elsevier Science Pub. Co., Inc.; 1AD. Available from: https://www.osti.gov/biblio/7181338.
[4] Richards RG, MacHunter DM, Gates PJ, et al. Gravity separation of ultra-fine (–0.1mm) minerals using spiral separators. Minerals Engineering. 2000;13(1):65–77.
[5] Arnold DJ, Stokes YM, Green JEF. Thin-film flow in helically-wound rectangular channels of arbitrary torsion and curvature. Journal of Fluid Mechanics. 2014/12/23 ed. 2015;764:76–94.
[6] Gao S, Wang Q, Zhou X, et al. Effect of the cross-sectional geometry of the mixed particle zone on the spiral separation process and its structural optimization. Minerals. 2024;14(12).
[7] Doheim MA, Abdel Gawad AF, Mahran GMA, et al. Computational prediction of water-flow characteristics in spiral separators: Part I. Flow depth and turbulence intensity. JES Journal of Engineering Sciences. 2008;36(4):935–950.
[8] Jain PK, Rayasam V. An analytical approach to explain the generation of secondary circulation in spiral concentrators. Powder Technology. 2017;308:165–177.
[9] Atasoy Y, Spottiswood DJ. A study of particle separation in a spiral concentrator. Minerals Engineering. 1995;8(10):1197–1208.
[10] Honaker RQ, Jain M, Parekh BK, et al. Ultrafine coal cleaning using spiral concentrators. Minerals Engineering. 2007;20(14):1315–1319.
[11] Holland-Batt AB, Hunter JL, Turner JH. The separation of coal fines using flowing-film gravity concentration. Powder Technology. 1984;40(1):129–145.
[12] Benusa MD, Klima MS. An evaluation of a two-stage spiral processing ultrafine bituminous coal. International Journal of Coal Preparation and Utilization. 2008;28(4):237–260.
[13] Hornsby DT, Watson SJ, Clarkson CJ. Fine coal cleaning by spiral and water washing cyclone. Coal Preparation. 1993;12(1–4):133–161.
[14] Bhima Rao R. Recovery of industrial silica sand from Subarnarekha meander placer deposit. Indian Chemical Engineer:1–12.
[15] Gulsoy OY, Kademli M. Effects of operational parameters of spiral concentrator on mica-feldspar separation. Mineral Processing and Extractive Metallurgy. 2006;115(2):80–84.
[16] Bazin C, Sadeghi M, Bourassa M, et al. Size recovery curves of minerals in industrial spirals for processing iron oxide ores. Minerals Engineering. 2014;65:115–123.

[17] Sadeghi M, Bazin C, Renaud M. Effect of wash water on the mineral size recovery curves in a spiral concentrator used for iron ore processing. International Journal of Mineral Processing. 2014;129:22–26.
[18] Pandiri S, Raju MV, Harikrishna M, et al. Enhanced iron recovery from ultrafine iron ore tailing through combined gravitational and magnetic separation process. Transactions of the Indian Institute of Metals. 2022;75(9):2435–2442.
[19] Izerdem D. Low-grade chromite beneficiation through spiral concentration: A review from the Turkish market perspective. Mineral Processing and Extractive Metallurgy Review. 2025;46(3):436–456.
[20] Bahrami A, Farajzadeh S, Kazemi F. The effect of particle size distribution and liberation degree on the separation performance of industrial spirals in low-grade chromite processing. Mining, Metallurgy & Exploration. 2021;38(1):277–287.
[21] Marion C, Grammatikopoulos T, Rudinsky S, et al. A mineralogical investigation into the preconcentration of the Nechalacho deposit by gravity separation. Minerals Engineering. 2018;121:1–13.
[22] Bhima Rao R, Rao RG. Development of alternate process flowsheets to recover an unconventional resource rare earth bearing placer garnet and sillimanite from Southeast Coast of India. Separation Science and Technology. 2024;59(4):697–708.
[23] Zhou M, Li L, Liu F, et al. Gravity-based preconcentration strategies for complex rare earth ore containing niobium and zirconium. Physicochem Probl Miner Process. 2024;60(1):183609.
[24] Haghighat E, Abouali S, Vaziri R. Constitutive model characterization and discovery using physics-informed deep learning. Engineering Applications of Artificial Intelligence. 2023;120:105828.
[25] Swamy M, Díez NG, Twerda A. Numerical modelling of the slurry flow in pipelines and prediction of flow regimes. València, Spain; 2015 [cited 2025 May 31]. p. 311–322. Available from: http://library.witpress.com/viewpaper.asp?pcode=MPF15-027-1.
[26] Subramaniam S. Lagrangian-Eulerian methods for multiphase flows. Progress in Energy and Combustion Science. 2013;39(2):215–245.
[27] Van den Akker HE. The Euler-Euler approach to dispersed two-phase flows in the turbulent regime. ERCOFTAC Bulletin. 1998;36:30–33.
[28] Chen G, Wang Q, He S. Assessment of an Eulerian multi-fluid VOF model for simulation of multiphase flow in an industrial Ruhrstahl-Heraeus degasser. Metallurgical Research & Technology [Internet]. 2019;116(6).
[29] Blazek J. Chapter 7—Turbulence modelling. In: Blazek J, editor. Computational fluid dynamics: Principles and applications (Second Edition) [Internet]. Oxford: Elsevier Science; 2005. p. 227–270. Available from: https://www.sciencedirect.com/science/article/pii/B9780080445069500096.
[30] Speziale CG, Thangam S. Analysis of an RNG based turbulence model for separated flows. International Journal of Engineering Science. 1992;30(10):1379–IN4.
[31] Baker C, Johnson T, Flynn D, et al. Chapter 4—Computational techniques. In: Baker C, Johnson T, Flynn D, et al., editors. Train aerodynamics [Internet]. Butterworth-Heinemann; 2019. p. 53–71. Available from: https://www.sciencedirect.com/science/article/pii/B9780128133101000046.
[32] Matthews BW, Fletcher CAJ, Partridge AC. Computational simulation of fluid and dilute particulate flows on spiral concentrators. Applied Mathematical Modelling. 1998;22(12):965–979.
[33] Ye G, Ma L, Alberini F, et al. Numerical studies of the effects of design parameters on flow fields in spiral concentrators. International Journal of Coal Preparation and Utilization. 2022;42(1):67–81.
[34] Sudikondala P, Mangadoddy N, Kumar M, et al. CFD modelling of spiral concentrator-prediction of comprehensive fluid flow field and particle segregation. Minerals Engineering. 2022;183:107570.
[35] Gao S, Meng L, Zhou X, et al. Design of partial cross-sectional geometry and prediction of separation performance in the spiral separator. Separation Science and Technology. 2022;57(13):2127–2144.

5 Assessment of Slurry Erosion in Pumps and Hydraulic Turbines

Virinder Kumar and Jashanpreet Singh

5.1 INTRODUCTION

Erosive wear is the term used to describe the plastic deformation or removal of material resulting from the impact of high-speed slurry particles on a surface, often driven by erosion or corrosion processes. The target surface experiences mechanical failure due to the mechanical interactions generated by the relative motion of fluid particles in relation to the solid surface. The surface is struck by solid particulates, which result in material failure through the mechanism of random collisions or directional impact [1]. Slurry erosion wear results from the velocity distribution of slurry particles and the erosion of the solid surface. This wear occurs in two distinct phases. In the first phase, slurry particles suspended in the fluid strike the target surface, causing plastic deformation. In the subsequent phase, the incessant erosive effect results in the removal of surface material. Corrosion-induced material failure results from an electrochemical process that may be regulated by suitable techniques [1–3]. Numerous researchers implemented mild steel [4–19] to examine erosion wear. They disclosed that the erosion attrition rate was contingent upon the operating conditions. Clark and Llewellyn [20] discovered that erosion is inversely related to hardness and is related to the square of the velocity. Desale et al. [16] investigated that erosion attrition decreased as sediment concentration increased. The erosion behaviour of the target surface is influenced by the shape, size, and mechanical properties of both the target material and the eroding elements [10,14,16,20–29]. Operational factors, including the velocity of erodent particles, flow rate, particle concentration, and the angle of particle impact on the surface, significantly influence erosion wear.

The mining process subjects minerals to the most severe conditions. In the accelerated context, these hazards are exacerbated. Hydraulic components are subjected to increased wear when impacted by high-velocity particles [30,31]. Figure 5.1(a–c) illustrates schematic representations of the damaged pump components. Erosion wear is common in oil and gas, mining, and thermal power plants handling slurries. The chemical, thermal, and mining industries convey a variety of slurries, such as coal, oil, charcoal, detritus, and metal residues.

All systems that are intended to transport liquid, sediment, or slurry are referred to as pumps. As a result, the slurry pump is vital to the efficiency of the slurry transportation system. Liquids and solid–liquid combinations are frequently transported by slurry pumps in the mining and chemical industries. Slurries are typically transported using centrifugal and progressive-cavity slurry pumps. Various parameters, including the slurry's concentration and viscosity, the particles' physical properties, and the slurry's corrosive nature, all contribute to the pump's performance. Numerous researchers have attempted to understand the flow dynamics within slurry pumps when handling both solid–liquid mixtures and clear water. Design parameters are essential for estimating pump efficacy. Slurry pumps are widely used to transport various slurries in power plants, mining, pharmaceuticals, and refineries. In thermal power facilities, the disposal and utilisation of ash are critical issues. Thermal power plants, mining, and industrial sectors have long contended with the erosion problem produced by slurry suspended particles. Consequently, it emerges as a vital consideration for researchers [32,33].

DOI: 10.1201/9781003561156-6

Many engineers, designers, and researchers recognise the importance of implementing protective measures to safeguard machinery, equipment, and other systems in hydropower plants. This necessitates a sophisticated level of process-oriented expertise, which encompasses a viable solution to such failure scenarios. In northern Indian hydropower facilities, sediment erosion in hydro-turbines and their parts is a substantial challenge. Quartz and feldspar are among the stony particulates that are present in the rivers of the Himalayas. During the rainy season, sediment is transported via runoff, resulting in considerable degradation of the hydro-turbines. Water turbine components are susceptible to wear when subjected to high sediment levels due to the exceptional hardness of the silt particles [34]. Silt with around 50% quartz content leads to various turbine issues [35–37]. Hydropower facilities employ two schemes: pumped storage and run-of-river (ROR). Erosion of sediment resulting from particle sedimentation is evident in each system [38]. Hydro-turbines are mainly classified as impulse, reaction, or gravity types [39]. The categorisation of turbines is essential for identifying turbine failures. The main failure modes in impulse and reaction turbines include cavitation, abrasion, corrosion, fatigue, and inherent defects [30,40]. Failures attributable to material fatigue and flaws may also depend on the operational circumstances of the power plant; however, erosion, corrosion, cavitation and abrasion are contingent upon the conditions of the water supply. Figure 5.1(d-f) illustrates the common surface wear patterns observed in impulse and reaction turbines. Francis turbine components are vulnerable to sediment erosion [35]. This degradation can lead to reduced power generation, leakage, and deformation of the runner and blades, as well as structural issues [36]. Studies show that runners and draft tube cones of Francis and Kaplan turbines undergo cavitation [39].

This chapter presents a literature overview of prominent computational fluid dynamics (CFD) analysis work related to turbines and pumps. This chapter is organised into sections on slurry kinds,

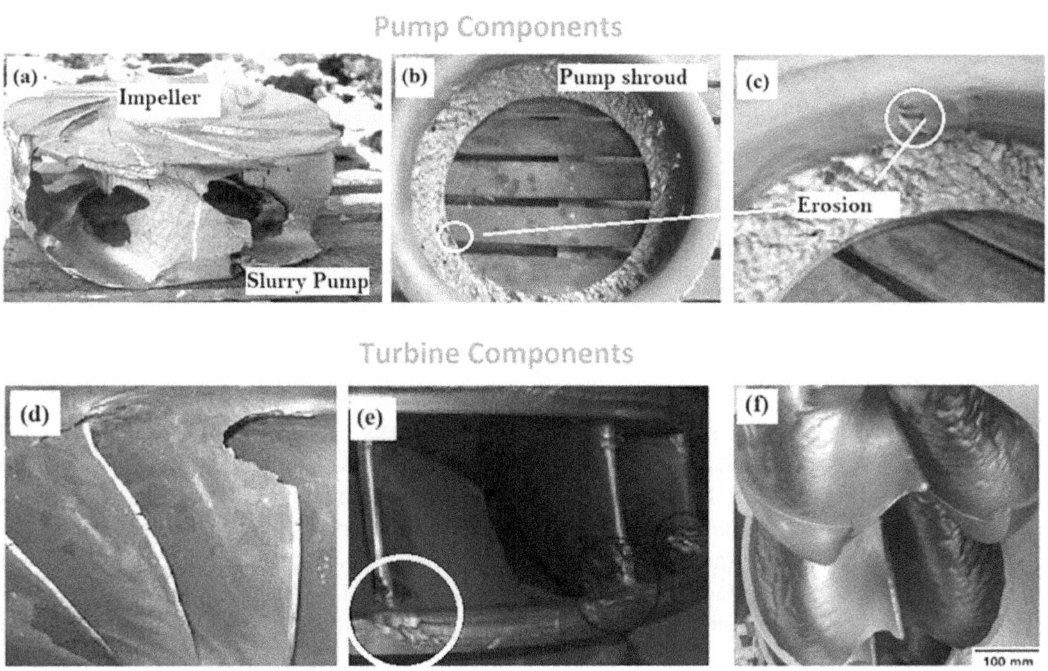

FIGURE 5.1 Eroded (a) impeller of slurry pump [41]; (b, c) pump shroud [42]; (d) runner at Jhimruk Hydropower plant, Nepal, after one year in service [43]; (e) inlet of runner at Cahua Hydropower plant, Peru [44,45]; (f) buckets of Pelton turbine of Toss Mini Hydel Project, Kullu, H.P., India [46].

(a–f: Permissions ©Elsevier)

fundamental erosion wear models, the erosion performance of hydraulic materials, and CFD related to turbines and pumps. The literature review concludes with an extensive analysis and potential inferences.

5.2 EROSIVE SLURRY PARTICLES LEAD TO EROSION

Slurry may exhibit characteristics of either homogeneity or heterogeneity. Typically, variously shaped particles undergo processing via hydraulic machinery. Usually, the particulates consist of alumina [26], silicon carbide [19], tailing [26], concrete [29], sand/silt [47], fly ash [47], coal [48], bottom ash [47], glass beads [49], quartz [20], ores [50], and others. In power plants, the majority of slurries are regarded as homogeneous. Nonetheless, the effective blending of slurries significantly exceeds the current methodologies employed. Power plants prepare slurries with specific concentration levels and particle size distributions to meet requirements for either disposal (such as ash) or combustion (such as coal). Two varieties of ash slurry mixtures are transported: the fine-size slurry, which is constituted of microscopic fly ash particles, and the coarse mix, which contains bottom ash. In hydropower facilities, silt is an undesirable substance accompanying flowing water in varying quantities. The classification of slurries is based on their flow behaviour at varying concentrations, a concept referred to as rheology. The rheological characteristics of slurry encompass several factors, including the size, density, shape, and concentration of the solid particles [48].

5.3 EROSION RESISTANCE OF HYDRAULIC MACHINERY MATERIALS

The material used in hydraulic apparatus and its parts is essential for the safe transportation of slurry. Turbines commonly employ 18Cr-8Ni austenitic steel, 13Cr-1Ni martensitic steel, and 13Cr4Ni or 16Cr5Ni martensitic–austenitic steels. Although austenitic steel with less than 0.2% molybdenum offers limited fatigue and creep resistance, it exhibits strong resistance to cavitation and corrosion. This steel was employed to construct the turbine runners at the hydroelectric facilities in Uttar Pradesh, Tamil Nadu, Pathri, and Mattur [51]. Turbine steels such as 13Cr4Ni and duplex steel SAF 2304 demonstrated consistently high erosion resistance, whereas 13Cr1Ni turbine steel showed the poorest performance. The 16Cr5Ni steel exhibited higher erosion rates than the 13Cr4Ni and the duplex steel. Martensitic steel 13Cr1Ni was substituted for 18Cr8Ni; however, its low hardness, ductility, and Ni content render it susceptible to sand erosion, cavitation, and corrosion. 13Cr6Ni steel was patented in the late 1960s; however, the patent rights' limitations led to the development of 13Cr4Ni and 17Cr4Ni steels and 16Cr5Ni. Among these, 13Cr4Ni and 16Cr5Ni are widely employed in turbine manufacturing [51].

Improving surface hardness may be an effective strategy to mitigate erosion wear in these components [16]. Stainless steel (SS) showed better erosion wear resistance than mild steel [19,52]. Adding chromium and nickel significantly improved the erosion resistance of cast iron. Rateick et al. [53] examined the impact of cementing of WC with 10% Co on SS 440C, utilising Al_2O_3 as the erodent material. SS is frequently used to construct components, including pipe connectors, elbows, pump casings, shaft sleeves, and many other components [54–70]. Erosion wear was examined in a variety of SS grades [19,71–73]. Burstein et al. [25] discovered that the erosion wear of SS304 intensified with impact angles between 5 and 22.5° and subsequently decreased until it approached 90°. The evaluation indicated that AA6063 exhibited the highest wear at a 15° impact angle.

Modi et al. [74] evaluated the slurry erosion resistance of high-carbon steel using coal and bottom ash as erodents. They applied surface treatments, hardening, and annealing processes to assess how microstructural changes affect erosion behaviour. The findings indicated that hardened steel exhibited approximately tenfold wear resistance compared to annealed steel, with both materials processed at 900°C for 1 hour. The Metture dam in India utilised SS304 for the first time in 1998, replacing grey cast iron pipes for raw water supply. SS exhibits significant hardness, impressive impact strength, and favourable welding characteristics, comprising a low carbon content of 0.3% by weight. SS304 and SS316 demonstrate superior erosion and wear resistance compared to low-carbon or mild steel [19,72,75]. Sharma [76] identified a direct correlation between cumulative erosion and exposure period in welded layers formed through the hard-facing deposition of 13Cr-4Ni-0.5Mo (13% Cr, 4% Ni, and 0.5% Mo). Various surface treatment techniques [73,77,78] were employed to enhance the mechanical characteristics of materials and resistance to erosion and abrasion. Coating deposition methods are viable for safeguarding materials used in turbomachinery [79–95]. Design optimisation plays a significant role in managing surface erosion [58].

5.4 EROSION ANALYSIS OF PUMP SURFACE BASED ON CFD

Slurry pumps are designed to convey liquids and solid–liquid mixes across several sectors [96]. Slurries are typically transported using centrifugal and lobe pumps. Centrifugal slurry pumps undergo erosion wear [26]. As previously mentioned, the pump's erosion performance is contingent upon unspecified characteristics. Erosion, cavitation, corrosion, abrasion, and scaling are the causes of surface failures in the pump. Wasp et al. [97] suggested that corrosion can be effectively controlled by adjusting the pH level and incorporating an oxygen inhibitor to remove dissolved oxygen. The slurry pump's design typically mirrors that of a regular pump; however, it features an increased spacing between vanes to accommodate the passage of thick slurry. The selection of materials is a crucial aspect of design for ensuring dependable functioning [98]. Wilson [99] delineated three distinct methods for categorising wear in pump components. Pump wear is primarily caused by solid particles striking the surface at normal, oblique, or perpendicular angles. Researchers [100,101] designed a slurry pump to reduce erosion and improve solid–liquid transport efficiency. Roudnev [102] presents a fundamental notion of the expulsion of the vanes. This design may facilitate the development of vortex flow, allowing abrasive particles to be directed away from the impeller shroud. Numerous studies have been documented using the particle image velocity (PIV) system for particle tracking, which proved advantageous for optimising pump component design (see Table 5.1) [103–105].

5.4.1 Experimental Studies on Erosion Wear in Pumps

Numerous studies have examined erosion wear in various pump parts utilising testers and loops. Evaluative testers for eroding wear are addressed in the preceding sections. Widenroth [112] employed the paint wear technique to evaluate erosion wear in a pump impeller and casing. He discovered that the erosion wear of ductile materials was minimal at elevated impact angles. As grain size increases, corrosion progressively transitions from the suction side to the pressure side of the blade. Widenroth [113] investigated impeller wear using a weight-loss approach and assessed casing wall thickness with an ultrasonic device. His findings revealed that vane thickness was reduced due to erosive wear, particularly in regions where solid particles impacted the surface at steep angles. Read [114] conducted experimental research utilising an ultrasonic instrument to determine the thickness of the casing wall. He noticed that erosion wear is most common in the border section of the dredge pump's casing. Deep cavities were observed on the impeller's peripheral vanes. Roco [96] identified three distinct factors contributing to pump wear: random, direct, and sliding impact from solid particulates on the surface. Severe erosion wear was observed next to the casing wall. A comparable assertion was articulated by Liu et al. [115]. Rayan and Shawky [116] evaluated the erosion behaviour of a cast steel impeller employing a weight-loss methodology. The experiments were conducted with a low-concentration solution (2–4% by volume). The degradation of the cast

TABLE 5.1
Erosion Wear Models Based on Various Influencing Factors

Researchers	Numerical Models for Erosion Wear	
Finnie [106,107]	$\varepsilon_{VP} = \begin{cases} \dfrac{m_p V^2{}_p}{P_{\psi k}}\left(\sin 2\alpha - \dfrac{6}{k}\sin^2\alpha\right) & \tan\alpha \le \dfrac{K}{6} \\ \varepsilon_{VP} = \dfrac{m_p V^2{}_p}{P_{\psi k}}\left(\dfrac{k\cos^2\alpha}{6}\right) & \tan\alpha \ge \dfrac{K}{6} \end{cases}$	(5.1)
Bitter [108]	$\varepsilon_{VP} = \dfrac{1}{2}\dfrac{M\left(V_p \sin\alpha - K\right)^2}{\delta}$	(5.2)
Bitter [109]	$\varepsilon_{VC} = \begin{cases} \dfrac{2M\left(V_p \sin\alpha - K\right)^2}{\sqrt{V_p \sin\alpha}}\left(V_p \cos\alpha - \dfrac{C\left(V_p \sin\alpha - K\right)}{\left(V_p \sin\alpha\right)^2}\chi\right); & \alpha \ge \alpha_{p0} \\ \dfrac{\frac{1}{2}M\left[V^2{}_p \cos^2\alpha - K_1\left(V_p \sin\alpha - K\right)^{3/2}\right]}{\chi}; & \alpha \le \alpha_{p0} \end{cases}$	(5.3)
Neilson and Gilchrist [105]	$\varepsilon_{VC} = \begin{cases} \dfrac{\frac{1}{2}M\left(V^2{}_p \cos^2\alpha - V^2{}_r\right)}{\chi} + \dfrac{1}{2}\dfrac{M\left(V_p \sin\alpha - K\right)^2}{\delta}; & \alpha \le \alpha_{p0} \\ \dfrac{\frac{1}{2}M\left(V^2{}_p \cos^2\alpha\right)}{\chi} + \dfrac{1}{2}\dfrac{M\left(V_p \sin\alpha - K\right)^2}{\delta}; & \alpha \ge \alpha_{p0} \end{cases}$	(5.4)
Oka et al. [110,111]	$E_T = KV_p{}^m d_p{}^n H_t{}^{K1} sin\alpha\left(1 + H_t\left(1 - sin\alpha\right)\right)^{n_2}$	(5.5)

steel impeller was reported to be comparable to the rate observed during cavitation, with a velocity exponent of 3 reported in relation to erosion wear.

5.4.2 Computational Study of Erosion Wear in the Pump and Its Parts

Numerical techniques are also employed to assess the pump wear caused by erosion. Numerous studies have employed the finite-element method (FEM) to evaluate wear in the pump and its parts. Minemura and Zhong [100] carried out a computational study on pump casing erosion utilising the FEM via the Lagrangian approach. The simulations employed the models suggested by Bitter [108,109]. They referenced two distinct mechanisms: cutting wear and deformation. Roco [117] utilised the finite-element approach to examine the progression of solid particle concentration across three distinct pump impeller blade profiles. The blade angle influences the formation of solid concentration. Dong et al. [118] conducted a computational analysis of wear in pumps utilising the FEM. They recorded the erosion wear progression through distinct phases: from steady to saturation wear, followed by a transition from saturation to sharp wear. The casing experienced pronounced erosion wear along the volute edge. Pagalthivarthi and Gupta [119] conducted a computational investigation of erosion wear in rotating channels utilising the FEM via the Lagrangian approach. They indicated that impact wear was significantly less than slide wear, namely by a factor of around

10 to 50 times. Pagalthivarthi and Visintainer [120] utilised a FEM approach to forecast the areas of maximum erosion on the casing wall. Additionally, Pagalthivarthi et al. [121,122] employed a 2D Eulerian–Eulerian model and a 3D FEM methodology to numerically analyse erosion wear on casings, shrouds, and impeller blades.

Numerous researchers have implemented finite-volume methods (FVMs) to forecast erosion wear in the pump and its parts. Roco and Addie [1] conducted a numerical analysis of pump deterioration many years ago. Utilising the energy approach, they implemented both FEM and FVM. They noted that slurry concentration and flow velocity strongly affect pump design optimisation, which can be controlled by modifying the thickness of the wetted walls. Roudnev et al. [102] similarly asserted this, performing 3D steady-state and transient simulations using the finite-volume method via ANSYS CFX 5. Pagalthivarthi et al. [124] also employed the finite-volume approach to forecast the impact and erosion wear on the pump casing. Using various impeller design configurations, Bross and Addie [125] conducted a numerical study on impeller vacuum seal wear. They found that the vacuum sealing region of the impeller is highly susceptible to severe erosion wear. Sellgren et al. [123] observed that at higher specific speeds, casing wear decreases, whereas at lower specific speeds, the impeller and suction liner experience reduced erosion. Furthermore, they found that the pressure face exhibited more degradation than the suction face, as shown in Figure 5.2. Shen et al. [126] investigated sediment-induced wear on the impeller of a double-suction centrifugal pump used in the Jingtai Yellow River Irrigation Project in China. They observed that sand particles caused grooving scratches and pitting on the blade surfaces. Zigzag-shaped erosion patterns were identified along the blade's trailing edge, illustrated in Figure 5.3 (a, b). Additionally, they found that the erosion rate density on the blade's suction (vacuum) side was higher than on the pressure side, as depicted in Figure 5.3 (c, d). Upon entering the impeller domain, particle velocity shifts from

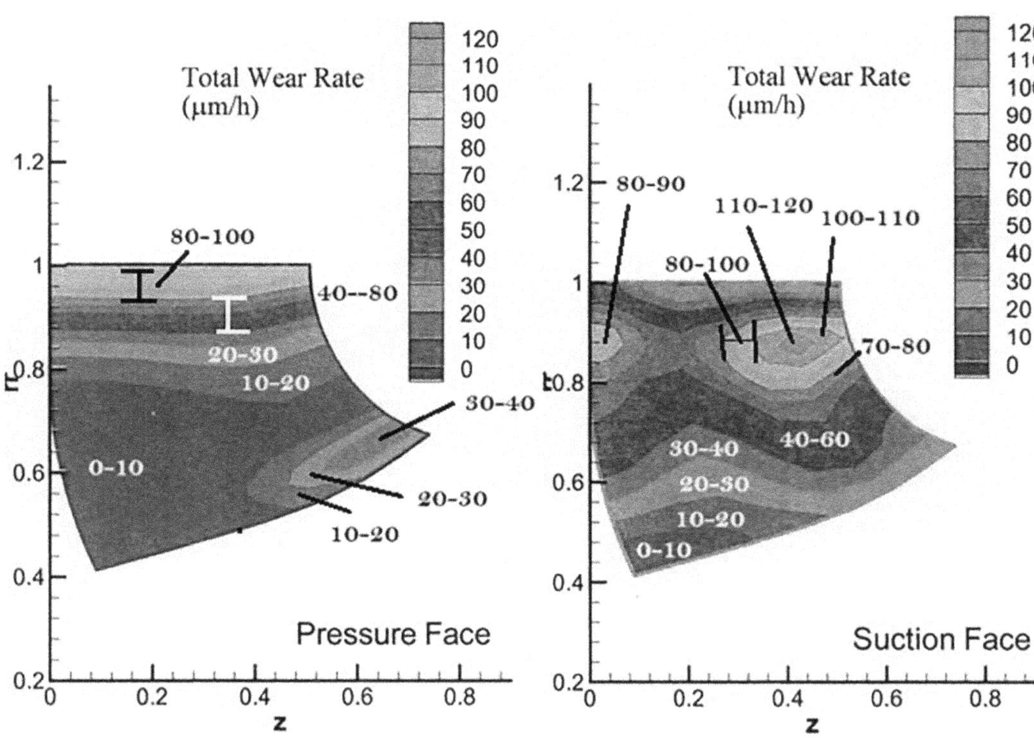

FIGURE 5.2 CFD-predicted erosion wear contour on the impeller blade [123].

(Open domain article)

axial to radial direction, evident from the comparison of experimental and numerical results at the impeller's leading edge in Figure 5.3 (e, f). As particles strike the rotating blades at steep angles upon entry, this area experiences intense and localised erosion.

Using the finite volume method, Li et al. [128] reported severe erosion concentrated at the impeller hub. They also found that reducing the impeller blade exit angle can help minimise

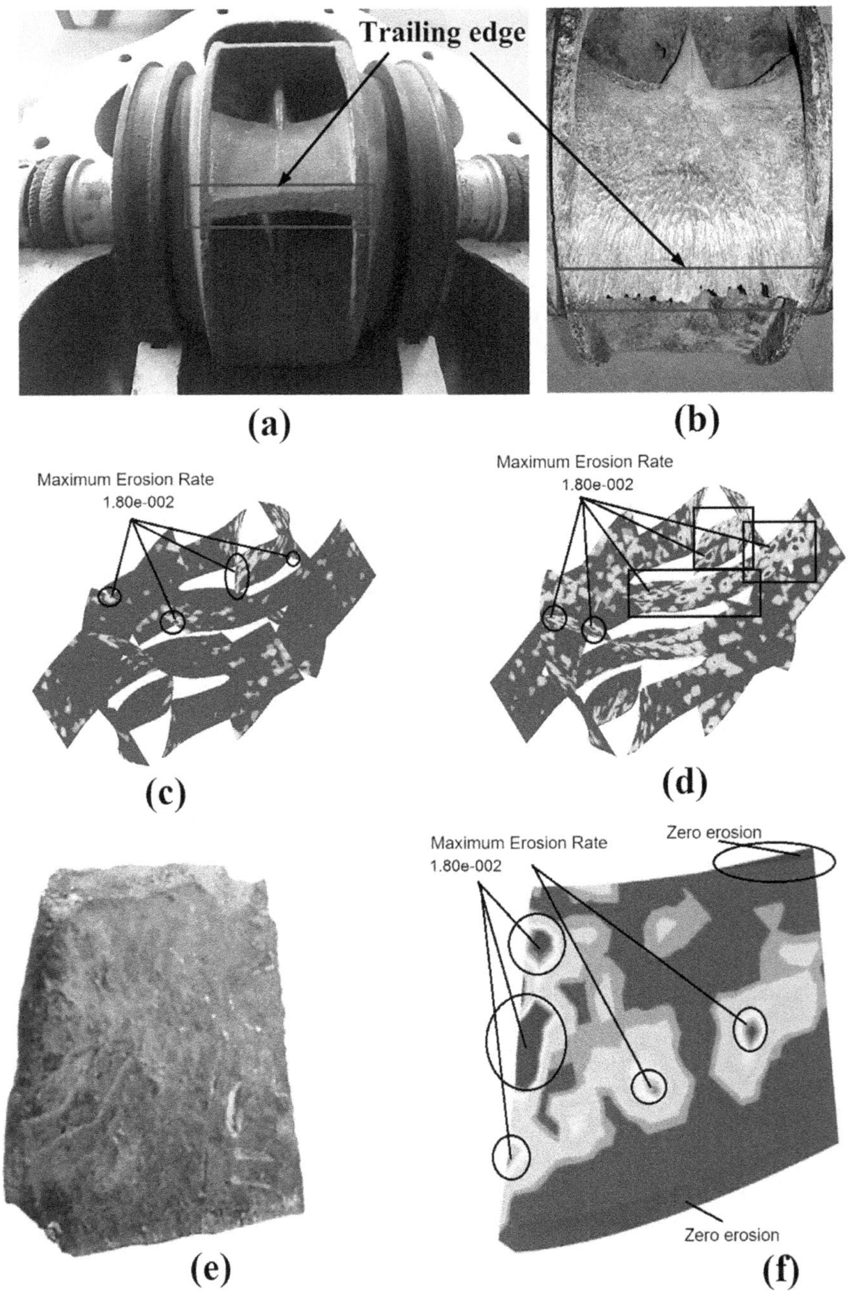

FIGURE 5.3 Pump impeller surfaces (a): before and (b): after 5000 hours of erosion; (c, d) erosion rate on pressure and suction sides; and (e) experimental vs (f) simulated wear on leading edge [126].

(Permission ©Elsevier)

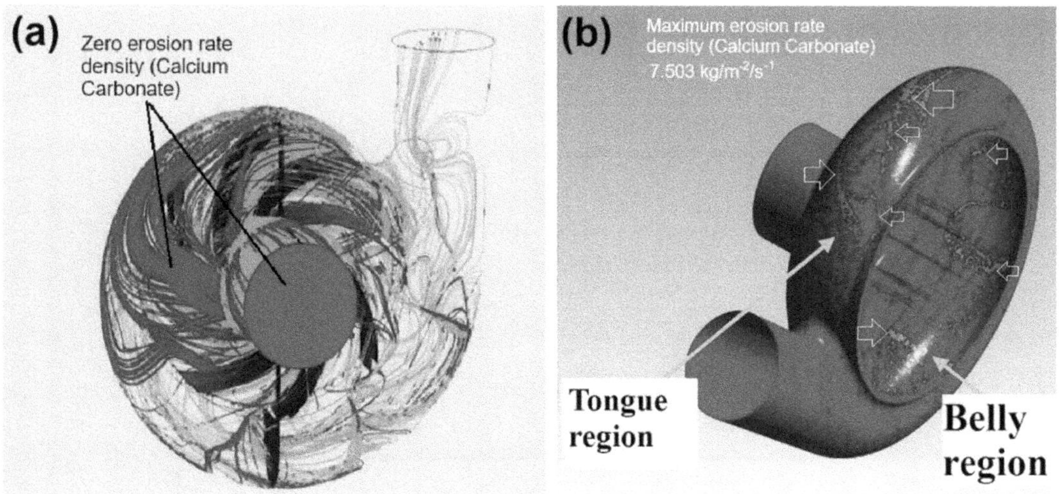

FIGURE 5.4 CFD results depicting (a) particle trajectories of lime slurry and (b) erosion rate density within the pump casing [127].

(Permission ©Elsevier)

erosion wear. Most studies have shown that the leading edge of the impeller blade undergoes the most severe erosion wear [129–131]. Using the discrete phase model, Pagalthivarthi et al. [132] used the CFD software Fluent to evaluate erosion wear on the pump casing. Many scholars have examined pump erosion wear [127,133–135]. Noon and Kim [127] conducted ANSYS-CFX simulations to investigate erosion rate density in a slurry pump. The study focused on lime slurry ($Ca(OH)_2$) with an average particle size of 1.20 mm, analysed within a temperature range of 70–90°C. The investigation revealed that the lime slurry particles consistently flow at the casing entry, initiating a swirling motion around the impeller blades, as seen in Figure 5.4 (a). They also observed peak erosion density at the volute casing's tongue–belly junction, as shown in Figure 5.4 (b). Figure 5.5 illustrates erosion rate density relative to velocity and particle concentration at 70°C.

5.5 SURFACE EROSION EVALUATION OF HYDRAULIC TURBINES BASED ON CFD

The deposition of sediment particulates causes surface failures in hydro-turbines. Quartz, alumina, and feldspar are present in high concentrations in the river water of the Asian highlands. Ongoing advancements have been noted in optimising turbine design to mitigate material degradation. Hydropower facilities employ two distinct systems: pumped storage and ROR. Nonetheless, erosion resulting from the sedimentation of particles is evident in both systems [136]. Researchers [137–139] indicated that ROR schemes experience hydro-abrasive erosion in turbine components. Researchers have reported the deterioration of turbine parts in Indian hydroelectric power projects [140,141]. Researchers [142–144] demonstrated the decline of turbine parts in Nepal's hydroelectric power facilities. Aslam-Noon and Kim [145] reported the degradation of turbine parts in Pakistani hydroelectric power projects. The erosion in hydroturbines is similarly contingent upon the previously mentioned characteristics [146–148]. Literature indicates that hydroturbines fail due to high fatigue, cavitation, erosion, and material defects. Francis and Kaplan turbines experience cavitation in runners and draft tube cones. Conversely, Pelton turbines experience cavitation in the runner buckets, nozzles, and needles. Diverse tester types were

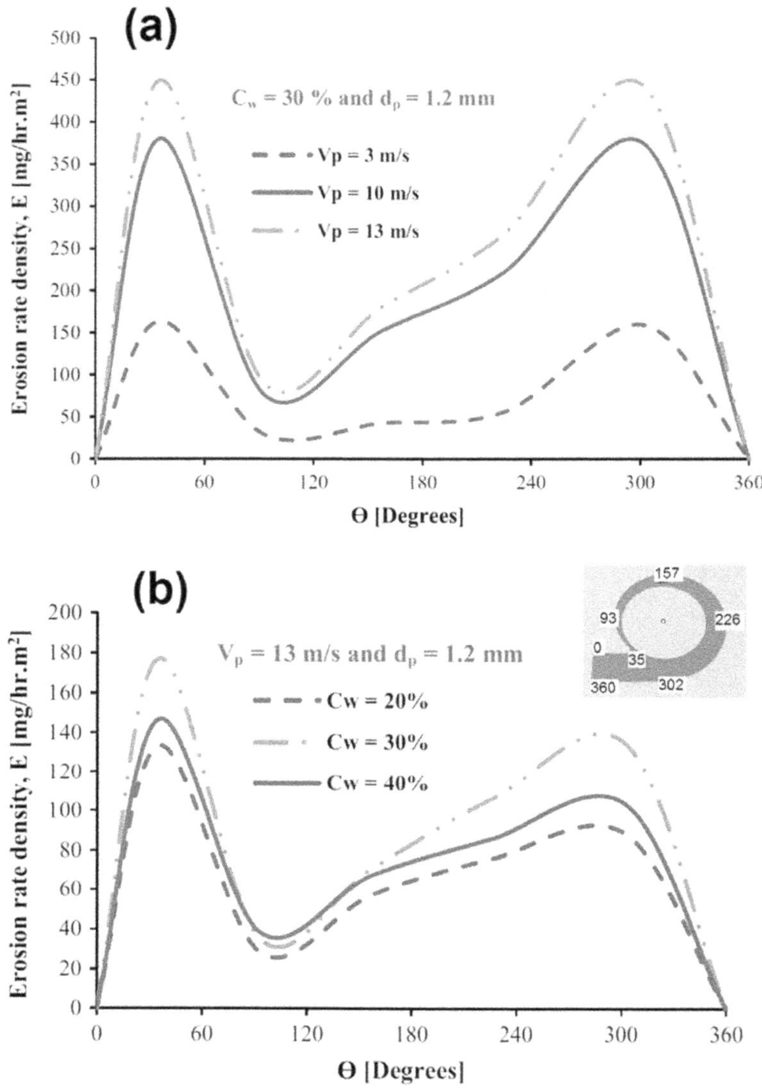

FIGURE 5.5 Density of erosion rate as a function of (a) velocity and (b) concentration at 70°C [127]. **(Permission ©Elsevier)**

employed to forecast turbine wear. Grewal et al. [149] developed a high-velocity jet apparatus that is particularly useful for evaluating erosion wear in Pelton and Turgo impulse turbines exposed to sediment flow [148,150,151]. Researchers have developed erosion tests that effectively predict the wear of Francis runner blades [152,153]. In addition, researchers [154,155] developed a tester to assess the erosion of various turbine materials.

The reviewed literature underscores the use of erosion wear correlations proposed by scholars such as Neilson and Gilchrist [105] to investigate the deformation wear in turbine components under normal impact angles. Krause and Grein [156] established a correlation to determine the abrasion wear rate (mm/h) on the surface of a Pelton runner constructed from conventional steel (X5CrNi13/4), as outlined in the following:

$$\delta = pqv^{3.4} f\left(d_{p50}\right) \tag{5.6}$$

In this context, the symbol p represents the constant of the equation. The symbols q, c, v, and $f(d_{p50})$ represent the constant, quartz content, mean sand concentration, relative jet velocity, and a function that defines particle size, respectively.

Naidu [157] reported an expression to estimate the rate of silt-induced erosion:

$$W = S_1 S_2 S_3 S_4 M_r \upsilon^x \tag{5.7}$$

S_1, S_2, S_3, and S_4 represent the coefficients for silt concentration, hardness, particle size, and particle morphology, respectively. The relative velocity of the flowing water is denoted by v, while the wear resistance of the material used in the component is indicated by the coefficient Mr. The exponent x is defined as follows:

- $x = 1.5$ (Pelton runner buckets)
- $x = 2.5$ (the pivot ring liner and guide vanes of the Francis and Pelton nozzle)
- $x = 3$ (Francis runner)

Asthana [158] proposed an expression for the prediction of turbine abrasion (*TA*):

$$TA = f\left(PE, \upsilon^z\right) \tag{5.8}$$

The symbol z represents the exponent of relative velocity. PE, which denotes the modified dispersed sediment content, and υ, which denotes the relative velocity between the turbine components and the streaming water, determine the function of TA. The following expression can be employed to calculate PE:

$$PE = P^{\alpha} a^{\beta} k_1 k_2 k_3 \tag{5.9}$$

In this context, k_1 represents the coefficient associated with shape features. The value of k_1 can be 0.75, 1.0, or 1.25, contingent upon the irregularities present in the particles. The symbol k_2 represents the coefficient of hardness, where $k_2 = 1$ indicates a hardness greater than 3 on Mohr's scale, and $k_2 = 0.5$ signifies a hardness less than 3 on Mohr's scale. Symbol k_3 equals 1, representing the coefficient of abrasion resistance for 13Cr4Ni steel. The symbol P represents the annual average of suspended sediment content (gm/l), indicating long-term sedimentation in the river. The value P equals 1, with a concentration of up to 5 g/l. The symbol *a* represents the coefficient associated with the average grain size of suspended sediment, specifically for a base measurement of 0.05 mm. The exponent β serves as the correction factor for the average particle size. The exponent value is $\beta = 1$, applicable for particle sizes up to 0.6 mm.

Schneider and Kachele [159] reported the following correlation to predict the wear rate (*W*):

$$W \sim cq\upsilon_f^n f\left(d_{p50}\right) \tag{5.10}$$

The nomenclature for the parameters mentioned previously is consistent with that used in the correlation by Krause and Grein [156]. The velocity exponent range was specified as 2.1 to 3. The fluctuation in the velocity exponent's value reflects its constraint.

Additionally, computational fluid dynamics instruments are implemented to investigate turbine erosion. The Finnie model [107] and the Tabakoff–Grant model [160] are the two most frequently employed computational models in the literature. Finnie's model [107] reported the erosive wear (E) in units of the kinetic energy of erodent particles:

$$E = k.V_P^3.f\left(\alpha\right) \tag{5.11}$$

where V_P represents the particle's impact velocity and $f(\alpha)$ is the function of the impingement angle at which the particles strike the target. The expression for the value of $f(\alpha)$ is given by:

$$f(\alpha) = \sin 2\alpha - 3\sin^2\alpha \qquad \{\tan\alpha \le 1/3\} \tag{5.12}$$

$$f(\alpha) = \frac{1}{3}\cos^2\alpha \qquad \{\tan\alpha > 1/3\} \tag{5.13}$$

Tabakoff and Grant [160] presented a comprehensive model for predicting erosion wear, which is as follows:

$$E = k_1.V_P^2.f(\alpha).\cos^2\alpha.\left(1 - R_T^2\right) + f\left(V_{pN}\right) \tag{5.14}$$

The impact angle function is defined as:

$$f(\alpha) = \left[1 + k_2 k_2 \sin\left(\alpha \frac{\pi/2}{\alpha_0}\right)\right]^2 \tag{5.15}$$

where k_1, k_2, k_3, and k_4 are the constants that are based on the properties of the erodent and the target. The value of $k_2 = 1$ if $\alpha \le 2\alpha_o$ and $k_2 = 0$ if $\alpha > 2\alpha_o$. The symbol R_T is represented by the following:

$$R_T = 1 - k_4 V_P.\sin\alpha \tag{5.16}$$

and $f\left(V_{PN}\right) = k_3\left(V_P \sin\alpha\right)^4$

This model yields results that are in excellent accord with the pattern identified by Lyczkowski and Bouillard [161].

The Francis turbine's stainless-steel vane was subjected to computational analysis by Hamed et al. [162]. The numerical simulations were conducted using Fluent 6.1 code. The turbulences were computed using the RNG k–ε model. The stainless-steel vane's irregularity was assessed within the 1.07–2.14 μm range. One million mesh points were utilised to calculate the solution. Quartz particles (150 μm) were introduced into the water. Figure 5.6 illustrates the outcomes reached during their investigation.

The computational analysis of the Francis turbine's rotor was conducted by Teran et al. [138] using the finite-volume methodology in the ANSYS code CFX. This investigation employed a Frozen rotor mesh with a multiple reference frame approach to study rotor-stator interactions. They employed water as a flowing medium, which contained 400,000 sediment particles. The steady-state condition was used to determine that the flow of media was turbulent and incompressible. Erosion was computed on the blades on the pressure and runner sides. They discovered that the blade experienced high erosion wear at a partial load of 3 MW, as illustrated in Figure 5.7 (a, b). Cavitation was also observed to be more pronounced at the optimum load (1.2 MW) and reduced at the partial load (3 MW), as illustrated in Figure 5.7 (c, d).

The turbine stage with contorted rotor blades was predicted by researchers [164–167] to have a variety of failure mechanisms. Chitrakar et al. [167] comprehensively analysed the sediment erosion in Francis turbines. Noon and Kim [145] performed a FEM using Finnie's model within the CFX software to predict the best efficiency point of the Francis turbine. The runner's rotation was determined to be 121.6 rpm. The pressure and velocity terms were coupled using a one-way coupling scheme. The mesh was divided into 15 layers, each with a thickness of 10% (length of the first layer: 0.030 m). Figure 5.8 illustrates the comparison between actual wear and CFD-predicted erosion contours across different components of the Francis turbine. The numerical investigation conducted by Fecarotta et al. [168] examined PAT's erosion due to solid particles' impingement. They created a CFD model by integrating the PAT with the in-house E-CODE. High PAT wear was observed in

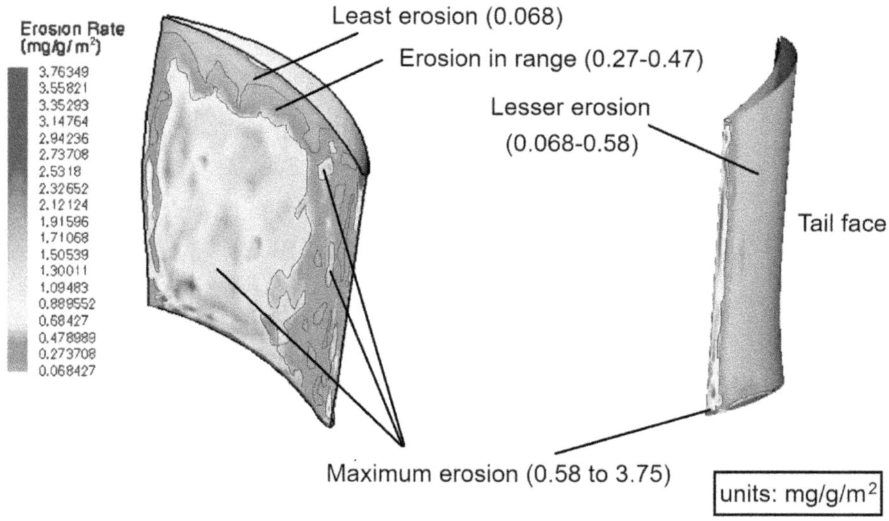

FIGURE 5.6 CFD simulation showing erosion rate distribution across the impeller blade surface [163].

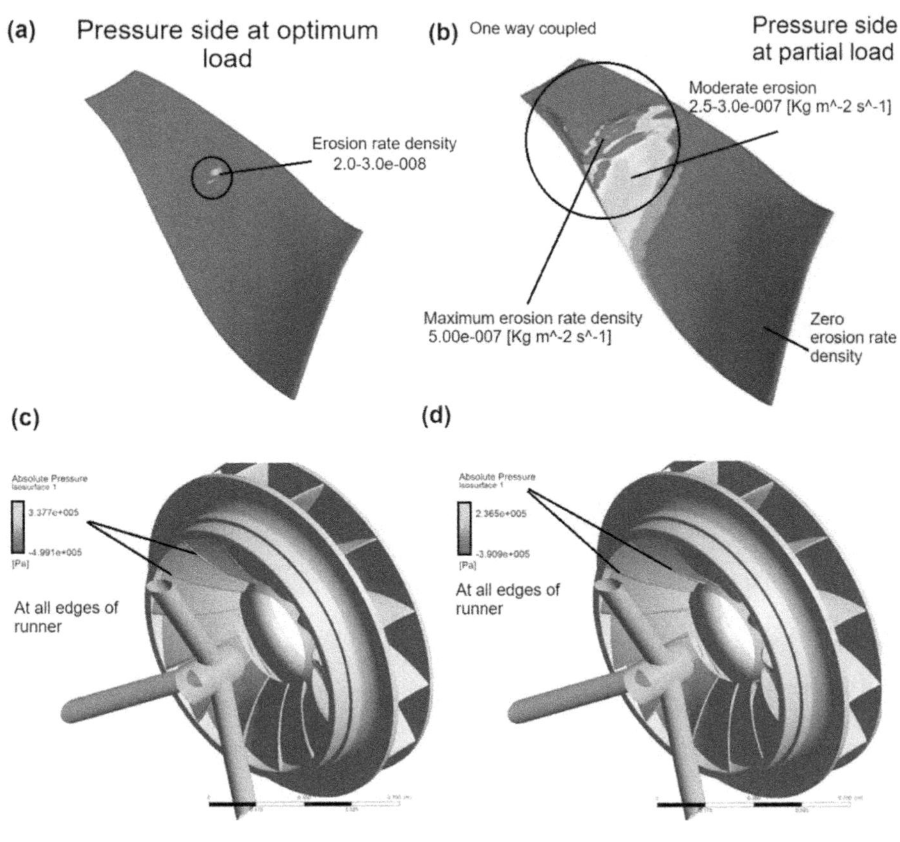

FIGURE 5.7 (a, b) Erosion on the pressure and runner blade sides. (c, d) Cavitation observed under optimum and partial load conditions [138].

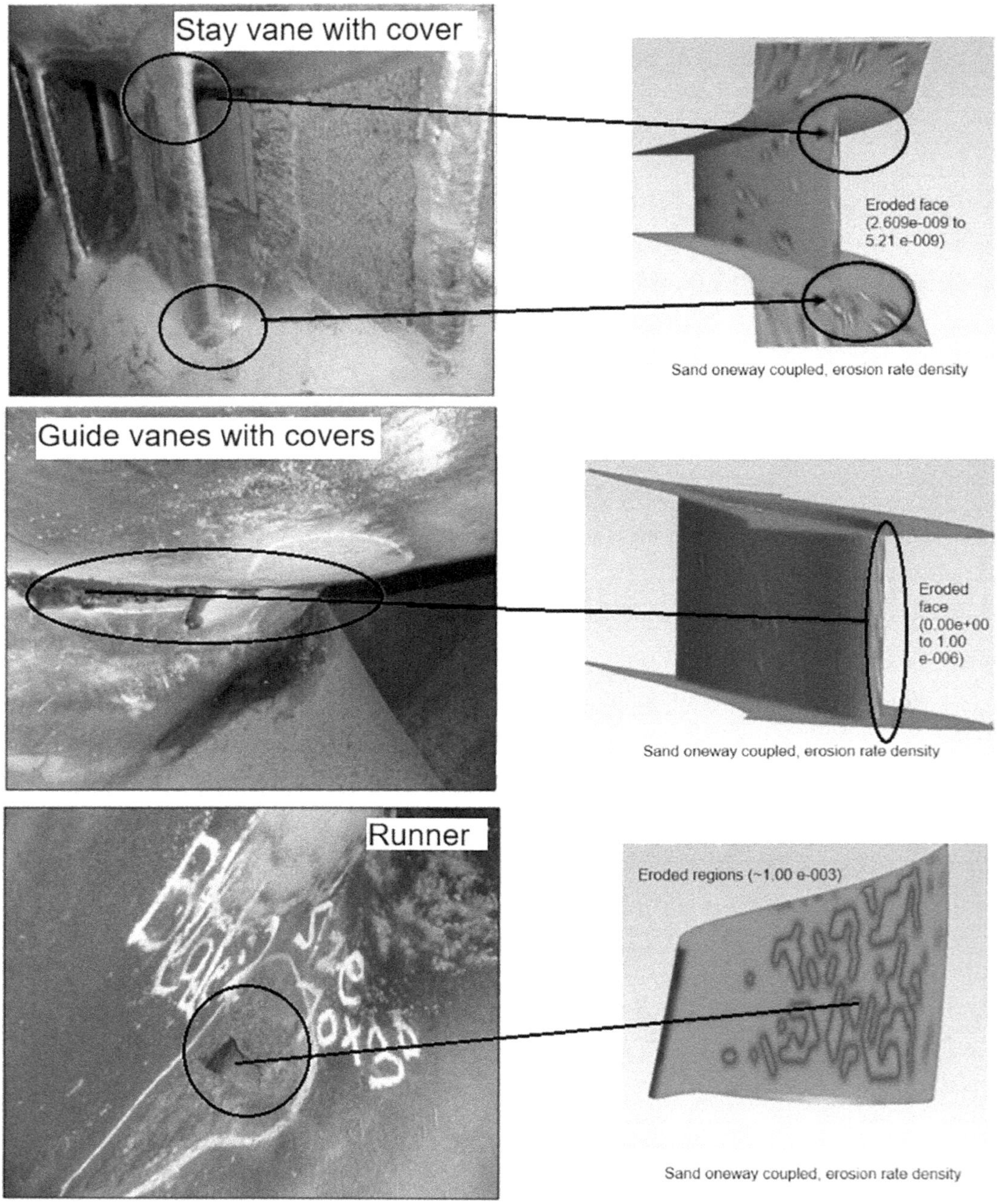

FIGURE 5.8 Observed erosion wear and CFD predicted erosion wear contours across different parts of a Francis turbine [145].

(Permission ©Elsevier)

the volute casing and outlet pipe, while the impeller experienced comparatively less attrition. The hydro-abrasive erosion in Pelton containers was assessed by Rai et al. [169] using a 3D scanner that consisted of a camera and a light-emitting diode. The cut-out section typically reduced the dividing height and roughness by 3% and 5% of the bucket breadth, respectively, as illustrated in Figure 5.9.

The performance of turbines in a hydro-abrasive erosion environment was extensively investigated by various researchers using the models mentioned earlier to optimise the turbine design [43,170–174]. Table 5.2 illustrates the various steady-state analyses conducted to investigate

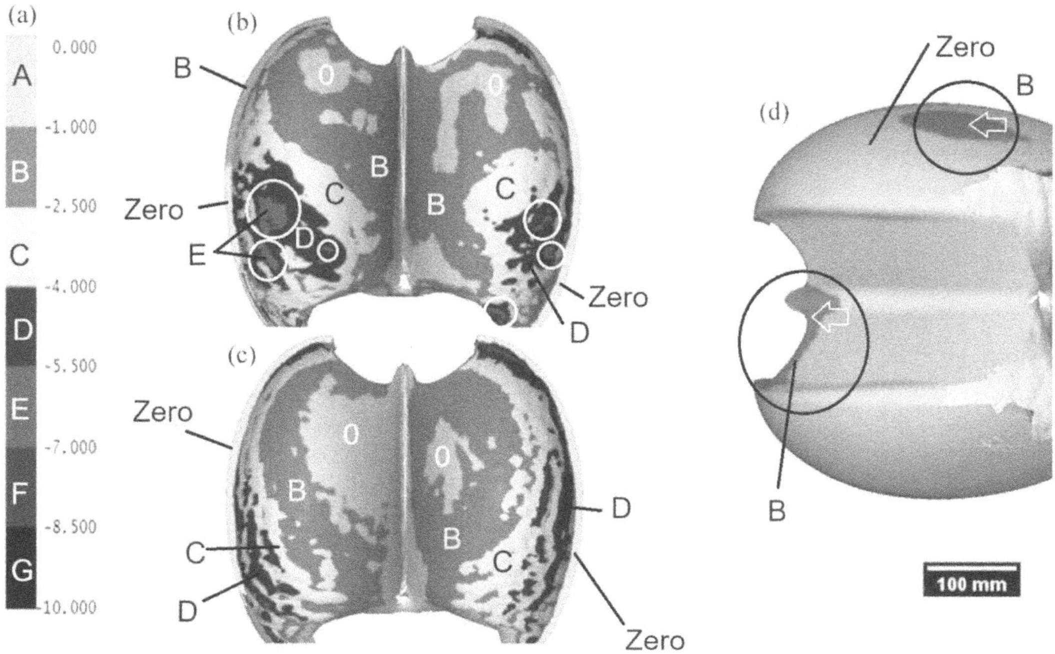

FIGURE 5.9 (a) Erosion extent, (b–c) variation in erosion depth due to internal and external hydro-abrasive processes, and (d) unprocessed Pelton buckets [169].

(Permission ©Elsevier)

TABLE 5.2
Numerical Simulation Investigations on Turbine Erosion Using CFD Tools

S. No.	Researchers	Turbine Type	Optimised Component	Grid	Turbulence	Advantages	Drawbacks
1	Thapa et al. [178]	Francis turbine	Blade	2,50,000	*SST k–ω*	• 33% reduction in erosion wear • Low erosion rate and improved efficiency of the blade profile	• 1–2% drop in efficiency
2	Eltvik et al. [179]	Francis turbine	Blade	3,01,000	*Standard k–ε*	• Erosion factor for five distinct turbine designs, dependent on their angles. • 20% decrease in erosion wear	–
3	Shrestha et al. [180]	Crossflow turbine	Guide van	7,18,215	*Standard k–ε*	• Optimised for 30%	• A high erosion rate is observed at the 20% and 40% guide vanes
4	Koirala et al. [144]	Francis turbine	Multiple runner blades	2,82,534	*SST k–ω*	• Thirteen blades show the best performance	• The efficacy of 11 blades is subpar when fully or partially loaded • The efficacy of 17 blades was 8% lower than that of 13 blades

(Continued)

TABLE 5.2 *(Continued)*
Numerical Simulation Investigations on Turbine Erosion Using CFD Tools

S. No.	Researchers	Turbine Type	Optimised Component	Grid	Turbulence	Advantages	Drawbacks
5	Khanal et al. [181]	Francis turbine	Blade	3,00,000	*SST k-ω*	• 31.5 times reduction in erosion wear	• High erosion wear at negative curvature
6	Baidar et al. [182]	Francis turbine	Guide van passage	13,79,922	*SST k-ω*	• The guide vane's angle significantly influences the localisation erosion wear	• Severe erosion occurs at the leading edge
7	Teran et al. [138]	Francis turbine	Rotor vane	*Domain G.G.I. algorithm*	—	• At a partial load of 3 MW, the blade experiences high erosion wear	• Cavitation was high at the optimal power level of 1.2 MW

hydro-abrasive erosion using the ANSYS CFX tool. The circulating media (quartz) were consistent across all investigations. According to Neopane et al. [175], erosion can be mitigated by optimising blade angle distribution and decreasing relative velocity in critical regions. ANSYS CFX and FLUENT solvers have been employed by researchers for numerical simulations. CFD tools have enabled researchers and designers to generate, modify, and visualise a prototype design of a component with greater flexibility. CFD modelling offers a cost-effective method for assessing turbine erosion. Conversely, it necessitates field data for validation [176,177]. The research demonstrates that CFD approaches are more time and cost efficient than experimental techniques, providing a more sustainable and environmentally friendly alternative.

5.6 CONCLUSIONS

This chapter has comprehensively explored the mechanisms, influencing factors, and predictive strategies for surface erosion in hydraulic turbines and slurry pumps operating in solid–liquid flow environments. Emphasis was placed on empirical and computational approaches to model erosion behaviour and the role of material selection, slurry characteristics, and flow dynamics. The review highlighted that the rate and nature of erosive wear are significantly influenced by particle size, shape, impact velocity, concentration, and impingement angle. The heterogeneity of industrial slurries, comprising materials like fly ash, bottom ash, coal, and quartz, contributes to the complexity of predicting wear patterns in hydraulic machinery. The rheological properties of the slurry and the operating conditions, particularly turbulent flow regimes, further complicate this interaction. Experimental investigations have established correlations between wear resistance and metallurgical properties of materials. SS such as SS304, SS316, and 13Cr4Ni have enhanced erosion resistance over traditional mild steels and cast iron, especially under high-impact and cavitating conditions. Coating techniques, surface hardening, and optimised heat treatment processes have emerged as viable solutions to improve component durability. CFD has proved a powerful tool in predicting erosion zones, optimising blade profiles, and evaluating the performance of different hydraulic components under abrasive conditions. Both steady-state and transient CFD simulations have provided valuable insights into flow patterns, erosion rate density, and areas of high stress concentration in pump impellers, casings, and turbine blades. Numerical models like those proposed by Finnie, Bitter, Oka, and Tabakoff–Grant have been widely applied in simulating erosion phenomena with varying degrees of accuracy. Despite advancements, several knowledge gaps remain. Most existing models do not fully account for particle shape irregularity, multiscale particle interactions, transient sediment concentration fluctuations, or non-Newtonian fluid behaviour.

Additionally, real-time validation of CFD results with high-fidelity experimental data remains limited. Future research should integrate experimental and numerical methodologies to develop unified, scalable, and adaptive models for erosion prediction. Understanding and mitigating particulate erosion in hydraulic machinery is vital for increasing operational lifespan, reducing maintenance costs, and ensuring energy efficiency. Incorporating advanced materials, improved surface engineering techniques, and robust CFD modelling will play a critical role in future design strategies. Collaborative research that bridges field data, laboratory tests, and simulation tools will be key to developing more resilient and efficient hydraulic systems across various industrial sectors.

REFERENCES

[1] Roco MC, Addie GR. Erosion wear in slurry pumps and pipes. Powder Technol. 1987;50:35–46.
[2] Lee DI, Lim HC. Erosion–corrosion damages of water-pump impeller. Int J Automot Technol [Internet]. 2009;10:629–634. Available from: https://doi.org/10.1007/s12239-009-0074-5.
[3] Li X, Lü T. Analysis of corrosion failure of petrochemical pipe elbow. J Cent South Univ Technol. 2005;12:119–123.
[4] Clark HM, Hawthorne HM, Xie Y. Wear rates and specific energies of some ceramic, cermet and metallic coatings determined in the Coriolis erosion tester. Wear. 1999;233:319–327.
[5] Clark HM. Specimen diameter, impact velocity, erosion rate and particle density in a slurry pot erosion tester. Wear. 1993;162:669–678.
[6] Hawthorne HM, Xie Y, Yick SK. A study of single particle-target surface interactions along a specimen in the Coriolis slurry erosion tester. Wear. 2002;253:403–410.
[7] Hawthorne HM. Some Coriolis slurry erosion test developments. Tribol Int. 2002;35:625–630.
[8] Islam MA, Alam T, Farhat ZN, et al. Effect of microstructure on the erosion behavior of carbon steel. Wear [Internet]. 2015;332–333:1080–1089. Available from: http://dx.doi.org/10.1016/j.wear.2014.12.004.
[9] Stack MM, Pungwiwat N. Slurry erosion of metallics, polymers, and ceramics: Particle size effects. Mater Sci Technol. 1999;15:337–344.
[10] Stack MM, El-badia TMA. Some comments on mapping the combined effects of slurry concentration, impact velocity and electrochemical potential on the erosion—corrosion of WC/Co—Cr coatings. Wear. 2008;264:826–837.
[11] Rao KVS, Ramesh CS, Girisha KG, et al. Slurry erosive wear behavior of plasma sprayed Cr2O3 coatings on steel substrates. Mater Today Proc [Internet]. 2017;4:10283–10287. Available from: https://doi.org/10.1016/j.matpr.2017.06.365.
[12] Clark HM. Particle velocity and size effects in laboratory slurry erosion measurements OR . . . do you know what your particles are doing? Tribol Int. 2002;35:617–624.
[13] Al-Bukhaiti MA, Abouel-Kasem A, Emara KM, et al. Particle shape and size effects on slurry erosion of AISI 5117 steels. J Tribol. 2016;138:24503.
[14] Al-Bukhaiti MA, Ahmed SM, Badran FMF, et al. Effect of impingement angle on slurry erosion behaviour and mechanisms of 1017 steel and high-chromium white cast iron. Wear. 2007;262:1187–1198.
[15] Al-Bukhaiti MA, Mohamad AAK, Emara KM, et al. Effect of slurry concentration on erosion wear behavior of AISI 5117 steel and high-chromium white cast iron. Ind Lubr Tribol. 2018;70:628–638.
[16] Desale GR, Gandhi BK, Jain SC. Effect of erodent properties on erosion wear of ductile type materials. Wear. 2006;261:914–921.
[17] Desale GR, Gandhi BK, Jain SC. Improvement in the design of a pot tester to simulate erosion wear due to solid-liquid mixture. Wear. 2005;259:196–202.
[18] Desale GR, Gandhi BK, Jain SC. Development of correlations for predicting the slurry erosion of ductile materials. J Tribol. 2011;133:1–10.
[19] Desale GR, Gandhi BK, Jain SC. Slurry erosion of ductile materials under normal impact condition. Wear. 2008;264:322–330.
[20] Clark HM, Llewellyn RJ. Assessment of the erosion resistance of steels used for slurry handling and transport in mineral processing applications. Wear. 2001;250–251:32–44.

[21] Xu X, Van der Zwaag S, Xu W. The effect of ferrite-martensite morphology on the scratch and abrasive wear behaviour of a dual phase construction steel. Wear [Internet]. 2016;348–349:148–157. Available from: http://dx.doi.org/10.1016/j.wear.2015.12.005.
[22] Li R, Zuo X, Hu Y, et al. Microstructure and properties of pipeline steel with a ferrite/martensite dual-phase microstructure. Mater Charact [Internet]. 2011;62:801–806. Available from: http://dx.doi.org/10.1016/j.matchar.2011.05.013.
[23] Hu Y, Zuo X, Li R, et al. Effect of initial microstructures on the properties of ferrite-martensite dual-phase pipeline steels with strain-based design. Mater Res. 2012;15:317–322.
[24] Cai F, Huang X, Yang Q. Mechanical properties, sliding wear and solid particle erosion behaviors of plasma enhanced magnetron sputtering CrSiCN coating systems. Wear [Internet]. 2015;324–325:27–35. Available from: http://dx.doi.org/10.1016/j.wear.2014.11.008.
[25] Burstein GT, Sasaki K. Effect of impact angle on the slurry erosion–corrosion of 304L stainless steel. Wear. 2000;240:80–94.
[26] Gandhi BK, Singh SN, Seshadri V. Study of the parametric dependence of erosion wear for the parallel flow of solid-liquid mixtures. Tribol Int. 1999;32:275–282.
[27] Gandhi BK, Singh SN, Seshadri V. Performance characteristics of centrifugal slurry pumps. J Fluid Eng. 2001;123:271–280.
[28] Clark HMI, Hartwich RB. A re-examination of the "particle size effect" in slurry erosion. Wear. 2001;248:147–161.
[29] Lindgren M, Perolainen J. Slurry pot investigation of the influence of erodent characteristics on the erosion resistance of austenitic and duplex stainless steel grades. Wear [Internet]. 2014;319:38–48. Available from: http://dx.doi.org/10.1016/j.wear.2014.07.006.
[30] Szala M, Hejwowski TJ, Lenart I. Cavitation erosion resistance of Ni-Co based coatings. Adv Sci Technol Res J. 2014;8:36–42.
[31] Szala M, Łukasik D. Cavitation wear of pump impellers. J Technol Exploit Mech Eng. 2016;2:40–44.
[32] Walker CI, Robbie P. Comparison of some laboratory wear tests and field wear in slurry pumps. Wear [Internet]. 2013;302:1026–1034. Available from: http://dx.doi.org/10.1016/j.wear.2012.11.053.
[33] Parent LL, Li DY. Wear of hydrotransport lines in Athabasca oil sands. Wear. 2013;301:477–482.
[34] Awal R, Chitrakar S, Thapa BS, et al. Analysis a general review on methods of sediment sampling and mineral content analysis. J Phys Conf Ser. 2019;1266:012005.
[35] Padhy MK, Saini RP. Effect of size and concentration of silt particles on erosion of Pelton turbine buckets. Energy [Internet]. 2009;34:1477–1483. Available from: http://dx.doi.org/10.1016/j.energy.2009.06.015.
[36] Chitrakar S, Dahlhaug OG, Neopane HP. Numerical investigation of the effect of leakage flow through erosion-induced clearance gaps of guide vanes on the performance of francis turbines. Eng Appl Comput Fluid Mech [Internet]. 2018;12:662–678. Available from: https://doi.org/10.1080/19942060.2018.1509806.
[37] Kapali A, Chitrakar S, Shrestha O, et al. A review on experimental study of sediment erosion in hydraulic turbines at laboratory conditions. J Phys Conf Ser. 2019;1266.
[38] Sangroula DP. Hydropower development and its sustainability with respect to sedimentation in Nepal. J Inst Eng. 1970;7:56–64.
[39] Kjolle A. No Title [Internet]. Trondheim, Norway; 2014. Available from: http://www.tev.ntnu.no/vk/publikasjoner/pdf/ArneKjolle/chapter14.pdfS.
[40] Padhy MK, Saini RP. Study of silt erosion mechanism in Pelton turbine buckets. Energy [Internet]. 2012;39:286–293. Available from: http://dx.doi.org/10.1016/j.energy.2012.01.015.
[41] Llewellyn RJ, Yick SK, Dolman KF. Scouring erosion resistance of metallic materials used in slurry pump service. Wear. 2004;256:592–599.
[42] Qiu N, Wang L, Wu S, et al. Research on cavitation erosion and wear resistance performance of coatings. Eng Fail Anal. 2015;55:208–223.
[43] Thapa BS, Thapa B, Dahlhaug OG. Current research in hydraulic turbines for handling sediments. Energy. 2012;47:62–69.
[44] Neopane, HP. Sediment Erosion in Hydro Turbines. Trondheim: Norwegian University of Science and Technology; 2010.
[45] Chitrakar S, Neopane HP, Dahlhaug OG. Study of the simultaneous effects of secondary flow and sediment erosion in Francis turbines. Renew Energy [Internet]. 2016;97:881–891. Available from: http://dx.doi.org/10.1016/j.renene.2016.06.007.
[46] Rai AK, Kumar A, Staubli T. Hydro-abrasive erosion in Pelton buckets: Classification and field study. Wear. 2017;392:8–20.

[47] Singh J, Kumar S, Mohapatra S. Study on role of particle shape in erosion wear of austenitic steel using image processing analysis technique. Proc Inst Mech Eng Part J J Eng Tribol [Internet]. 2019;233:712–725. Available from: http://journals.sagepub.com/doi/10.1177/1350650118794698.
[48] Singh JP, Kumar S, Mohapatra SK. Head loss investigations inside 90° pipe bend for conveying of fine coal-water slurry suspension. Int J Coal Prep Util [Internet]. 2017;00:1–17. Available from: https://doi.org/10.1080/19392699.2017.1415892.
[49] Arabnejad H, Mansouri A, Shirazi SA, et al. Development of mechanistic erosion equation for solid particles. Wear [Internet]. 2015;332–333:1044–1050. Available from: https://linkinghub.elsevier.com/retrieve/pii/S0043164815000381.
[50] Singh M, Kumar S, Kumar S, et al. Characterisation of iron-ore suspension at in-situ conditions. Mater Today Proc [Internet]. 2018;5:17845–17851. Available from: https://doi.org/10.1016/j.matpr.2018.06.110.
[51] Goel DB, Sharma MK. Present state of damages and their repair welding in Indian hydroelectric projects. Proc Silt Damages to Equip Hydro Power Station Remedial Meas. New Delhi, India; 1996. p. 137–152.
[52] Wood RJK, Puget Y, Trethewey KR, et al. The performance of marine coatings and pipe materials under fluid-borne sand erosion. Wear. 1998;219:46–59.
[53] Rateick Jr RG, Karasek KR, Cunningham AJ, et al. Solid-particle erosion of tungsten carbide/cobalt cermet and hardened 440C stainless steel—a comparison. Wear. 2006;261:773–778.
[54] Singh J, Singh S, Pal Singh J. Investigation on wall thickness reduction of hydropower pipeline underwent to erosion–corrosion process. Eng Fail Anal [Internet]. 2021;127:105504. Available from: https://doi.org/10.1016/j.engfailanal.2021.105504.
[55] Kumar P, Kumar S, Singh J. Rheological and computational analysis of crude oil transportation. Int J Mech Aerospace, Ind Mechatron Manuf Eng. 2017;11:429–432.
[56] Singh J. Slurry erosion performance analysis and characterisation of high-velocity oxy-fuel sprayed Ni and Co hardsurfacing alloy coatings. J King Saud Univ—Eng Sci [Internet]. 2023;35:415–429. Available from: https://doi.org/10.1016/j.jksues.2021.06.009.
[57] Singh J, Kumar S, Mohapatra SK. Study on solid particle erosion of pump materials by fly ash slurry using Taguchi's orthogonal array. Tribol—Finnish J Tribol [Internet]. 2021;38:31–38. Available from: https://journal.fi/tribologia/article/view/97530.
[58] Singh J. Investigation on Slurry Erosion of Different Pumping Materials and Coatings. Patiala, India: Thapar Institute of Engineering and Technology; 2019.
[59] Singh J, Mohapatra SK, Kumar S. Performance analysis of pump materials employed in bottom ash slurry erosion conditions. J Tribol. 2021;30:73–89.
[60] Singh J. A review on mechanisms and testing of wear in slurry pumps, pipeline circuits and hydraulic turbines. J Tribol. 2021;143:1–83.
[61] Singh J, Singh JP. Performance analysis of erosion resistant Mo2C reinforced WC-CoCr coating for pump impeller with Taguchi's method. Ind Lubr Tribol. 2022;74:431–441.
[62] Singh J, Singh S. Neural network prediction of slurry erosion of heavy-duty pump impeller/casing materials 18Cr-8Ni, 16Cr-10Ni-2Mo, super duplex 24Cr-6Ni-3Mo-N, and grey cast iron. Wear [Internet]. 2021;476:203741. Available from: https://doi.org/10.1016/j.wear.2021.203741.
[63] Singh J, Singh JP. Numerical analysis on solid particle erosion in elbow of a slurry conveying circuit. J Pipeline Syst Eng Pract. 2021;12:04020070.
[64] Singh J, Singh S. Neural network supported study on erosive wear performance analysis of Y2O3/WC-10Co4Cr HVOF coating. J King Saud Univ—Eng Sci [Internet]. 2024;36:662–676. Available from: https://doi.org/10.1016/j.jksues.2021.12.005.
[65] Kumar S, Singh J, Mohapatra SK. Role of particle size in assessment of physico-chemical properties and trace elements of Indian fly ash. Waste Manag Res. 2018;36:1016–1022.
[66] Kumar S, Singh J, Mohapatra SK. Influence of particle size on leaching characteristic of fly ash. 15th Int Conf Environ Sci Technol 31 August to 2 Sept 2017. Rhodes, Greece; 2017. p. 01243.
[67] Kumar P, Kumar S, Singh J. Effect of natural surfactant on the rheological characteristics of heavy crude oil. Mater Today Proc [Internet]. 2018;5:23881–23887. Available from: https://doi.org/10.1016/j.matpr.2018.10.180.
[68] Kumar P, Singh J. Computational study on effect of Mahua natural surfactant on the flow properties of heavy crude oil in a 90° bend. Mater Today Proc [Internet]. 2021;43:682–688. Available from: https://doi.org/10.1016/j.matpr.2020.12.612.

[69] Singh J, Kumar S, Mohapatra SK, et al. Shape simulation of solid particles by digital interpretations of scanning electron micrographs using IPA technique. Mater Today Proc [Internet]. 2018;5:17786–17791. Available from: https://doi.org/10.1016/j.matpr.2018.06.103.
[70] Singh J. Application of thermal spray coatings for protection against erosion, abrasion, and corrosion in hydropower plants and offshore industry. In: Thakur L, Vasudev H, editors. Therm Spray Coatings [Internet]. 1st ed. Boca Raton: CRC Press; 2021. p. 243–283. Available from: https://www.taylorfrancis.com/books/9781003213185/chapters/10.1201/9781003213185-10.
[71] Grewal HS, Agrawal A, Singh H. Design and development of high-velocity slurry erosion test rig using CFD. J Mater Eng Perform. 2013;22:152–161.
[72] Divakar M, Agarwal VK, Singh SN. Effect of the material surface hardness on the erosion of AISI316. Wear. 2005;259:110–117.
[73] Santa JF, Baena JC, Toro A. Slurry erosion of thermal spray coatings and stainless steels for hydraulic machinery. Wear. 2007;263:258–264.
[74] Modi OP, Pandit P, Mondal DP, et al. High-stress abrasive wear response of 0.2% carbon dual phase steel: Effects of microstructural features and experimental conditions. Mater Sci Eng A. 2007;458:303–311.
[75] Lindgren M, Suihkonen R, Vuorinen J. Erosive wear of various stainless steel grades used as impeller blade materials in high temperature aqueous slurry. Wear. 2015;328–329:391–400.
[76] Sharma MK. Problems in repair welding of 13/4 stainless steel and their remedial measures. Proc Silt Damages to Equip Hydro Power Station Remedial Meas. New Delhi; 1996. p. 166–181.
[77] Manisekaran T, Kamaraj M, Sharrif SM, et al. Slurry erosion studies on surface modified 13Cr-4Ni steels: Effect of angle of impingement and particle size. J Mater Eng Perform. 2007;16:567–572.
[78] Shivamurthy RC, Kamaraj M, Nagarajan R, et al. Influence of microstructure on slurry erosive wear characteristics of laser surface alloyed 13Cr-4Ni steel. Wear. 2009;267:204–212.
[79] Vasudev H. Wear characteristics of Ni-WC powder deposited by using a microwave route on mild steel. Int J Surf Eng Interdiscip Mater Sci [Internet]. 2020;8:44–54. Available from: http://services.igi-global.com/resolvedoi/resolve.aspx?doi=10.4018/IJSEIMS.2020010104.
[80] Singh G, Vasudev H, Bansal A, et al. Microwave cladding of Inconel-625 on mild steel substrate for corrosion protection. Mater Res Express [Internet]. 2020;7:026512. Available from: https://iopscience.iop.org/article/10.1088/2053-1591/ab6fa3.
[81] Singh M, Vasudev H, Singh M. Surface protection of SS-316L with boron nitride based thin films using radio frequency magnetron sputtering technique. J Electrochem Sci Eng [Internet]. 2022;00:1–13. Available from: https://www.pub.iapchem.org/ojs/index.php/JESE/article/view/1247.
[82] Prashar G, Vasudev H. High temperature erosion behavior of plasma sprayed Al2 O3 coating on AISI-304 stainless steel. World J Eng [Internet]. 2021;18:760–766. Available from: https://www.emerald.com/insight/content/doi/10.1108/WJE-10-2020-0476/full/html.
[83] Vasudev H, Thakur L, Bansal A, et al. High temperature oxidation and erosion behaviour of HVOF sprayed bi-layer Alloy-718/NiCrAlY coating. Surf Coatings Technol [Internet]. 2019;362:366–380. Available from: https://doi.org/10.1016/j.surfcoat.2019.02.012.
[84] Vasudev H, Thakur L, Singh H, et al. A study on processing and hot corrosion behaviour of HVOF sprayed Inconel718-nano Al2O3 coatings. Mater Today Commun [Internet]. 2020;25:101626. Available from: https://doi.org/10.1016/j.mtcomm.2020.101626.
[85] Vasudev H, Thakur L, Singh H, et al. Erosion behaviour of HVOF sprayed Alloy718-nano Al2 O3 composite coatings on grey cast iron at elevated temperature conditions. Surf Topogr Metrol Prop [Internet]. 2021;9:035022. Available from: https://iopscience.iop.org/article/10.1088/2051-672X/ac1c80.
[86] Vasudev H, Prashar G, Thakur L, et al. Electrochemical corrosion behavior and microstructural characterisation of hvof sprayed Inconel718-Al2 O3 composite coatings. Surf Rev Lett [Internet]. 2022;29:2250017. Available from: https://www.worldscientific.com/doi/abs/10.1142/S0218625X22500172.
[87] Prashar G, Vasudev H. A review on the influence of process parameters and heat treatment on the corrosion performance of Ni-based thermal spray coatings. Surf Rev Lett [Internet]. 2022;29:1–18. Available from: https://www.worldscientific.com/doi/abs/10.1142/S0218625X22300015.
[88] Singh P, Vasudev H, Bansal A. Effect of post-heat treatment on the microstructural, mechanical, and bioactivity behavior of the microwave-assisted alumina-reinforced hydroxyapatite cladding. Proc Inst Mech Eng Part E J Process Mech Eng [Internet]. 2022;095440892211161. Available from: http://journals.sagepub.com/doi/10.1177/09544089221116168.

[89] Singh G, Vasudev H, Bansal A, et al. Influence of heat treatment on the microstructure and corrosion properties of the Inconel-625 clad deposited by microwave heating. Surf Topogr Metrol Prop [Internet]. 2021;9:025019. Available from: https://iopscience.iop.org/article/10.1088/2051-672X/abfc61.
[90] Singh P, Bansal A, Vasudev H, et al. In situ surface modification of stainless steel with hydroxyapatite using microwave heating. Surf Topogr Metrol Prop [Internet]. 2021;9:035053. Available from: http://dx.doi.org/10.1088/2051-672X/ac28a9.
[91] Prashar G, Vasudev H. A comprehensive review on sustainable cold spray additive manufacturing: State of the art, challenges and future challenges. J Clean Prod [Internet]. 2021;310:127606. Available from: https://doi.org/10.1016/j.jclepro.2021.127606.
[92] Vasudev H, Prashar G, Thakur L, et al. Electrochemical corrosion behavior and microstructural characterization of HVOF sprayed Inconel-718 coating on gray cast iron. J Fail Anal Prev [Internet]. 2021;21:250–260. Available from: https://doi.org/10.1007/s11668-020-01057-8.
[93] Vasudev H, Singh P, Thakur L, et al. Mechanical and microstructural characterisation of microwave post processed Alloy-718 coating. Mater Res Express [Internet]. 2020;6:1265f5. Available from: https://iopscience.iop.org/article/10.1088/2053-1591/ab66fb.
[94] Prashar G, Vasudev H, Thakur L. Performance of different coating materials against slurry erosion failure in hydrodynamic turbines: A review. Eng Fail Anal [Internet]. 2020;115:104622. Available from: https://doi.org/10.1016/j.engfailanal.2020.104622.
[95] Singh M, Vasudev H, Kumar R. Corrosion and tribological behaviour of BN thin films deposited using magnetron sputtering. Int J Surf Eng Interdiscip Mater Sci [Internet]. 2021;9:24–39. Available from: https://services.igi-global.com/resolvedoi/resolve.aspx?doi=10.4018/IJSEIMS.2021070102.
[96] Roco MC. Modeling erosion wear in centrifugal slurry pumps. Hydrotransport. 1984;9:291–316.
[97] Wasp EJ, Kenny JP, Gandhi RL. Solid—liquid flow: Slurry pipeline transportation [Pumps, valves, mechanical equipment, economics]. Ser Bulk Mater Handl; (United States). 1977;1.
[98] Neville A, Hodgkiess T, Dallas JT. A study of the erosion–corrosion behaviour of engineering steels for marine pumping applications. Wear. 1995;186:497–507.
[99] Wilson G. The design aspects of centrifugal pumps for abrasive slurries. Can Min Metall Bull. 1973;66(733):91–102.
[100] Minemura K, Zhong Y, Uchiyama T. Numerical prediction of erosion wear on pump casing under solid-water two-phase flow conditions. Trans Japan Soc Mech Eng Ser B [Internet]. 1995;61:2571–2578. Available from: http://www.jstage.jst.go.jp/article/kikaib1979/61/587/61_587_2571/_article/-char/ja/.
[101] Zhong Y, Minemura K. Measurement of erosion due to particle impingement and numerical prediction of wear in pump casing. Wear [Internet]. 1996;199:36–44. Available from: https://linkinghub.elsevier.com/retrieve/pii/0043164896069748.
[102] Roudnev AS. Some aspects of slurry pump design. World Pumps [Internet]. 1999;1999:58–61. Available from: https://linkinghub.elsevier.com/retrieve/pii/S0262176200876816.
[103] Kadambi JR, Charoenngam P, Subramanian A, et al. Investigations of particle velocities in a slurry pump using PIV: Part 1. The tongue and adjacent channel flow. J Energy Resour Technol [Internet]. 2004;126:271–278. Available from: https://asmedigitalcollection.asme.org/energyresources/article/126/4/271/461274/Investigations-of-Particle-Velocities-in-a-Slurry.
[104] Kadambi JR, Mehta M, Sankovic J, et al. Velocity measurements of particles in the impeller of a centrifugal slurry pump. 12th Int Symp Calouste Gulbenkian Fund Lisbon. 2004. p. 1–11.
[105] Neilson JH, Gilchrist A. Erosion by a stream of solid particles. Wear [Internet]. 1968;11:111–122. Available from: https://linkinghub.elsevier.com/retrieve/pii/0043164868905917.
[106] Sheldon GL, Finnie I. On the ductile behaviour of nominally brittle material cutting. J Eng Ind. 1966;88:387–392.
[107] Finnie I. Some observations on the erosion of ductile metals. Wear. 1972;19:81–90.
[108] Bitter JGA. A study of erosion phenomena: Part I. Wear. 1963;6:169–190.
[109] Bitter JGA. A study of erosion phenomena: Part II. Wear. 1963;6:169–190.
[110] Oka YI, Okamura K, Yoshida T. Practical estimation of erosion damage caused by solid particle impact: Part 1. Effects of impact parameters on a predictive equation. Wear. 2005;259:95–101.
[111] Oka YI, Yoshida T. Practical estimation of erosion damage caused by solid particle impact: Part 2. Mechanical properties of materials directly associated with erosion damage. Wear. 2005;259:102–109.
[112] Wiedenroth W. Wear tests executed with a 125mm ID loop and a model dredge pump. Proc 9th Int Conf Hydraul Transp Solids Pipes; 1984. p. 317–330.

[113] Widenroth W. The influence of sand and gravel on the characteristics of centrifugal pimps, some aspects of wear in hydraulic transportation installations. Proc Hydrotrnasport 1; 1970.
[114] Read EN. Experience gained with large dredge pumps in a sand mining operation. Hydrotransport. Volume 8, BHRA Fluid Engineering, Johannesburg (South Africa), Paper F1, (1982).
[115] Liu J, Xu H, Qi L, et al. Study on erosive wear and novel wear-resistant materials for centrifugal slurry pumps. Heat Transf Summer Conf; 2004. p. 101–104.
[116] Rayan MA, Shawky M. Evaluation of wear in a centrifugal slurry pump. Proc Inst Mech Eng Part A J Power Eng. 1989;203:19–23.
[117] Roco M. Calculation of solid particles concentration in centrifugal pump impellers using finite element technique. Hydrotransport. 1980;7:359–376.
[118] Xing D, Hai-lu Z, Xin-yong W. Finite element analysis of wear for centrifugal slurry pump. Procedia Earth Planet Sci. 2009;1:1532–1538.
[119] Pagalthivarthi KV, Gupta PK. Prediction of erosion wear in multi-size particulate flow through a rotating channel. Fluid Dyn Mater Process. 2009;5:93–122.
[120] Pagalthivarthi K, Visintainer R. Finite element prediction of multi-size particulate flow through three-dimensional channel: Code validation. J Comput Multiph Flows. 2013;5:57–72.
[121] Pagalthivarthi KV, Furlan JM, Visintainer RJ. Comparison of 2D and 3D predictions of erosion wear in centrifugal slurry pump casings. Fluids Eng Div Summer Meet. American Society of Mechanical Engineers; 2013. p. V01CT20A004.
[122] Pagalthivarthi KV, Furlan JM, Visintainer RJ. Wear rate prediction in multi-size particulate flow through impellers. Fluids Eng Div Summer Meet. American Society of Mechanical Engineers; 2013. p. V01CT20A002.
[123] Sellgren A, Addie G, Visintainer R, et al. Prediction of slurry pump component wear and cost. Annu Texas A&M Dredg Semin 19/06/2005-22/06/2005. Western Dredging Association; 2005.
[124] Pagalthivarthi KV, Helmly FW. Applications of materials wear testing to solids transport via centrifugal slurry pumps. Wear Test Adv Mater. 1992;1167:114–126.
[125] Bross S, Addie G. Prediction of impeller nose wear behaviour in centrifugal slurry pumps. Exp Therm fluid Sci. 2002;26:841–849.
[126] Shen Z, Chu W, Li X, et al. Sediment erosion in the impeller of a double-suction centrifugal pump—a case study of the Jingtai Yellow River Irrigation Project, China. Wear [Internet]. 2019;422–423:269–279. Available from: https://linkinghub.elsevier.com/retrieve/pii/S0043164818311657.
[127] Noon AA, Kim M-H. Erosion wear on centrifugal pump casing due to slurry flow. Wear. 2016;364:103–111.
[128] Li Y, Zhu Z, He Z, et al. Abrasion characteristic analyses of solid-liquid two-phase centrifugal pump. J Therm Sci. 2011;20:283–287.
[129] Ahmad K, Baker RC, Goulas A. Computation and experimental results of wear in a slurry pump impeller. Proc Inst Mech Eng Part C J Mech Eng Sci. 1986;200:439–445.
[130] Addie GR, Visintainer RJ. Experiences with a numerical method of calculating slurry pump casing wear. Proc 4th Int Pump Users Symp. Turbomachinery Laboratories, Department of Mechanical Engineering, Texas, A&M . . .; 1987.
[131] Krüger S, Martin N, Dupont P. Assessment of wear erosion in pump impellers. Proc 26th Int Pump Users Symp. Turbomachinery Laboratory, Texas, A&M University; 2010.
[132] Pagalthivarthi KV, Gupta PK, Tyagi V, et al. CFD prediction of erosion wear in centrifugal slurry pumps for dilute slurry flows. J Comput Multiph flows. 2011;3:225–245.
[133] Madadnia J, Amjad M, Kusnan J. Analysis of severe erosion in industrial centrifugal slurry pumps. Fluids Eng Div Summer Meet. American Society of Mechanical Engineers; 2013. p. V002T06A007.
[134] Shen ZJ, Li RN, Han W, et al. The research on particle trajectory of solid-liquid two-phase flow and erosion predicting in screw centrifugal pump. IOP Conf Ser Mater Sci Eng. IOP Publishing; 2016. p. 12052.
[135] Adamkowski A, Henke A, Lewandowski M. Resonance of torsional vibrations of centrifugal pump shafts due to cavitation erosion of pump impellers. Eng Fail Anal. 2016;70:56–72.
[136] Sangroula DP. Hydropower development and its sustainability with respect to sedimentation in Nepal. J Inst Eng. 2009;7:56–64.
[137] Dorji U, Ghomashchi R. Hydro turbine failure mechanisms: An overview. Eng Fail Anal. 2014;44:136–147.
[138] Teran LA, Aponte RD, Munoz-Cubillos J, et al. Analysis of economic impact from erosive wear by hard particles in a run-of-the-river hydroelectric plant. Energy. 2016;113:1188–1201.

[139] Felix D, Albayrak I, Abgottspon A, et al. Hydro-abrasive erosion of hydraulic turbines caused by sediment—a century of research and development. IOP Conf Ser Earth Environ Sci [Internet]. 2016;49:122001. Available from: https://iopscience.iop.org/article/10.1088/1755-1315/49/12/122001.
[140] Singh M, Banerjee J, Patel PL, et al. Effect of silt erosion on Francis turbine: A case study of Maneri Bhali Stage-II, Uttarakhand, India. ISH J Hydraul Eng [Internet]. 2013;19:1–10. Available from: http://www.tandfonline.com/doi/abs/10.1080/09715010.2012.738507.
[141] Rai AK, Kumar A, Hies T, et al. Field application of a multi-frequency acoustic instrument to monitor sediment for silt erosion study in Pelton turbine in Himalayan region, India. IOP Conf Ser Earth Environ Sci [Internet]. 2016;49:122004. Available from: https://iopscience.iop.org/article/10.1088/1755-1315/49/12/122004.
[142] Sujakhu S, Neopane HP. Mineral analysis of sediments in nepalese hydropower plants: A case study of Sunkoshi and Sundarijal hydropower plant. Int J Adv Renew Energy Res. 2013;2.
[143] Neopane HP, Sujakhu S. Particle size distribution and mineral analysis of sediments in nepalese hydropower plant: A case study of Jhimruk hydropower plant. Kathmandu Univ J Sci Eng Technol. 2013;9:29–36.
[144] Koirala R, Thapa B, Neopane HP, et al. Sediment erosion in guide vanes of Francis turbine: A case study of Kaligandaki Hydropower Plant, Nepal. Wear [Internet]. 2016;362–363:53–60. Available from: https://linkinghub.elsevier.com/retrieve/pii/S0043164816300898.
[145] Aslam Noon A, Kim M-H. Erosion wear on Francis turbine components due to sediment flow. Wear [Internet]. 2017;378–379:126–135. Available from: https://linkinghub.elsevier.com/retrieve/pii/S0043164817303587.
[146] Wills DJ, Truscott GF. Solids-handling pumps a survey of current practice, problems and developments. Fifth Int Conf Hydraul Transp Solids Pipes, BHRA; 1978. p. F6.
[147] Neopane HP, Dahlhaug OG, Cervantes M. Sediment erosion in hydraulic turbines. Glob J Res Eng Mech Mech Eng. 2011;11:17–26.
[148] Padhy MK, Saini RP. Study of silt erosion mechanism in Pelton turbine buckets. Energy. 2012;39:286–293.
[149] Grewal HS, Agrawal A, Singh H. Identifying erosion mechanism: A novel approach. Tribol Lett [Internet]. 2013;51:1–7. Available from: http://link.springer.com/10.1007/s11249-013-0156-4.
[150] Rai AK, Kumar A, Staubli T. Developing a test rig to measure hydro-abrasive erosion in Pelton turbine. Proc Intl Conf Hydropower Sustain Dev. Roorkee, India; 2015. p. 535–547.
[151] Khurana S, Varun, Kumar A. Silt erosion study on the performance of an impulse turbine in small hydropower. Int J Ambient Energy. 2016;37:520–527.
[152] Thapa B, Brekke H. Effect of sand particle size and surface curvature in erosion of hydraulic turbine. IAHR Symp Hydraul Mach Syst. Stockholm; 2004.
[153] Tuzson JJ. Laboratory slurry erosion tests and pump wear rate calculations. J Fluids Eng. 2015;106:135–140.
[154] Chattopadhyay R. Case study: High silt wear of hydroturbine runners. Wear. 1993;162–164:1040–1044.
[155] Roman JM, Xin LY, Hui WM, et al. Dealing with abrasive erosion in hydro turbine. Hydropower Dams. 1997;3:67–71.
[156] Krause M, Grein H. Abrasion research and prevention. Int J Hydropower Dams. 1996;3:17–20.
[157] Naidu BS. Addressing the problems of silt erosion at hydro plants. Int J Hydropower Dams. 1997;4:72–77.
[158] Asthana BN. Determination of optimal sediment size to be excluded for run-of-river project—a case study. Roorkee, India; 1997.
[159] Schneider C, Kachele T. Recent research results on predicting and preventing silt erosion. In: Varma CVJ, Naidu BSK, Rao ARG, editors. 1st Int Conf Silting Probl Hydro Power Plants. New Delhi, India and Brookfield, USA: A.A. Balkema Publishers; 1999. p. III.
[160] Grant G, Tabakoff W. Erosion prediction in turbomachinery due to environmental solid particles. J Aircr. 1974;12:471–478.
[161] Lyczkowski RW, Bouillard JX. Corrigendum to “state-of-the-art review of erosion modeling in fluid/solids systems” [Progress in Energy and Combustion Science 28 (2002) 543–602]☆. Prog Energy Combust Sci [Internet]. 2003;29:85. Available from: https://linkinghub.elsevier.com/retrieve/pii/S036012850200045X.
[162] Hamed AA, Tabakoff W, Rivir RB, et al. Turbine blade surface deterioration by erosion. J Turbomach [Internet]. 2005;127:445–452. Available from: https://asmedigitalcollection.asme.org/turbomachinery/article/127/3/445/471258/Turbine-Blade-Surface-Deterioration-by-Erosion.

[163] Wood RJK, Lu P. Leading edge topography of blades—a critical review. Surf Topogr Metrol Prop [Internet]. 2021;9:023001. Available from: https://iopscience.iop.org/article/10.1088/2051-672X/abf81f.
[164] Masoodi JH, Harmain GA. A methodology for assessment of erosive wear on a Francis turbine runner. Energy [Internet]. 2017;118:644–657. Available from: https://linkinghub.elsevier.com/retrieve/pii/S0360544216315407.
[165] Ilieva GI. Erosion failure mechanisms in turbine stage with twisted rotor blade. Eng Fail Anal [Internet]. 2016;70:90–104. Available from: https://linkinghub.elsevier.com/retrieve/pii/S1350630716306276.
[166] Ilieva G. Mechanisms of eater droplets deposition on turbine blade surfaces and erosion wear effects. J Appl Fluid Mech [Internet]. 2017;10:551–567. Available from: http://jafmonline.net/JournalArchive/download?file_ID=42108&issue_ID=239.
[167] Chitrakar S, Neopane HP, Dahlhaug OG. Study of the simultaneous effects of secondary flow and sediment erosion in Francis turbines. Renew Energy [Internet]. 2016;97:881–891. Available from: https://linkinghub.elsevier.com/retrieve/pii/S0960148116305195.
[168] Fecarotta O, Messa GV, Pugliese F, et al. Preliminary development of a method for impact erosion prediction in pumps running as turbines. EWaS3 2018 [Internet]. Basel Switzerland: MDPI; 2018. p. 680. Available from: https://www.mdpi.com/2504-3900/2/11/680.
[169] Rai AK, Kumar A, Staubli T. Hydro-abrasive erosion in Pelton buckets: Classification and field study. Wear [Internet]. 2017;392–393:8–20. Available from: https://linkinghub.elsevier.com/retrieve/pii/S0043164817306506.
[170] Chávez JC, Valencia JA, Jaramillo GA, et al. Failure analysis of a Pelton impeller. Eng Fail Anal [Internet]. 2015;48:297–307. Available from: https://linkinghub.elsevier.com/retrieve/pii/S1350630714002738.
[171] Sangal S, Singhal MK, Saini RP. CFD based analysis of silt erosion in Kaplan hydraulic turbine. 2016 Int Conf Signal Process Commun Power Embed Syst. IEEE; 2016. p. 1765–1770.
[172] Neopane HP, Dahlhaug OG, Cervantes MJ. The effect of sediment characteristics for predicting erosion on Francis turbines blades. Int J Hydropower Dams. 2012;19:79–83.
[173] Mangla M. Innovative approach to minimise silt erosion in hydro turbines. Int Conf Hydropower Sustain Dev Dehradun; 2015. p. 158–170.
[174] Thapa BS, Dahlhaug OG, Thapa B. Sediment erosion in hydro turbines and its effect on the flow around guide vanes of Francis turbine. Renew Sustain Energy Rev. 2015;49:1100–1113.
[175] Neopane HP, Dahlhaug OG, Thapa B. Alternative design of a Francis turbine for sand laden water. Int Conf Small Hydropower-Hydro Sri Lanka; 2007. p. 24.
[176] Barstad LF. CFD analysis of a Pelton turbine. Institutt for energi-og prosessteknikk; 2012.
[177] Sadlo F, Peikert R, Parkinson E. Vorticity based flow analysis and visualisation for pelton turbine design optimisation. IEEE Vis 2004. IEEE; 2004. p. 179–186.
[178] Thapa BS, Thapa B, Eltvik M, et al. Optimising runner blade profile of Francis turbine to minimise sediment erosion. IOP Conf Ser Earth Environ Sci. IOP Publishing; 2012. p. 32052.
[179] Eltvik M, Thapa BS, Dahlhaug OG, et al. Numerical analysis of effect of design parameters and sediment erosion on a Francis runner. Fourth Int Conf Water Resour Renew Energy Dev Asia, Thail. 2012.
[180] Shrestha S, Hodgkiess T, Neville A. Erosion–corrosion behaviour of high-velocity oxy-fuel Ni-Cr-Mo-Si-B coatings under high-velocity seawater jet impingement. Wear. 2005;259:208–218.
[181] Khanal K, Neopane HP, Rai S, et al. A methodology for designing Francis runner blade to find minimum sediment erosion using CFD. Renew Energy. 2016;87:307–316.
[182] Baidar B, Chitrakar S, Koirala R, et al. Selection of optimal number of francis runner blades for a sediment laden micro hydropower plant in Nepal. Int J Fluid Mach Syst. 2015;8:294–303.

Section II

Protection Methods against Slurry Flows and Erosion

6 Erosion Wear Behavior of Composite Materials for Slurry and Solid Particle Erosion Applications

Praveen Kumar Verma, Hitesh Vasudev, Sunitha K, and Yanhai Cheng

6.1 INTRODUCTION

Erosion wear is a serious problem in many industrial applications where materials are subjected to high-velocity solid particles or slurry conditions. Aerospace, oil and gas, power generation, and marine engineering industries frequently face erosion wear, which causes material loss, equipment failure, and maintenance costs. The effect of erosion wear on industrial parts, including turbine blades, pipelines, heat exchangers, and pumps, calls for the design of advanced materials with improved erosion resistance. Erosion wear happens when surfaces experience repetitive impacts from solid or liquid particles, resulting in material loss over time. This is commonly experienced in settings where abrasive particles are carried by a carrier medium, like water or air. In slurry erosion, suspended solid particles in a fluid strike surfaces, wearing away the material over time. Solid particle erosion, on the other hand, is direct particle impact without a liquid phase, causing surface damage through several mechanisms such as micro-cutting, fatigue wear, and brittle fracture [1–2].

Material loss owing to erosion can compromise the performance and lifespan of components, thus requiring frequent replacements and maintenance. Erosion-resistant materials could go a long way in mitigating operational expenditure and enhancing the reliability of a system. In the various methodologies adopted to counter erosion, composite materials have shown promise, as they possess superior wear-resistant and mechanical properties [3].

Composite materials have been identified as a potential solution to mitigate erosion wear because of their superior mechanical properties, high wear resistance, and capacity to endure harsh environments. Composites differ from traditional monolithic materials in that they contain a matrix material with high-strength fillers, which improves their toughness, hardness, and impact resistance. Different composite systems, such as polymer matrix composites (PMCs), metal matrix composites (MMCs), and ceramic matrix composites (CMCs), have been widely investigated to solve erosion wear problems in severe industrial applications. These materials exhibit a balance between hardness and toughness, which are important parameters in withstanding particle impact and abrasive forces. Moreover, the development of nanotechnology and hybrid composites has further expanded the range of erosion-resistant materials with enhanced durability and flexibility to various operating conditions [4–5].

The primary objective of this review is to provide an in-depth analysis of erosion wear mechanisms and the role of composite materials in mitigating this phenomenon. The chapter aims to explore the fundamental principles governing erosion wear, including the influence of material properties, impact conditions, and environmental factors. A detailed discussion on various composite systems, their fabrication techniques, and their effectiveness in erosion resistance will be

DOI: 10.1201/9781003561156-8

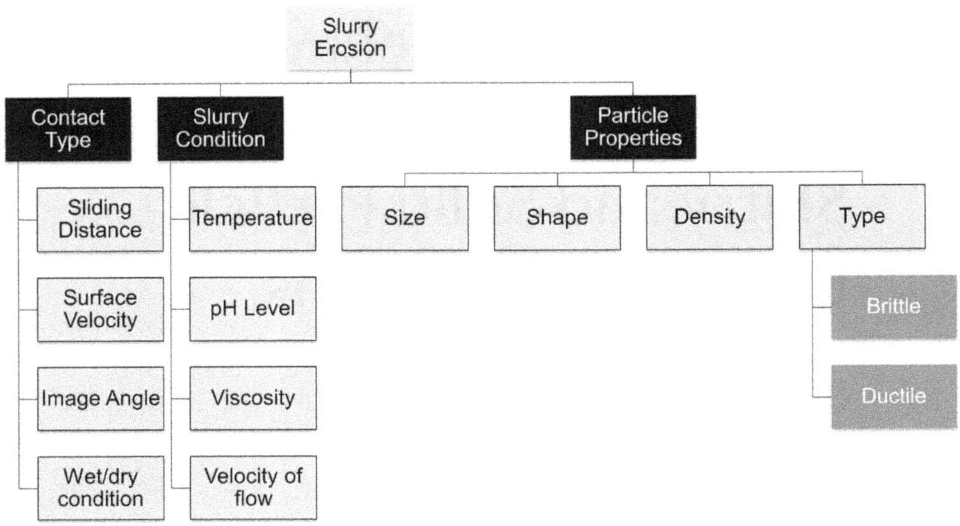

FIGURE 6.1 Different parameters involved in slurry erosion [9].

presented. Furthermore, the review will highlight recent advancements in composite materials, emphasizing innovative approaches such as nanocomposites, functionally graded materials, and bio-inspired structures. By consolidating findings from existing literature, this review seeks to offer valuable insights into material selection, design strategies, and potential research directions for improving erosion-resistant composites [6–7].

The scope of this review encompasses a wide range of industrial applications where erosion wear poses a significant challenge. This includes the aerospace industry, where turbine blades, engine components, and aircraft surfaces are subjected to high-velocity particle impacts. The power generation sector, particularly thermal and hydroelectric plants, also experiences severe erosion wear in boiler tubes, steam turbines, and water turbines due to high-speed fluid flow and solid particle entrainment. Additionally, in the mining and mineral processing industries, equipment such as slurry pumps, pipelines, and crushers face extensive material degradation due to abrasive particle interactions. The review will also address erosion wear concerns in the oil and gas industry, focusing on drilling tools, pipelines, and offshore structures exposed to harsh environmental conditions [8]. Figure 6.1 shows the various parameters involved in slurry erosion.

To provide a structured analysis, this chapter is organized into several sections. Following this introduction, the second section will delve into the fundamental mechanisms of erosion wear, categorizing different types of erosion, such as solid particle erosion, liquid impingement erosion, and cavitation erosion. This section will also discuss the key parameters influencing erosion wear, including impact velocity, angle of incidence, particle properties, and material characteristics. The third section will focus on the role of composite materials in enhancing erosion resistance, detailing various composite types, reinforcement materials, and matrix combinations. It will also examine the fabrication techniques and processing methods that influence the wear behavior of composites [10].

6.2 FUNDAMENTALS OF EROSION WEAR

Erosion wear is a process of material degradation due to repeated contact of solid particles, liquid droplets, or gas streams with a surface, resulting in progressive material loss. It is a critical problem in aerospace, energy, mining, and biomedical applications, where parts are subjected to hostile

TABLE.6.1
Erosion Wear Behavior of Composite Materials

Parameter	Slurry Erosion	Solid Particle Erosion
Mechanism	Material loss due to continuous impact of liquid-solid mixture	Surface degradation from high-velocity solid particle impact
Dominant Wear Modes	Hydrodynamic impact, cavitation, fatigue	Abrasion, brittle fracture, plastic deformation
Impact Angle Effect	High erosion at intermediate angles (20°–40°)	Ductile materials erode more at 30°; brittle at 90°
Material Dependence	MMCs and CMCs show higher resistance than PMCs	Harder and tougher composites perform better
Reinforcement Type	Ceramics (SiC, Al_2O_3), fibers (carbon, glass), nanoparticles	Hard ceramic phases (WC, TiC, B_4C), hybrid fillers
Coating Solutions	Thermal spray coatings, polymeric overlays	HVOF, plasma spray, carbide-based coatings
Experimental Techniques	Slurry pot test, rotating disc test, jet erosion test	ASTM G76, G65, high-speed particle impact tests
Microstructural Analysis	SEM (wear pattern), EDS (composition), XRD (phase analysis)	SEM (erosion morphology), EDS (elemental study), XRD
Industrial Applications	Hydropower, pipelines, turbines, marine structures	Aerospace, mining, oil and gas, power plants
Computational Modeling	CFD for slurry flow dynamics, FEM for impact analysis	DEM and FEM for particle-material interaction
Challenges	Material degradation over time, erosion–corrosion synergy	Rapid surface deterioration, difficulty in real-world prediction

environments. Erosion wear can be categorized according to the type of eroding medium. Solid particle erosion takes place when hard particles are suspended within a fluid or gas and hit a surface, which is typically found in pipelines and turbines. Liquid droplet erosion happens when high-speed liquid droplets hit solid surfaces, which can occur in components such as heat exchangers and steam turbines. Cavitation erosion arises from the collapse of vapor bubbles within a liquid in close proximity to a solid surface, producing high-pressure shock waves that wear away material. Slurry erosion happens when suspended solid particles in a liquid medium impact surfaces and cause material loss in hydraulic and mining processes [11–16]. Table 6.1 present the erosion wear behavior of composite materials.

The mechanisms of erosion wear are influenced by material characteristics and operating conditions. Abrasion is one of the major mechanisms, wherein hard particles cut or plow through the surface and result in micro-grooving. This mechanism prevails at lower impact angles (parallel to the surface). Fatigue is another mechanism that arises from repeated cyclic stresses resulting from incessant particle impacts, resulting in crack growth and propagation. Subsurface cracking and delamination cause material loss over time, hence fatigue being a major consideration in ductile materials. Brittle fracture is prevalent in hard and brittle materials, where cracks grow quickly on impact, resulting in material chipping or spalling. Brittle materials show greater erosion rates at regular impact angles (90°) because there is a greater concentration of stress at the impact point [17].

The rate and severity of erosion wear are determined by several factors. The wear mechanism depends significantly on the impact angle. Ductile materials are subjected to maximum erosion at moderate impact angles (20°–40°), whereas brittle materials bear maximum erosion at regular

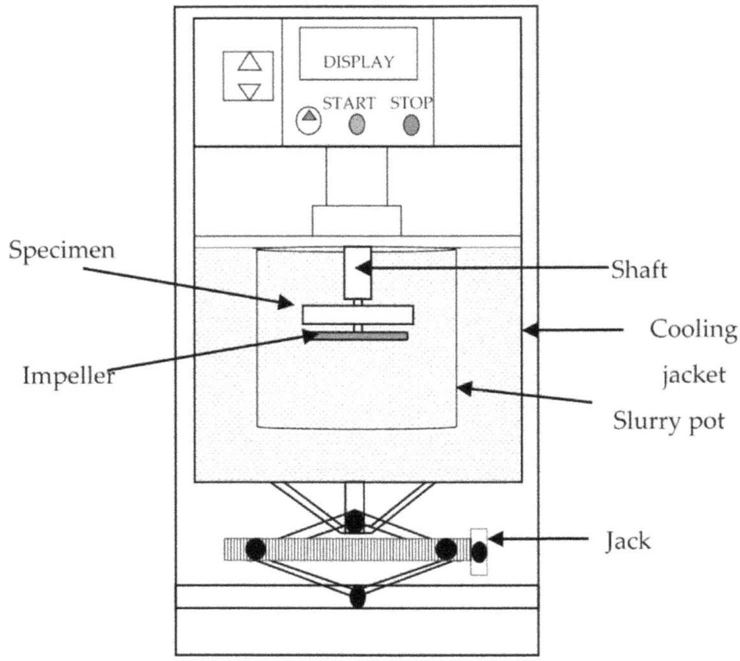

FIGURE 6.2 Schematic diagram of the erosion wear pot tester [20].
(CC BY 4.0 attribution)

impact angles (90°). Particle velocity is also an important parameter, with greater velocity causing increased kinetic energy of the impacting particles and thus more material removal. Erosion rate usually has a power-law dependence on velocity, which implies that even a small change in velocity can dramatically increase wear. Hardness of a material is another important parameter for erosion resistance; hard materials respond better to wear, while soft materials deform and wear faster. Also, the material's properties, including ductility, fracture toughness, and microstructure, affect how it is affected by erosion. Ductile materials develop fatigue-induced erosion, whereas brittle materials suffer from fracture and chipping. Surface coatings and treatments may be used to improve erosion resistance by enhancing hardness and toughness, and these are viable options for increasing the lifespan of components subject to erosive conditions [18–19]. Figure 6.2 shows a schematic diagram of the erosion wear pot tester.

Understanding of the mechanisms and influencing variables of erosion wear is important for material and protective coating design to withstand aggressive conditions. Optimization of material selection, surface treatment, and operating conditions helps designers produce better erosion-resistant components, decreasing maintenance and increasing performance in applications where erosion wear is an important issue [21].

6.3 COMPOSITE MATERIALS FOR EROSION RESISTANCE

Composite materials have become essential in erosion-resistant applications due to their ability to withstand harsh conditions while maintaining structural integrity. Different types of composites are used depending on the operating environment, including polymer matrix composites, metal matrix composites, ceramic matrix composites, and hybrid or functionally graded composites. These composites possess properties that are individually characteristic and that make them favorable for

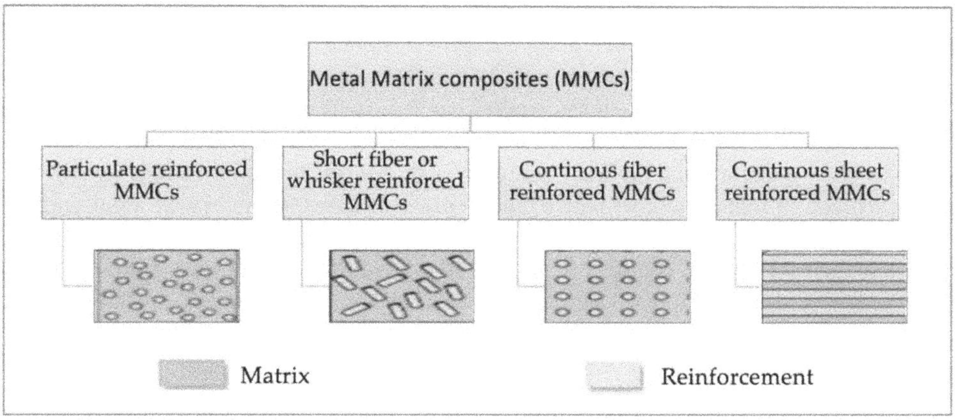

FIGURE 6.3 Metal matrix composite reinforcements [23].

different applications prone to erosion, including aerospace parts, turbine blades, heat exchangers, and industrial pipes. Their efficacy in eroding resistance is also highly dependent on the reinforcing materials used, including fibers, ceramics, and nanoparticles, that improve their strength and durability [22]. Figure 6.3 presents the various metal matrix composite reinforcements.

Polymer matrix composites are extensively utilized in erosion applications because they are light in weight, easy to process, and possess good impact resistance. PMCs are made up of a polymer-based matrix, which is generally composed of epoxy, polyester, or thermoplastic resins, reinforced with particles or fibers. These composites are widely utilized in applications where weight reduction is significant, for example, in aerospace and automotive applications. Although they have their benefits, PMCs are typically softer than metal and ceramic-based composites and thus more vulnerable to high-velocity particle impacts. Nevertheless, the addition of reinforcing agents like carbon fibers, glass fibers, and aramid fibers greatly enhances their wear resistance. Carbon fiber-reinforced PMCs (CFRPs) have high strength-to-weight ratios and good fatigue resistance and are therefore well suited for high-performance applications. GFRPs are reinforced PMCs which offer better impact resistance and thus are widely applied in protective coatings for turbine blades and industrial equipment. Aramid fibers such as Kevlar provide good toughness and resist micro-cracking and so are ideal for impact protection and ballistic applications [24].

MMC is a class of material used in the erosion application where high temperature as well as stress is involved. MMCs are a metal matrix, for example, aluminum, titanium, or nickel, reinforced with ceramic whiskers, particles, or fibers. They possess the toughness and ductility of metals and the hardness and wear resistance of ceramics and thus are very effective in withstanding erosion wear. Aluminum composites with SiC or boron carbide (B_4C) particulate reinforcements, which possess remarkable hardness and resistance to erosion, are the most widely applied MMCs. Titanium-based MMCs find use in the fields of aerospace as well as bio-medical devices because they demonstrate higher strength coupled with good resistance to corrosion. Nickel-based MMCs, reinforced with hard ceramic phases like tungsten carbide (WC) or chromium carbide (Cr_3C_2), are usually applied in severe environments, such as gas turbines and power plants, where erosion due to high temperatures is a serious issue. Reinforcement selection in MMCs plays an important role in deciding the wear characteristics, as it increases hardness, decreases material loss, and increases high-velocity particle impact resistance [25–26].

Ceramic matrix composites are engineered for high-temperature, erosive, and corrosive conditions. CMCs are made from a ceramic matrix, for example, silicon carbide (SiC), aluminum oxide

(Al_2O_3), or zirconium dioxide (ZrO_2), reinforced with ceramic fibers or ceramic particles. CMCs exhibit high hardness, thermal stability, and oxidation resistance, which can be applied in aerospace parts, gas turbine engines, and heat exchangers. In contrast to brittle, catastrophically failing monolithic ceramics, CMCs have enhanced toughness that results from their reinforcement with fibers. Silicon carbide fiber-reinforced SiC (SiC/SiC) composites, for example, offer outstanding thermal-shock and erosive resistance and are thus suited for high-temperature applications. Alumina-based CMCs are utilized in industrial coatings to shield metal surfaces from abrasive wear and erosion. In addition, high-fracture-toughness zirconia-based composites find applications in biomedical implants and thermal barrier coatings. CMCs' ability to preserve their mechanical integrity at extreme conditions also makes them an ideal material for applications involving sustained erosion resistance [27–29].

Hybrid and functionally graded composites (FGCs) are a high-tech group of materials developed to maximize erosion resistance through various combinations of reinforcement phases or by smoothly changing material composition. Hybrid composites combine two or more forms of reinforcements, for example, a mixture of fibers, nanoparticles, and ceramic particles, to improve mechanical properties. Hybrid PMCs reinforced with a mixture of carbon fibers and ceramic nanoparticles, for instance, have better hardness, toughness, and impact resistance than traditional PMCs. Hybrid MMCs, with a combination of hard ceramic reinforcements and ductile metallic phases, offer a compromise between toughness and wear resistance. Functionally graded composites, however, exhibit a progressive change in composition or microstructure between one surface and another, enabling the tailoring of material properties in different regions. For erosion applications, FGCs can be designed to possess a hard, erosion-resistant surface layer combined with a tough, damage-tolerant core. This structure is especially useful for turbine blades, aerospace components, and biomedical implants, where localized erosion resistance is essential for service life extension [30–32].

The function of reinforcement materials in composite systems is central to the improvement of erosion resistance and general mechanical performance. Fibers, including carbon, glass, and aramid, contribute strength and toughness to polymer and ceramic matrices, lowering the rate of material loss due to erosion. Carbon fibers, especially, possess high stiffness and resistance to fatigue and thus find use in structural applications where there is a need for long durability. Glass fibers also have good impact resistance and are commonly employed in coatings and protective coatings for industrial machinery. Aramid fibers, having superior toughness, find use in high-speed particle impact and ballistic protection applications [33].

Ceramic reinforcements, such as silicon carbide, aluminum oxide, and zirconium dioxide, are important to enhance the wear resistance and hardness of metal and ceramic matrices. These reinforcements work as hard obstacles against eroding particles, decreasing material loss and increasing component lifetime. SiC and Al_2O_3 particles are widely used in MMCs and CMCs to resist erosion in severe environments. In addition, zirconia-based reinforcements help in enhancing fracture toughness, which makes them important in applications where crack propagation resistance is critical [34–35].

Nanoparticles are increasingly being used in composite materials to improve their mechanical properties at the microstructural level. Nanoparticle reinforcements like carbon nanotubes (CNTs), graphene, and nano-sized ceramics enhance hardness, strength, and wear resistance considerably by improving the composite microstructure. CNT-reinforced PMCs are stiff and thermally stable and have applications in aerospace as well as other high-performance engineering applications. Graphene-based reinforcements also exhibit excellent erosion resistance due to their superior surface area and mechanical properties. Nano-ceramic reinforcements add enhanced grain refinement, plus hardness, in MMCs as well as CMCs, providing an overall enhanced erosion resistance [36–37].

6.4 SLURRY AND SOLID PARTICLE EROSION OF COMPOSITE MATERIALS

Erosion wear is a major problem in several industries where composite materials are subjected to abrasive particles in liquid suspension (slurry erosion) or as solid particles in a gas or air environment (solid particle erosion). Both erosion processes are characterized by gradual loss of material by repeated particle impact, but with different mechanisms and factors influencing them. Knowledge of these processes, the effects of composite material composition, and experimental and computational methods for the analysis of erosion is necessary in order to optimize material performance under severe environments [38–39].

Slurry erosion takes place when suspended particles in a fluid medium impinge on a surface, resulting in material degradation after a period. The severity of slurry erosion is a function of particle size, velocity, concentration, angle of impact, and surface material properties. The main erosive mechanisms of slurry erosion are cutting, plastic deformation, fatigue, and brittle fracture. In ductile materials like polymer matrix composites and metal matrix composites, erosion generally results from plastic deformation and fatigue. Recurring particle impact causes surface work-hardening, crack formation, and material removal. Conversely, brittle materials, including ceramic matrix composites, suffer erosion predominantly by crack propagation and brittle fracture, with the removal of material in small fragments as a result of the lack of impact energy absorption [40].

It is one of the key problems in slurry erosion analysis that the interaction of mechanical and hydrodynamic factors is intricate, making it challenging to accurately predict the behavior of material. Particle size and shape variations, fluid turbulence, and temperature changes also make erosion more difficult to predict. Further, though harder materials are more resistant to erosion, higher hardness will result in brittleness, which decreases impact resistance. Therefore, composite material design with a desired balance of hardness, toughness, and impact energy absorption is important for achieving better slurry erosion resistance [41].

The structure of composite materials plays a major role in determining their resistance to slurry erosion. PMCs, being light and corrosion-resistant, tend to be deficient in hardness to resist extended exposure to slurry. Nevertheless, reinforcement with ceramic particles like silicon carbide (SiC) or aluminum oxide (Al_2O_3) can enhance their erosion resistance. MMCs, consisting of metallic matrices like aluminum, titanium, or nickel reinforced with hard ceramic particles, offer increased hardness and wear resistance. MMCs containing SiC, tungsten carbide (WC), or boron carbide (B_4C) have superior slurry performance because of their high hardness and heat stability [42].

CMCs like silicon carbide (SiC)-reinforced silicon carbide (SiC/SiC) composites possess great erosion resistance because of their hardness and resistance to oxidation. Yet, they are susceptible to impact damage because of their brittleness, particularly at glancing angles. Hybrid and functionally graded composites maximize erosion resistance through the use of multiple reinforcement phases, which include a hard outer layer for erosion protection and a tough core to dissipate impact energy. These high-performance composite designs are especially beneficial in uses such as hydropower turbines, oil pipelines, and aerospace components subjected to high-velocity slurry environments [43].

Slurry erosion testing is a necessity in analyzing material performance in controlled environments. The most prominent techniques utilized include slurry pot testing, jet impingement testing, and recirculating slurry loop testing. In the slurry pot testing technique, composite samples are put into a spinning slurry mix, where the particle concentration, velocity, and impact angle may be controlled and varied. It is helpful to compare the rates of material erosion under varying conditions. In jet impingement testing, a high-velocity jet of slurry is impinged against a test specimen, replicating field erosion conditions in pipelines, turbines, and industrial systems. Varying jet velocity and impingement angle, researchers can examine various erosion mechanisms and material behaviors [44]. Figure 6.4 shows the solid erosion test setup.

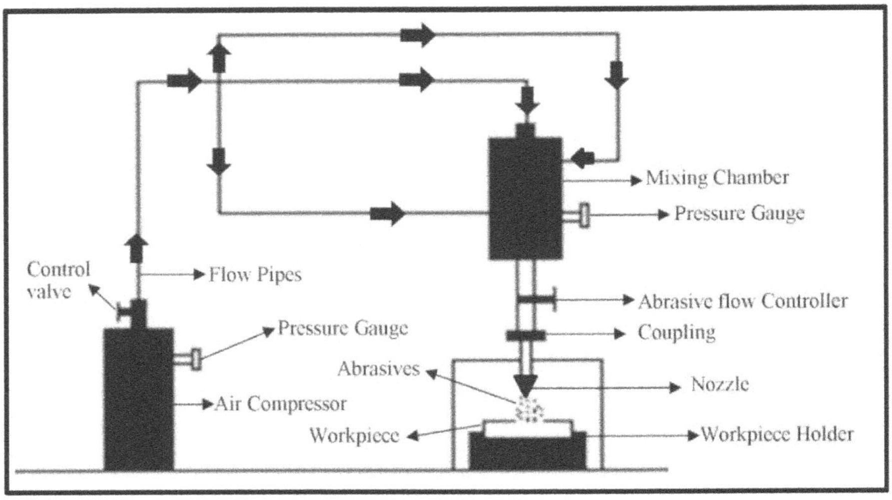

FIGURE 6.4 Schematic of the developed solid erosion test setup [45].

Recirculating slurry loop testing offers continuous erosion evaluation under industrially applicable flow regimes, allowing long-term material performance testing. Sophisticated surface analysis methods including scanning electron microscopy (SEM) and atomic force microscopy (AFM) are employed to study eroded surfaces, determine wear mechanisms, and assess material deterioration. Weight loss measurements prior to and following tests provide quantitative information on erosion rates, which forms the basis of comparison of various composite materials [46].

Several studies have explored the slurry erosion behavior of composite materials. For instance, investigations on SiC- and WC-reinforced aluminum-based MMCs concluded that composites based on WC showed significantly reduced erosion rates owing to increased hardness and toughness. Analogously, experiments with polymer composites reinforced by graphene nanoplatelets revealed enhanced slurry erosion resistance, limiting material loss in abrasive conditions. In the use of hydropower, ceramic-coated turbine blades have been observed to increase service life by reducing material degradation from erosive waterborne particles. This brings about the significance of reinforcing and material selection strategies in order to improve slurry erosion resistance [47].

Solid particle erosion, in which high-velocity particles strike a surface within a gaseous or air environment, is affected by impact angle, particle velocity, and particle type. The impact angle dictates the prevailing wear mode; ductile materials see greatest erosion at moderate angles (20°–40°) due to plastic deformation, whereas brittle materials have higher erosion rates for normal impact angles (90°) because of crack propagation. Particle speed influences erosion rates considerably, as greater speed results in higher transfer of kinetic energy and hence higher material removal. Erosion rate is frequently in a power-law function of speed such that even minor increases in speed can increase wear immensely. Particle properties like hardness, shape, and size also impact erosion severity. Pointy particles produce more aggressive cutting and plowing compared to round particles, and larger particles impart more energy per impact, with a subsequent rise in wear rates [48].

Under solid particle erosion, PMCs with hard filler reinforcements like carbon fibers or SiC particles exhibit better resistance than unreinforced polymers. MMCs, especially aluminum- and titanium-based composites with ceramic reinforcements, exhibit better performance owing to their high hardness and impact resistance. CMCs, though naturally hard, experience brittle fracture, and hence their performance is highly sensitive to impact conditions. Functionally graded materials and hybrid composites present an effective alternative by presenting a hard, wear-resistant outer layer and an impact energy-absorbing tough core and thus increasing durability in erosion-high-risk environments [49].

Computational fluid dynamics and finite-element analysis are key numerical simulation techniques employed in erosion prediction and material design. CFD simulations model particle flow behavior and impact conditions to enable researchers to better comprehend erosion patterns and design material properties for improved performance. FEA is employed to simulate stress distribution and crack growth in composites, which facilitates designing more erosion-resistant materials. Coupling experimental testing with numerical models provides a precise estimation of erosion behavior and material life, influencing development of high-performance erosion-resistant composites [50–51].

6.5 EXPERIMENTAL METHODS FOR EROSION WEAR EVALUATION

Erosion testing is critical in assessing the wear resistance of materials subjected to solid particle or slurry environments. Several standard testing procedures have been established to provide consistency and reliability in erosion evaluation. Among them, ASTM G65 is one of the most popular standards for determining the abrasion resistance of metallic and non-metallic materials. It involves a dry sand-rubber wheel test, in which particles of abrasives are pushed against a rotating specimen under controlled conditions. The technique offers a standardized manner in comparing the wear resistance of materials and coatings. Other ASTM standards like ASTM G76 for solid particle erosion testing and ASTM G119 for slurry erosion testing assist in assessing material behavior under given erosion conditions. These standard procedures are guaranteed to be reproducible and are of great benefit in selecting materials appropriate for erosive conditions [52].

Besides standard procedures, some of the most commonly used experimental procedures for erosion testing are the jet erosion test, slurry pot test, and rotating disc test. The jet erosion test reproduces actual field erosion conditions by firing a high-velocity slurry or solid particle jet onto the test sample. By having control over impact angles, velocity, and particle concentration, the test method assesses material resistance to dynamic impact well. Slurry pot testing is another common method, in which a sample is submerged in a rotating slurry blend, enabling controlled variation of particle velocity, concentration, and impact angle. This type of test is especially effective for erosion simulation of pipeline, turbines, and pump conditions. Rotating disc test consists of a high-speed rotating disc that subjects the test material to erosive particles and yields information on wear mechanisms and degradation of material. All these methods have their merits, and the choice will be based on the application and operating condition [53–55].

Advanced characterization methods including scanning electron microscopy, energy dispersive spectroscopy (EDS), and X-ray diffraction (XRD), are frequently employed to understand mechanisms of erosion at the microstructural level. SEM is a robust imaging instrument that generates high-resolution images of eroded surfaces, allowing researchers to analyze wear patterns, crack initiation, and particle embedding within materials. SEM analysis aids in the differentiation of various erosion mechanisms like plastic deformation, brittle fracture, and fatigue wear, providing information on material failure modes. EDS, which is commonly coupled with SEM, enables elemental analysis of eroded surfaces. It aids in the detection of compositional changes, contamination, and the occurrence of wear debris, which may be of significance in determining material degradation under erosive conditions. XRD is another key method employed to examine the phase composition and crystallographic structure of materials prior to and following erosion testing. Through the detection of phase transformations, oxidation, and residual stress formation, XRD offers important information regarding material stability and structural changes caused by erosion [56–58].

6.6 RECENT ADVANCES IN COMPOSITE MATERIALS FOR EROSION RESISTANCE

Nanocomposite development has enhanced wear resistance in erosive environment-exposed materials considerably. Nanocomposites involve nanoscale reinforcements like carbon nanotubes, graphene, nanoceramics, and metallic nanoparticles into polymer, metal, or ceramic matrices. The reinforcements increase mechanical properties like hardness, toughness, and fracture resistance, hence minimizing material loss under erosive conditions. For example, polymer matrix composites reinforced with graphene nanoplatelets exhibit improved erosion resistance due to their high strength-to-weight ratio and self-lubricating properties. Similarly, metal matrix composites reinforced with nano-sized aluminum oxide (Al_2O_3) or silicon carbide (SiC) show superior hardness and thermal stability, making them ideal for high-temperature erosion applications. The homogeneous dispersion of the nanoparticles in the matrix is key to improvement in resistance by inhibiting localized concentration of stress and crack propagation [59]. Figure 6.5 shows the atom structure of carbon nanotubes, graphene nanoplatelets, and fullerene-C60 particles.

Surface treatments and coatings provide another useful method of reducing erosion damage. Some coating technologies, including thermal spray coatings, chemical vapor deposition (CVD), physical vapor deposition (PVD), and laser cladding, improve the wear resistance of structural materials (Figure 6.6). Harsh ceramic-based coatings such as tungsten carbide (WC), chromium carbide (CrC), and titanium nitride (TiN) offer a protective shield against erosive particles, greatly increasing the longevity of aerospace, power generation, and oil and gas equipment components. Further, newer multifunctional coatings that contain nanostructured layers or functionally graded materials exhibit better toughness without compromising high hardness. Plasma-sprayed HA and alumina, for instance, exhibit outstanding erosion resistance in biomedical and marine environments [61].

The introduction of self-healing composites and smart materials has made novel erosion-resistant application possibilities a reality. Self-healing composites contain microcapsules or vascular networks infused with healing agents that trigger healing when damaged, restoring material integrity and extending service life. Polymer-based self-healing materials with the embedded liquid-phase healing agents have shown promise for aerospace and automobile applications, which are costly in terms of maintenance. Shape-memory alloys (SMAs) and piezoelectric materials are also capable of actively responding to external stimuli through changes in microstructure, lowering erosion-related degradation. Smart materials are especially useful in dynamic situations where erosion conditions change drastically. Computational modeling and simulations are essential in predicting erosion

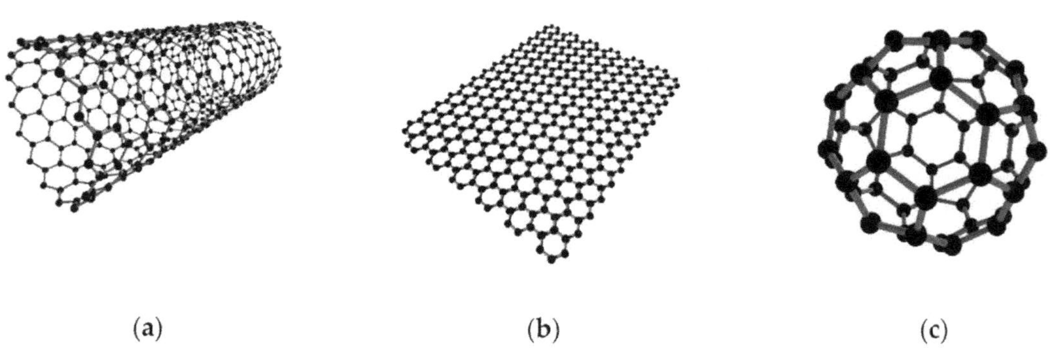

FIGURE 6.5 Atom structure of (a) carbon nanotubes, (b) graphene nanoplatelets, and (c) fullerene-C60 particles [60].

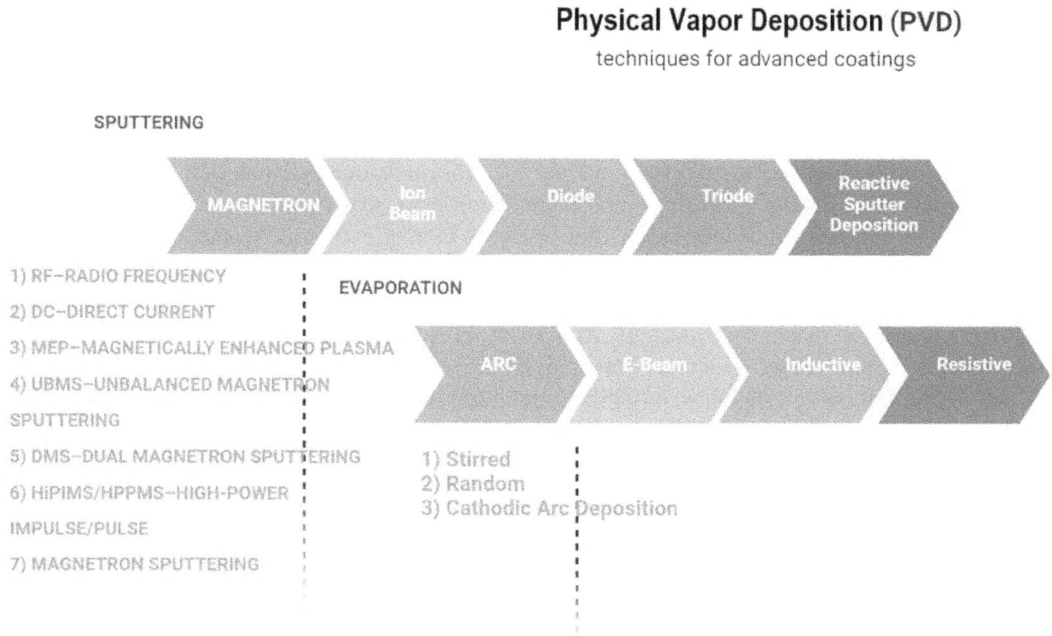

FIGURE 6.6 Physical vapor deposition (PVD) methods used for improved coatings [62].
(CC BY 4.0 attribution)

wear behavior and optimizing material design. Computational fluid dynamics, finite-element analysis, and other techniques enable simulating particle impact dynamics and material response [63].

6.7 INDUSTRIAL APPLICATIONS AND CASE STUDIES

Erosion-resistant materials are of prime importance in applications where parts are exposed to high-velocity particles, slurry erosion, and harsh environmental conditions. In the defense and aerospace industries, erosion wear is a significant problem for aircraft turbine blades, leading edges, and missile parts exposed to high-speed dust, sand, and rain erosion. Composite materials with erosion-resistant coatings, including thermal barrier coatings and nanocomposite reinforcements, contribute to the increased service life of aerospace components. Functionally graded materials are also utilized to deliver a hard external surface while retaining toughness in the core, minimizing material loss at harsh operating conditions. The defense sector applies the same principles to armor coatings and missile systems, where erosion resistance is essential to ensure structural integrity and performance.

In the petroleum and gas sector, erosion wear is a major problem in pumps, pipelines, drilling equipment, and valve parts subject to sand-laden crude oil and abrasive slurries. Materials like tungsten carbide-strengthened metal matrix composites and ceramic-based coatings reduce erosion damage in high-pressure flow regimes. High-velocity oxygen fuel (HVOF) and plasma-sprayed coatings are common for increasing surface hardness and wear resistance, maximizing the operating life of vital equipment. Furthermore, polymer-based composite liners and corrosion-resistant alloys (CRAs) are being used more and more in pipeline infrastructure to avoid both erosion and corrosion, making oil and gas transport systems efficient and safe [64].

In power plants, especially in gas and steam turbines, boiler tubes, and coal-fired plant equipment, erosion due to high-speed ash particles and steam flows causes material degradation over time. In response to this, thermal spray coatings such as carbide-based and ceramic-based coatings are deposited onto turbine blades and boiler tubes for improving wear resistance. Nickel-based superalloys with protective coats are widely used in high-temperature applications for resisting erosion as well as oxidation. New-generation materials such as self-healing composites are also being developed to improve the life of power plant components by healing microdamage resulting from particle impact [65].

In applications involving hydropower and marine, erosion caused by cavitation, waterborne particulates, and slurry flow is a formidable problem for offshore structures, hydro-turbines, and ship propellers. For withstanding combined erosion and corrosion effects, downstream and hard-coating alloys of stainless steel like WC-Co and SiC-based ceramics are applied. Nanocomposite coatings combined with hybrid coatings enhance the erosion resistance of hydropower turbines without compromising their high mechanical strength. Moreover, self-lubricating polymeric coatings are being investigated to minimize wear in underwater parts. Progress in advanced erosion-resistant materials is continually improving efficiency, reliability, and durability across these sectors, with longer life and lower maintenance costs [66].

6.8 CONCLUSION

The erosion wear performance of composite materials in slurry and solid particle erosion conditions depends on several factors such as material composition, impact parameters, and environmental conditions. Composite materials like polymer matrix composites, metal matrix composites, and ceramic matrix composites provide different levels of erosion resistance based on their reinforcement phase and structure. The addition of hard ceramic particles, fibers, and nanoparticles significantly improves wear resistance, and thus composites can be used in high-wear conditions. Hybrid and functionally graded composites also optimize erosion performance through a balance of hardness, toughness, and impact energy absorption.

Surface treatments, coatings, and novel manufacturing technologies like thermal spray coatings, nanocomposite reinforcement, and self-healing materials have been extensively investigated to minimize erosion damage. Experimental methods, such as jet erosion tests, slurry pot tests, and rotating disc tests, assist in determining material performance under controlled conditions, whereas characterization techniques like SEM, EDS, and XRD assist in understanding erosion mechanisms and material degradation. Computational modeling and simulations also assist in predicting wear behavior and material design optimization for erosion-susceptible applications.

Fields like aerospace, oil and gas, power plants, and hydropower depend on erosion-resistant composite materials to maximize the longevity and reliability of critical components. Continued innovation in new composites, surface engineering technologies, and predictive modeling will further enhance the discipline, resulting in more efficient and durable materials for erosion-intensive applications. Future work should be on multifunctional composites with enhanced mechanical properties and self-adaptive features to resist extreme erosion conditions.

REFERENCES

1. Zhou M, Xie Y, Li Y, et al. Erosion wear behavior of composite materials in slurry and solid particle erosion environments. Wear. 2019;432:49–59.
2. Li X, Zhang Q, Liu X, et al. Influence of particle size and impact angle on erosion behavior of composite materials. Tribol Int. 2018;122:99–108.
3. Fadaei H, Ghasemi M, Zarrebini M. Study on erosion wear performance of composite coatings under slurry and solid particle erosion. Surf Coat Technol. 2020;387:125535.
4. Singh G, Ranjan P, Pandey R, et al. Erosion behavior of particulate reinforced composite materials: A review. J Mater Sci. 2019;54:8129–8147.
5. Ashrafi B, Sadeghi M. Influence of fiber reinforcement on the erosion resistance of polymer matrix composites. Polym Compos. 2018;39:3285–3294.

6. Wang X, Li Z, Xu X, et al. Effect of reinforcement particles on the slurry erosion behavior of composite materials. Compos Part B Eng. 2019;175:107131.
7. Gupta R, Mishra R. Erosion wear resistance of composite materials for slurry applications. J Mater Process Technol. 2017;244:101–108.
8. Zhao Y, Liu J, Wang D, et al. Erosion wear behavior of composite coatings under solid particle impact. J Therm Spray Technol. 2020;29:1187–1197.
9. Kalácska Á, Nagarajan R, Tóth LF, et al. Surface damage analysis on the application of abrasion and slurry erosion in targeted steels using an erosion test rig. Lubricants. 2022 Nov;10(11):316.
10. Jiang H, Chen X, Guo Z. High performance slurry erosion behavior of polymer matrix composites. Wear. 2018;414–415:1–10.
11. Zhang Z, Liu Y, Zhang H, et al. Solid particle erosion behavior of metal matrix composites: A review. Tribol Int. 2017;111:228–240.
12. Kumar R, Yadava V, Verma A. Erosion behavior of composite materials: Influence of matrix and reinforcement. J Reinforced Plast Compos. 2020;39:1469–1483.
13. Huang Z, Wu X, Chen D, et al. Erosion characteristics of composite materials: A comparative study. Wear. 2019;428–429:84–93.
14. Ranjan P, Singh G, Yadav V, et al. Composite materials for erosion wear resistance: A systematic review. Mater Manuf Process. 2020;35:1–15.
15. Li C, Zhang H, Xue Y, et al. Erosion wear of ceramic matrix composites under solid particle impact. Ceram Int. 2018;44:4290–4297.
16. Zang X, Chen X, Wang Y, et al. Erosion resistance of metal matrix composite coatings for high-temperature applications. Surf Coat Technol. 2021;406:126535.
17. Ma Z, Tang L, Zhang Z, et al. The effect of particle size and velocity on erosion of composite materials in slurry applications. J Appl Polym Sci. 2020;137:48528.
18. Choi J, Lee H, Park Y, et al. Erosion wear behavior of reinforced polymer composites in high-velocity solid particle impact. Compos Sci Technol. 2020;192:108131.
19. Banerjee S, Rao G. Erosion wear characteristics of metal matrix composites in solid particle impacts. Wear. 2021;462:204422.
20. Kumar K, Kumar S, Singh G, et al. Erosion wear investigation of HVOF sprayed WC-10Co4Cr coating on slurry pipeline materials. Coatings. 2017 Apr 12;7(4):54.
21. Yang X, Wang J, Sun J, et al. Comparative study of slurry erosion resistance of metal matrix composites reinforced with different particles. Wear. 2019;430–431:75–83.
22. Lin Z, Zhang L, Jiang P, et al. Solid particle erosion behavior of composite materials for aerospace applications. J Aerosp Eng. 2020;33:04020044.
23. Gutema EM, Lemu HG. Conventional machining of metal matrix composites towards sustainable manufacturing—present scenario and future prospects. J Compos Sci. 2024 Sep 12;8(9):356.
24. Shankar M, Venkatraman B, Srinivasan N. A study on the slurry erosion behavior of composite materials. Wear. 2021;461:203463.
25. Nayak S, Mishra P. Experimental study on slurry erosion behavior of composite materials: Effect of particle concentration. Tribol Lett. 2020;68:1–10.
26. Ghosh S, Rani R, Gupta R, et al. Erosion wear behavior of aluminum-based composite materials under solid particle impact. Tribol Int. 2020;145:106147.
27. Verma PK, Minhas AS, Singh P, Kumar S. Slurry erosion behaviour of thermal sprayed Al2O3 and Cr2O3 coatings for turbine steels. In AIP Conference Proceedings 2024 Feb 20 (Vol. 2986, No. 1). AIP Publishing.
28. Cao L, Liu Y, Zhang H. Effect of particle impact angle on erosion wear of polymer matrix composites. Tribol Lett. 2019;67:1–8.
29. Ganjian M, Eskandari S, Barati R, et al. A review on slurry erosion behavior of composite materials: Mechanisms and influencing factors. J Mater Eng Perform. 2021;30:2294–2312.
30. Kumar S, Singh G, Ranjan P. Erosion wear resistance of polymer composites: Effect of fillers and additives. Compos Part B Eng. 2020;185:107733.
31. Shi Y, Wang Y, Ma X, et al. Erosion behavior of composites: Effect of matrix composition and fiber reinforcement. Wear. 2020;448:203226.
32. Kang M, Jeon S, Park J. High-temperature slurry erosion behavior of composite materials. J Therm Spray Technol. 2019;28:986–994.

33. Cheng J, Zhang L, Xu H, et al. Investigation of erosion resistance of composite materials under solid particle impact. Wear. 2020;458:203531.
34. Liu X, Xu S, Liang Y. Erosion resistance of hybrid composite coatings under solid particle impact. J Therm Spray Technol. 2021;30:278–289.
35. Pati N, Sahoo S, Sahoo P, et al. Erosion wear behavior of composite materials reinforced with ceramic particles. Mater Sci Eng A. 2021;800:140385.
36. Mishra P, Patil A. A review on slurry erosion behavior of composites: Mechanisms and modeling. Wear. 2020;442–443:203135.
37. Zhou T, Jiang H, Li Z, et al. Erosion resistance of composite coatings: A study on wear mechanisms. J Mater Sci. 2019;54:8659–8674.
38. Zhang C, Liu Q, Zhang L, et al. Effect of matrix reinforcement on the erosion wear resistance of composites in slurry applications. Compos Part A Appl Sci Manuf. 2020;136:105967.
39. Li X, Wang L, Huang Y, et al. Erosion behavior of polymer matrix composites under solid particle impact at various velocities. Tribol Lett. 2019;67:47.
40. Rao P, Tiwari A, Roy M. Influence of particulate reinforcement on erosion wear resistance of composites. Mater Sci Eng B. 2020;262:114595.
41. Tang J, Yang Y, Li C, et al. Erosion behavior of metal matrix composites in slurry erosion environments. J Mater Sci. 2019;54:3479–3491.
42. Zhang Z, Zhang H, Liu F, et al. Investigation of erosion resistance of composite materials under high-velocity solid particle impact. Compos Part B Eng. 2019;165:356–363.
43. Sahoo S, Patil P, Raj B. Erosion wear of composite materials: Influence of particle impact angle and velocity. Tribol Int. 2020;142:105986.
44. Kumar S, Chaurasia M. Influence of composite matrix materials on solid particle erosion behavior. Wear. 2021;468–469:204661.
45. Antil SK, Antil P, Singh S, et al. Artificial neural network and response surface methodology based analysis on solid particle erosion behavior of polymer matrix composites. Materials. 2020 Mar 18;13(6):1381.
46. Xie Y, Zhang X, Li Z. Erosion behavior of composite materials for industrial applications. Tribol Lett. 2021;69:1–9.
47. Mohan M, Ravi S. Erosion wear behavior of metal matrix composite coatings. Surf Coat Technol. 2020;403:126393.
48. Tiwari S, Mishra S. Erosion wear of composites under slurry and solid particle impacts: A comparison study. Wear. 2021;472–473:204711.
49. Wang J, Choi W, Zhang H, et al. Wear behavior of composite coatings under slurry erosion conditions. Wear. 2018;394–395:29–37.
50. Xu J, Liu Q, Chen Y. Erosion behavior of reinforced composite materials in high-velocity solid particle impact. Mater Des. 2019;174:107755.
51. Li Z, Zhang Y, Guo H, et al. Erosion wear resistance of composite coatings under liquid-solid impact. J Mater Process Technol. 2020;276:116354.
52. Liu Y, Zhang L, Wang Y, et al. Erosion behavior of particulate reinforced composite materials under solid particle impact. J Compos Mater. 2020;54:2541–2552.
53. Cheng X, Zhang C, Gao H. Effect of fiber reinforcement on the slurry erosion behavior of polymer composites. Wear. 2021;478:203764.
54. Rao G, Banerjee S. Erosion behavior of particulate reinforced composites: Effect of matrix hardness. Tribol Int. 2021;149:105617.
55. Nanda A, Nair S. Erosion wear analysis of composite coatings: Influence of particle concentration and size. Wear. 2019;426–427:65–73.
56. Li W, Chen Z, Liu S, et al. Solid particle erosion behavior of polymer matrix composites: Role of particle shape. Tribol Int. 2020;151:106521.
57. Verma A, Yadav V. Wear and erosion behavior of composite coatings under high-velocity solid particle impact. J Mater Eng Perform. 2020;29:3299–3310.
58. Das S, Kumar A, Sarma P. Effect of fiber reinforcement on erosion wear of polymer composites. J Reinf Plast Compos. 2020;39:3199–3209.
59. Tang L, Li H, Zhang Z, et al. Study on the slurry erosion behavior of composite materials in harsh environments. Wear. 2020;447:203092.

60. Wang X, Tang F, Cao Q, et al. Comparative study of three carbon additives: Carbon nanotubes, graphene, and fullerene-c60, for synthesizing enhanced polymer nanocomposites. Nanomaterials. 2020 Apr 27;10(5):838.
61. Ma J, Lee K, Liao Y. High-temperature erosion of composite materials for aerospace applications. J Therm Spray Technol. 2021;30:217–227.
62. Dabees S, Mirzaei S, Kaspar P, et al. Characterization and evaluation of engineered coating techniques for different cutting tools. Materials. 2022 Aug 16;15(16):5633.
63. Mishra S, Reddy A, Kumar P, et al. Effect of particle impact velocity on erosion of composite materials: A review. Tribol Int. 2019;138:1–8.
64. Li Q, Wang D, Liu Y, et al. Erosion wear behavior of ceramic matrix composites: Experimental and modeling approaches. Ceram Int. 2020;46:10489–10497.
65. Rani R, Gupta M, Sharma S. Erosion wear of composite materials for slurry and solid particle applications. Wear. 2021;470:203587.
66. Zhou J, Yang S, Zhou Y. Slurry erosion wear mechanisms of composite materials for industrial applications. J Mater Sci. 2020;55:7842–7857.

7 Fabrication of Metallic Materials for Erosion Resistance

Sunitha K, Praveen Kumar Verma, Hitesh Vasudev, and Che Hua Yang

7.1 INTRODUCTION

Erosion wear is a critical issue in industries that involve high-velocity solid particle impact, liquid droplet erosion, or cavitation erosion of metallic components. The impact of these erosion mechanisms can lead to material degradation, loss of structure integrity, and reduced operational efficiency. This becomes particularly critical in aerospace, marine, power generation, and mining industries where components are subjected to severe erosive environments. The choice of appropriate materials and manufacturing techniques is vital to maximizing erosion resistance, extending service life, and lowering maintenance costs. Erosion wear results when repeated impacts of solid particles, liquid droplets, and cavitation bubbles attack a metallic surface, removing material progressively [1–2]. Solid particle erosion is a type of material degradation due to the impact of high-velocity solid particles on a surface, leading to progressive wear and loss of material.

It is a highly common phenomenon in applications such as gas turbines, pipelines, and mining equipment, where abrasive particles are carried in fluid or gas streams. The impact of such particles wears away tiny fragments from the material surface, reducing the structural strength of the components gradually. Particle velocity, size, hardness, and incidence angle are the parameters that play a significant role in erosion rate. Hard and abrasive materials such as ceramic coatings or high-temperature alloys are used to mitigate the effect of solid particle erosion in process uses [3–4]. Liquid droplet erosion is another kind of erosion that occurs as a result of high-speed liquid droplets impacting a solid surface again and again, causing gradual material loss. It is particularly prevalent in steam turbines and aircraft engines, in which liquid drops of water suspended in high-velocity steam or air strike metal surfaces. Continuous action of such drops creates local stress and surface fatigue and results in the formation of cracks and subsequent material loss. Liquid droplet erosion is a function of the size of the droplet, the impact velocity, and material properties. To counter the impact, engineers turn to coatings, surface treatments, and erosiogenic materials to strengthen the concerned components [4–6]. Cavitation erosion, however, results from the formation and collapse of vapor bubbles in a liquid on the metal surface. The process generates high-energy shock waves that result in serious material degradation over a long period. Cavitation erosion is particularly difficult to handle in ship propellers, hydraulic turbines, and pumps, where pressure fluctuations form vapor bubbles in the fluid. Upon implosion of these bubbles, they form localized high-pressure zones, which erode the underlying material, leading to pitting and surface degradation. The severity of cavitation erosion is influenced by fluid velocity, pressure oscillations, and material characteristics. To reduce cavitation damage, engineers employ advanced materials, surface coatings, and optimized design methods that reduce pressure oscillations and turbulence [7–8]. Figure 7.1 shows surface material behavior under two different types of impacts. In Figure 7.1(a), a normal impact results in material deformation at the point of contact directly underneath, tending to create craters.

DOI: 10.1201/9781003561156-9

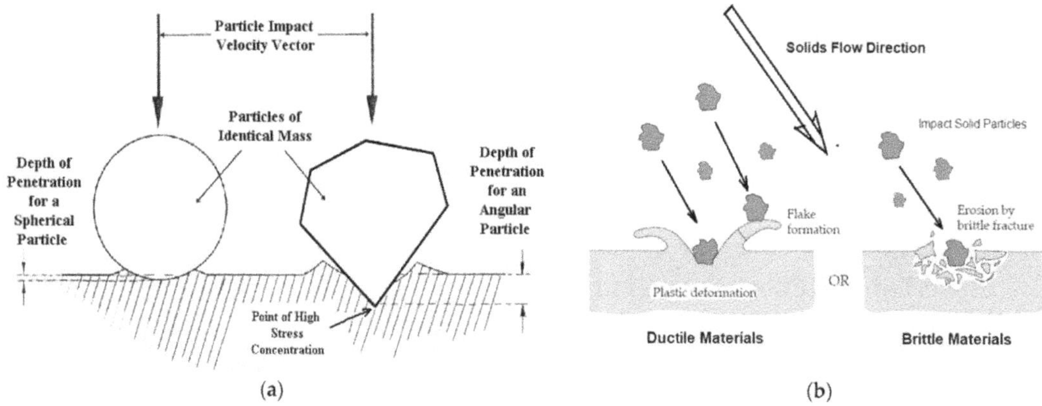

FIGURE 7.1 Surface material under impacts: (a) normal impact; (b) oblique impact [9]. **(CC BY 4.0 attribution)**

In Figure 7.1(b), an oblique impact leads to asymmetric deformation and material displacement along the impact path, with potential surface shearing.

Each of these mechanisms of erosion, namely solid particle erosion, liquid droplet erosion, and cavitation erosion, is a substantial concern in many industrial applications. The correct choice of materials, protective coatings, and improvement in design are all important in reducing erosion-induced failures and increasing the life of critical hardware. Knowledge of the mechanisms and the factors that affect erosion enables industries to come up with efficient strategies to reduce material degradation and guarantee sustained performance under hostile environments. The choice of erosion-resistant materials is influenced by hardness, toughness, corrosion resistance, and microstructural stability. Metallic materials with excellent mechanical characteristics and advanced coatings are usually used to resist erosion damage. Nevertheless, the ongoing need for evolving industrial requirements demands the discovery of new fabrication methods and material compositions providing enhanced erosion resistance. To combat erosion wear, various manufacturing methods have been explored, including alloying, surface hardening, thermal spray coatings, and additive manufacturing. Alloying introduces elements such as chromium, nickel, cobalt, and titanium into the base metal that contribute hardness, toughness, and corrosion resistance. Surface hardening methods including carburizing, nitriding, and laser hardening involve introducing surface properties with improved hardness without compromising the overall ductility of the material. Thermal spray coatings, including high-velocity oxy-fuel (HVOF) spraying, plasma spraying, and cold spraying, allow the deposition of erosion-resistant metal coatings on metal substrates. Additive manufacturing (AM) technologies like selective laser melting (SLM) and electron beam melting (EBM) allow parts with engineered microstructures and intricate material combinations to be produced with improved erosion resistance [10–13].

Recent studies have focused on clarifying material microstructure–erosion behavior relationships. Grain size, phase composition, and hardness distribution play important roles in the erosion resistance of a material. Fine-grained microstructures and dispersion-strengthened materials exhibit better erosion resistance due to their capacity to resist alternating impact forces. The development of functionally graded materials (FGMs) and multilayer coatings has also contributed to improving erosive performance. In addition, computational modeling and simulation methodologies have emerged as essential tools to predict erosion behavior and optimize material design. Finite element analysis (FEA), computational fluid dynamics (CFD), and machine learning-based prediction models are gaining prominence to analyze material response to erosive conditions and inform the creation of novel erosion-resistant materials [14].

This review focuses on presenting an overview of the latest developments in the fabrication of metallic materials with enhanced erosion resistance. It reviews different fabrication processes, microstructural changes, and performance testing methodologies to identify the current trends in the field. The discussion also covers the role of processing conditions on erosion resistance and the likely future trends towards improving the erosion resistance of metallic components in high-erosion environments [15].

7.2 FABRICATION TECHNIQUES FOR EROSION-RESISTANT METALLIC MATERIALS

7.2.1 Alloying and Compositional Optimization

Alloying is one of the basic methods of improving the erosion resistance of metals through the alteration of their microstructure and mechanical behavior. The introduction of alloying elements enhances hardness, toughness, and corrosion resistance, enabling metals to be stronger in harsh conditions. High-strength alloys like stainless steel, nickel-based superalloys, and titanium alloys find extensive applications in those industries that require high levels of erosion resistance because they can handle critical conditions [16].

Stainless steel is one of the most widely used erosion-resistant alloys because of its strength, toughness, and corrosion resistance. The availability of elements like chromium, nickel, and molybdenum provides it with superior resistance against loss of material owing to solid particle erosion, liquid droplet erosion, and cavitation. The passive oxide layer formed by chromium is especially good at covering the base metal and preventing wear and corrosion. This renders stainless steel highly appropriate for application in steam turbines, marine water conditions, and power facilities where components are required to withstand high-speed fluids and abrasive particles. Nickel-based superalloys such as Inconel and Hastelloy exhibit excellent erosion resistance due to their heat strength and high-temperature oxidation resistance. These are designed to withstand extreme heat and mechanical stress, and thus they are employed for gas turbines, aerospace components, and chemical processing plant equipment. Nickel enhances the strength and thermal stability of the alloy, while chromium, cobalt, and aluminum improve its oxidation and corrosion resistance. These elements also contribute to the formation of protective surface oxides, which minimize material degradation by erosion. The ability of nickel-based superalloys to maintain mechanical properties at elevated temperatures renders them very efficient in withstanding erosion due to solid particles and liquid droplets. Titanium alloys are another class of high-performance materials possessing excellent erosion resistance due to their low weight, high strength-to-weight ratio, and improved corrosion resistance. The addition of vanadium and aluminum improves the strength and hardness of titanium alloys, making them suitable for aerospace, marine, and biomedical applications. These alloys have high cavitation erosion resistance in ship propellers and hydraulic turbines, where frequent changes in pressure lead to loss of material. The natural buildup of a protective oxide layer makes titanium alloys more durable and resists frequent repair and replacement [17–20].

7.2.2 Thermal Spray Coatings

Thermal spray methods are extensively used to improve the erosion resistance of metal parts by applying protective coatings that reduce material loss. Such methods involve high-velocity oxy-fuel, plasma spraying, and cold spraying, each with its own set of benefits regarding coating adhesion, density, and microstructural properties. By using erosion-resistant materials like WC-Co (tungsten carbide-cobalt), Cr_3C_2-NiCr (chromium carbide-nickel chromium), and TiC-based (titanium carbide) coatings, industry can extend the life of vital components exposed to high-velocity particle impact, liquid droplet erosion, and cavitation by a considerable amount [21]. HVOF spraying is among the most efficient techniques for the deposition of dense, well-adhered coatings with superior

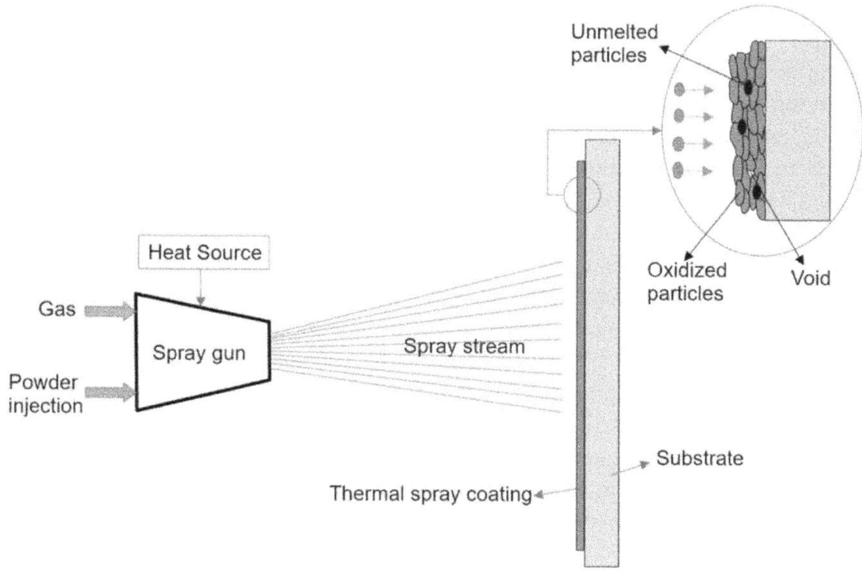

FIGURE 7.2 Schematic illustration of thermal spray coating [24].
(CC BY 4.0 attribution)

hardness and wear properties. The technique uses the burning of oxygen and fuel to create a supersonic gas jet to push powder particles onto a substrate with high velocity. The coatings produced have excellent mechanical properties such as higher hardness, low porosity, and high bond strength. WC-Co coatings, typically deposited using HVOF, are highly resistant to solid particle erosion because they have high hardness and toughness. These coatings are extensively used in aerospace, mining, and power generation applications, where equipment is subjected to abrasive particles and harsh operating conditions [22–23]. Figure 7.2 shows the setup diagram of thermal spraying.

Plasma spraying, another commonly employed thermal spray process, utilizes a high-temperature plasma jet to melt and propel coating materials onto a substrate. This process is ideally suited for the deposition of ceramic-based coatings, like Cr_3C_2-NiCr, which provide good resistance to high-temperature erosion and oxidation. Cr_3C_2-NiCr coatings are generally used on steam turbines, gas turbines, and process equipment, where exposure to liquid flow and high-speed gas causes degradation of material. Chromium carbide gives hardness and wear resistance, whereas nickel-chromium increases toughness and resistance to oxidation, resulting in long-term operation under severe conditions [25–26]. Cold spraying, which is a fairly new thermal spray process, is particularly suited to depositing coatings with little or no thermal degradation. In this technique, metal or composite particles are accelerated at supersonic velocities and strike the substrate in a solid condition, creating solid and closely adherent coatings. TiC coatings, deposited by cold spraying, have superior erosion resistance because of their high hardness and chemical stability. They are extremely useful in marine, aerospace, and automotive applications where cavitation and high-velocity fluid impact create problems [27].

7.2.3 Surface Hardening Techniques

Surface processes like carburizing, nitriding, and laser hardening are commonly employed to improve the erosion and wear resistance of metallic materials. These processes alter the metal surface properties by enhancing hardness while maintaining the core material's ductility and toughness. This blend is critical in erosive environment applications where material loss must be resisted by high hardness, while the core material should maintain its mechanical integrity during impact and stress [28].

Carburizing is a heat treatment where a metal, commonly low-carbon steel, is subjected to a carbon-containing atmosphere at high temperatures. The treatment enables carbon atoms to diffuse into the surface, creating a hardened outer layer while leaving a ductile core. The hardened surface enhances wear and erosion resistance, making carburized parts well suited for use in aerospace, automotive, and industrial machinery applications. Gears, bearings, and shafts are particularly benefited by carburizing since they are repeatedly subjected to erosive and contact forces. The process can be carried out with gas, liquid, or solid mediums, but gas carburizing is the most common because it can deliver uniform case depth and controlled carbon diffusion [29]. Figure 7.3 shows schematic diagrams of carburizing. In Figure 7.3(a), the traditional carburizing process brings carbon into the surface at elevated temperatures through gaseous or solid media. In Figure 7.3(b), the new solid-solution carburizing process increases diffusion efficiency and carbon incorporation, providing better surface hardness and wear resistance with better control.

Nitriding is another useful surface treatment that increases erosion resistance by adding nitrogen into the metal surface at comparatively low temperatures than carburizing. Nitriding, unlike carburizing, does not involve quenching, which minimizes the chances of distortion

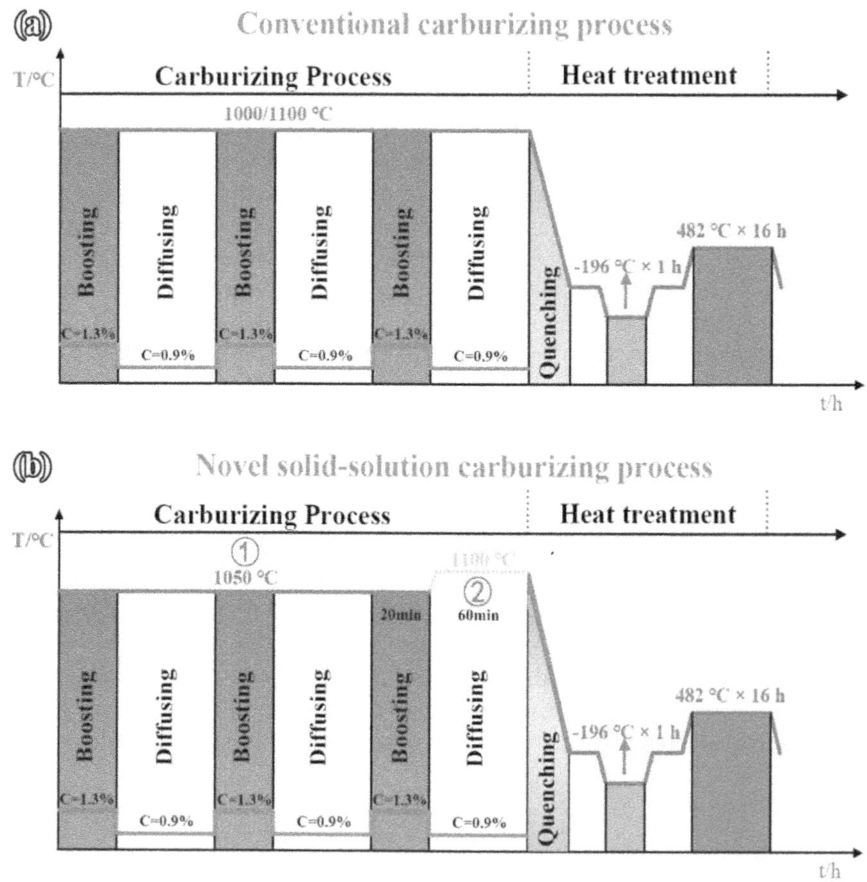

FIGURE 7.3 Schematic diagram of carburizing process: (a) conventional carburizing process; (b) novel solid-solution carburizing process [30].

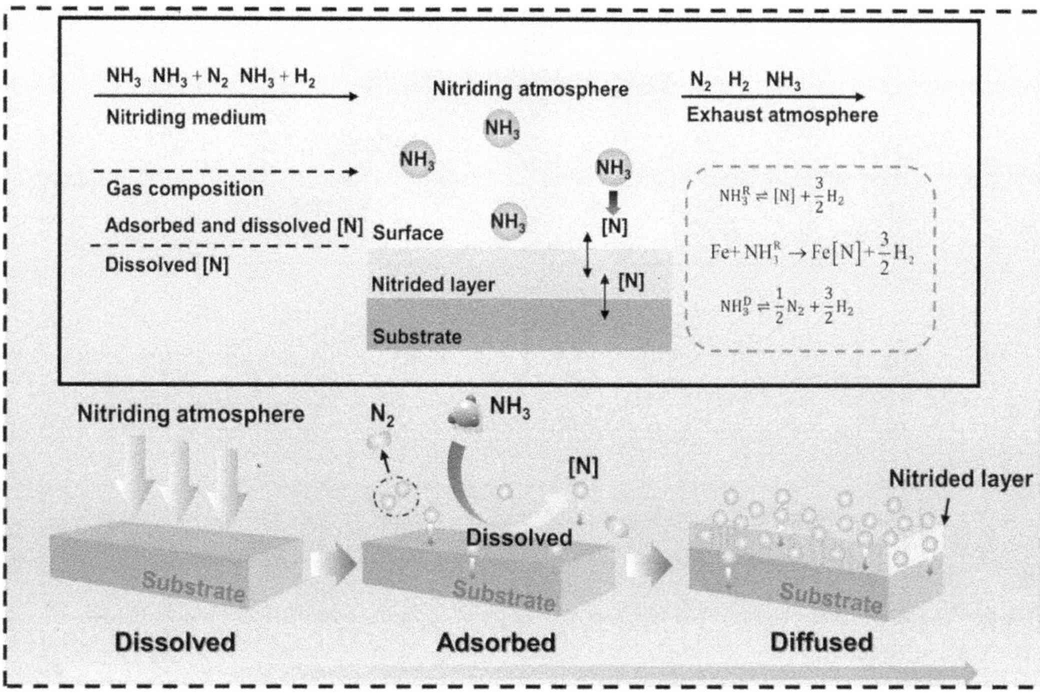

FIGURE 7.4 Interface reaction process of gas nitriding [32].
(CC BY 4.0 attribution)

in parts with complex shapes. The treatment is often used on steels with alloying elements like chromium, molybdenum, and aluminum, which have stable nitrides that improve surface hardness. Nitrided layers exhibit outstanding wear resistance, corrosion protection, and fatigue strength and are appropriate for use in turbine blades, crankshafts, and hydraulic cylinders. The diffusion-controlled process of nitriding permits hardness depth control with high precision so that parts show optimal performance in erosion-critical environments [31]. Gas nitriding is a surface hardening technology that allows the nitriding atmosphere to penetrate to the surface of metal or alloy material and diffuse to a certain depth in the conditions of high pressure and high temperature. Nitrogen atoms chemically combine with metal surface atoms during gas nitriding to produce a hard, wear resistant, corrosion resistant, and metal fatigue resistant nitrided layer, presented in Figure 7.4.

Laser hardening is a new method of surface treatment that involves heating and cooling rapidly the surface of a metal by using a concentrated high-energy beam of laser and thereby producing a hardened layer with no need to introduce additional alloying elements. It is beneficial in localized hardening, by which particular locations of a part can be hardened without making changes in other unwanted areas. Laser hardening creates a fine-grained microstructure that improves erosion resistance by minimizing material loss caused by solid particle and liquid droplet impact. Laser hardening is widely applied in automotive, aerospace, and tooling industries where precision hardening is necessary to increase the service life of vital components [33]. Figure 7.5 shows a schematic diagram of energy transfer of the laser during surface hardening using a laser. The laser beam provides intense heat energy to the surface of the material, rapidly heating it over the austenitizing temperature. Heat conducted into the substrate provides self-quenching and creates a hard martensitic layer without surface melting, and this improves wear resistance.

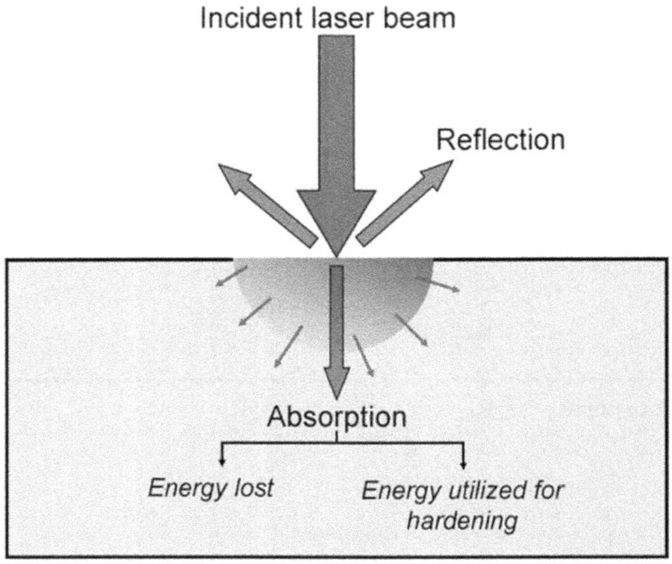

FIGURE 7.5 Schematic diagram showing laser energy transfer during laser surface hardening [34]. **(CC BY 4.0 attribution)**

Each of these surface treating processes has specific benefits based on the requirements of the application. Carburizing achieves deep surface hardness with great wear resistance and is good for high-load applications. Nitriding gives outstanding hardness and corrosion resistance without distortion and is suitable for precision-engineered components. Laser hardening gives precise hardening with little thermal impact and is a preferred choice where localized wear protection is needed. Surface processes like carburizing, nitriding, and laser hardening are crucial in enhancing the erosion resistance of metal materials. Through the increase in surface hardness without compromising core ductility, these processes considerably prolong the service life of parts subjected to erosive conditions, lowering maintenance expenses and enhancing overall system reliability [35].

7.3 ADDITIVE MANUFACTURING FOR EROSION-RESISTANT MATERIALS

New developments in additive manufacturing (AM), specifically selective laser melting (SLM) and electron beam melting (EBM), have transformed the manufacturing of metallic parts with engineered microstructures and improved erosion resistance. The AM processes have superior design freedom, allowing complex geometries and functional graded materials to be fabricated that are inaccessible by traditional manufacturing. Through optimal microstructure, porosity, and composition, AM provides the opportunity to create materials having higher wear and erosion resistance, and thus they are most appropriately used in the applications of aerospace, energy, mining, and biomedicine [36].

SLM is a powder-bed fusion AM process where a high-power laser selectively melts and bonds metal powder in a layer-by-layer fashion, forming a completely dense metallic part. One of the main strengths of SLM is microstructure control at the fine scale, tailoring grain orientation and phase content to enhance hardness and erosion resistance. Materials like nickel-based superalloys, titanium alloys, and cobalt-chromium alloys were successfully produced with SLM for high-performance uses. The enhanced mechanical properties resulting from SLM processing are due to the rapid

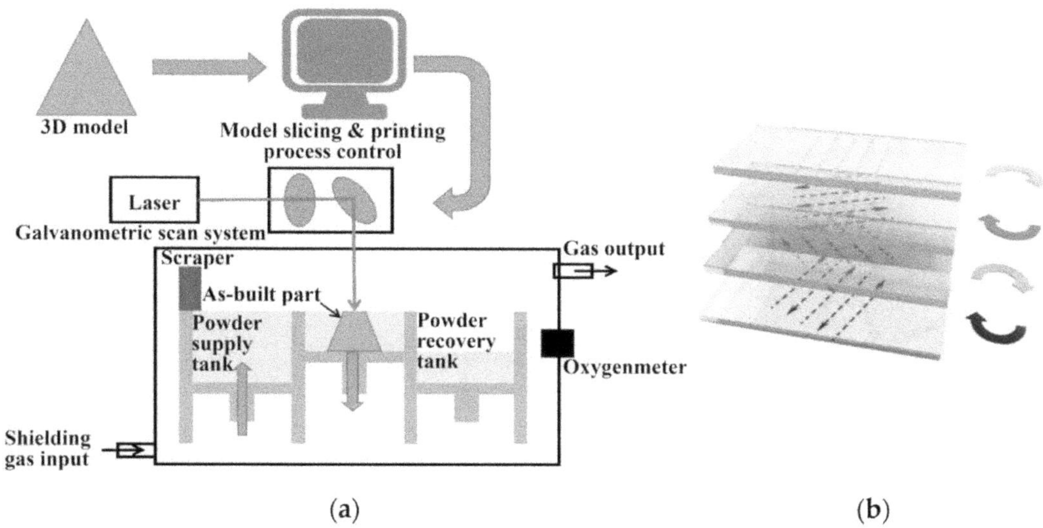

FIGURE 7.6 Schematic diagrams of SLM device (a) and laser scanning strategy (b) [39].

solidification and fine-grained microstructure, which improve wear resistance under erosive conditions. Moreover, post-processing methods like heat treatment and hot isostatic pressing (HIP) can also finish the microstructure and alleviate residual stress, enhancing overall durability. SLM also makes possible the printing of lattice structures and complex cooling channels that contribute to improved erosion resistance through diminished surface impact stress and energy dissipation. It is especially valuable for turbine blades, nozzles, and high-speed fluid-handling hardware, where solid particle or liquid droplet-induced erosion is critical [37–38]. Figure 7.6 presents schematic diagrams pertinent to the selective laser melting process. In Figure 7.6(a), the SLM machine consists of a laser source, powder delivery system, and build platform. In Figure 7.6(b), the laser scanning strategy depicts the path and pattern of the laser beam employed to selectively melt powder layers for optimal fusion and part quality.

EBM is a second powder-bed fusion method but with a high-energy electron beam rather than a laser used to melt and fuse the metal powder. EBM has the distinct benefit of being used to process high-melting-point alloys like titanium- and nickel-based alloys in vacuum environments, which reduces oxidation and maximizes mechanical properties. One of the most important advantages of EBM is its capacity to create parts with controlled porosity and grain structures, which lead to outstanding erosion resistance at high temperatures. This makes it particularly suitable for aerospace and power generation applications, where high-speed gas flow and particulate impact can lead to severe material degradation. The columnar grain structure created in EBM-processed parts offers greater toughness and crack resistance, ensuring extended service life in erosive environments. In addition, EBM allows the manufacture of tailor-made implantable biomedical devices with the surface roughness and porosity designed to improve their wear and degradation resistance in biological environments. This can be especially important for orthopedic and dental implants which need to be highly durable under cyclic loading conditions [40–41]. Figure 7.7 shows the setup of Freemelt ONE equipment employed for measuring smoke phenomena on electron beam melting. The e-beam starts to ramp on a ramping plate prior to its interaction with the powder bed. A fast camera (Sony RX100 V) observes the formation of smoke through a viewport. A tailor-made start plate is used for repeatable initiation of smoke at a specified height of powder. The specifications of the e-beam diameter and start plate are indicated to the right in the Figure 7.7.

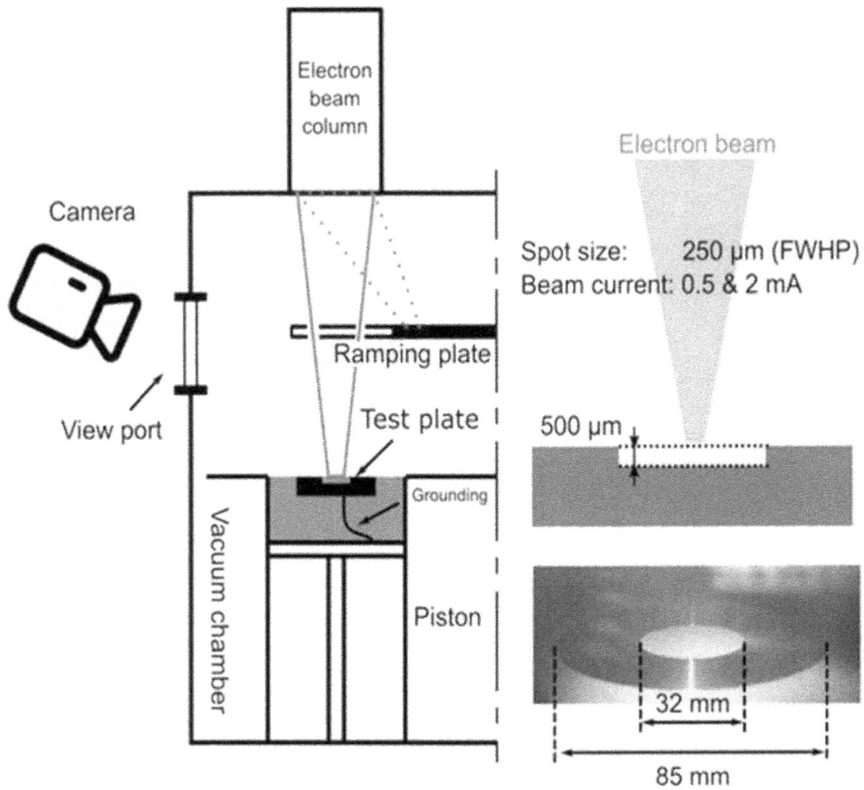

FIGURE 7.7 Freemelt ONE machine setup. The e-beam ramped up on a ramping plate prior to hitting the target powder bed. A high-speed camera (Sony rx100 v) was positioned alongside the viewport to observe the Smoke. A specially fabricated start plate guaranteed reproducible smoke triggering on a loose powder bed at a set powder height. Information regarding the e-beam diameter and the start plate can be seen on the right side of the picture [42].

One of the most revolutionary capabilities of AM is the capacity to produce functionally graded materials and gradient coatings that possess better erosion resistance. FGMs enable gradual change in composition and microstructure throughout a component, tailoring material properties to specific operating needs. For instance, an AM-produced turbine blade may possess a wear-resistant ceramic top coat, a strong metal intermediate layer, and a ductile metal core. Such a configuration minimizes erosion-induced material loss while sustaining general mechanical integrity. Likewise, functionally graded coatings like WC-Co, Cr_3C_2-NiCr, and TiC-based coatings may be directly coated onto substrates using AM techniques, promoting optimal adhesion and durable performance under erosive conditions. AM also facilitates the integration of nano-reinforcements like ceramic particles, graphene, or carbon nanotubes in metal matrices, providing further erosion resistance. The reinforced composites possess increased hardness, toughness, and impact resistance and have significantly lowered material wear during high-velocity solid particle or liquid droplet erosion [43].

7.4 INFLUENCE OF MICROSTRUCTURE ON EROSION RESISTANCE

Microstructural properties are key in determining the erosion resistance of metallic materials. Grain size, phase distribution, and precipitation hardening affect directly the hardness, toughness, and wear resistance of the material and consequently the ability of the material to resist erosive conditions. Fine-grained microstructures, dispersion-strengthened alloys, and intermetallic phase structures effectively improve erosion resistance by enhancing mechanical properties and minimizing material loss under impact condition [44].

Grain size is one of the most significant microstructural characteristics controlling erosion resistance. Fine-grained materials tend to have higher strength and hardness as a result of higher grain boundary density, which serves as an obstacle to dislocation movement. This finer microstructure improves the material's capacity to withstand plastic deformation and fracture under high-velocity particle or liquid droplet impact. For example, materials subjected to severe plastic deformation or rapid solidification processes are prone to having improved erosion resistance due to their fine-grained microstructure. Additionally, thermomechanical treatment involving heat treatment and recrystallization also maximizes wear performance by controlling grain size. Phase distribution within a material also significantly impacts its erosion resistance. A properly distributed and homogeneous secondary phase, such as carbides, nitrides, or oxides, can enhance hardness and prevent erosion-induced material removal at localized points. The incorporation of strengthening phases such as tungsten carbide (WC) or chromium carbide (Cr_3C_2) dispersed within a ductile matrix provides a tough wear balance. These strengthening phases are obstacles to the erosive particle penetration and reduce rates of material losses. A finely dispersed intermetallic phase, such as NiAl or TiAl, can improve erosion resistance at high temperatures in aerospace and power generation environments by maintaining structural integrity under adverse operating conditions [45–46]. Age hardening, or precipitation hardening, is another significant microstructural mechanism for enhancing erosion resistance. The process involves the formation of finely dispersed precipitates within the metal matrix, which inhibits dislocation movement and increases hardness.

Alloys such as precipitation-hardened stainless steel, nickel-based superalloys, and aluminum alloys are preferred by this mechanism, exhibiting improved erosive wear resistance. Precipitation of hard phases, such as γ' (gamma prime) in the case of nickel-based superalloys, under controlled conditions improves high-temperature erosion resistance substantially by preventing material degradation under severe conditions [47–48]. Oxide dispersion and intermetallic phase are responsible for improved erosion resistance by developing a protective surface film that is resistant to wear. Integration with oxide dispersion-strengthened (ODS) materials, such as Al_2O_3, Y_2O_3, or ZrO_2, provides better thermal stability and wear resistance. These finely dispersed oxides are strengthening particles that add mechanical properties and prevent the initiation of microcracks under erosive impact. Furthermore, oxide layers formed through high-temperature oxidation form a passive barrier that reduces direct material loss by erosion. Similarly, intermetallic phases such as Fe_3Al and TiAl are extremely resistant to hardness and oxidation and hence used in aerospace and industrial applications where components are exposed to high-speed erosion [49–50].

Finally, microstructural features like grain size, phase structure, precipitation hardening, oxide dispersion, and intermetallic phase formation are crucial in increasing erosion resistance. By controlling these parameters through alloy design, heat treatment, and sophisticated manufacturing methods, materials can be tailored to resist hostile erosive environments and provide durability and long-term performance in demanding applications [51].

7.5 PERFORMANCE EVALUATION OF EROSION-RESISTANT METALLIC MATERIALS

The resistance of metal materials and coatings to erosion is measured using several standardized testing techniques that mimic actual erosive environments. Tests supply excellent information on the behavior of materials in various forms of erosive wear, such as solid particle erosion, liquid droplet erosion, and cavitation erosion. Some typical test methods are jet impingement testing, cavitation erosion testing, and slurry erosion testing, each mimicking particular modes of erosion.

Comparative testing of various metallic materials and coatings by these test methods enables researchers to determine the best materials for high-performance applications in aerospace, power generation, mining, and biomedical industries [52].

Jet impingement testing is often used to measure the erosion resistance of materials to solid particles and liquid droplets. In the method, a high-velocity jet of solid particles or liquid droplets is directed onto the surface of the material at a selected angle. The test conditions, such as velocity, impact angle, particle size, and composition of the medium, can be changed to simulate actual operating conditions in gas turbines, steam turbines, and aircraft engines. The erosion rate can typically be measured in terms of weight loss, surface roughness change, or microstructural analysis by scanning electron microscopy. Comparative studies have shown that hard coatings such as WC-Co and Cr_3C_2-NiCr are more erosion resistant than bare metals due to their high hardness and effective dissipation of impact energy. Moreover, functionally graded materials developed with additive manufacturing have demonstrated higher functionality by realizing a seamless transition in hardness and toughness along the material surface [53].

Cavitation erosion testing is done to quantify material resistance to damage caused by collapsing vapor bubbles within a fluid. It is of vital concern for equipment such as ship propellers, hydraulic turbines, and pumps, where material loss resulting from cavitation is a considerable factor. The test typically involves subjecting a sample to ultrasonic cavitation or to high-speed rotating equipment immersed in a liquid environment, causing pressure oscillations that create and collapse vapor bubbles. The shock waves resulting from bubble collapse cause surface fatigue, microcracking, and material loss over time. Comparative research has shown that nitrided and carburized surfaces have better cavitation resistance because of their hardened layers, which inhibit crack initiation and propagation. Furthermore, oxide dispersion-strengthened (ODS) alloys, including Y_2O_3-reinforced steels, have shown excellent resistance because of their capacity for absorbing and distributing stress waves [54–55]. Slurry erosion testing evaluates the performance of a material in a condition where solid particles are carried in a liquid carrier, such as pipelines, pumps, and valves that handle abrasive slurries. In this test, the samples are subjected to a flowing slurry of abrasive particles, and the erosion rate is measured by weight loss and surface degradation analysis. Research has shown that harder materials with strong intermetallic phases like TiAl and Ni-based superalloys are better under slurry erosion conditions. Thermal spray coatings utilizing HVOF and plasma spraying have proved highly effective in minimizing material loss through a wear-resistant coating [56–57].

Jet impingement, cavitation erosion, and slurry erosion testing play an important role in assessing metallic material and coating durability in erosive conditions. Comparative tests show that novel materials, such as thermal spray coatings, additive-structured frameworks, and surface-treated alloys, possess greatly enhanced erosion resistance. The test procedures continue to inform the creation of next-generation erosion-resistant materials for high-performance applications [58].

7.6 APPLICATIONS OF EROSION-RESISTANT METALLIC MATERIALS

Erosion-resistant metallic materials are critical to many high-performance industries, such as aerospace, marine, power generation, and mining. The materials are engineered to resist harsh erosive environments from high-velocity solid particles, liquid droplets, cavitation, and slurry flow. The choice of suitable materials depends on the prevalent erosion mechanism of an application so that optimal performance and durability can be achieved. In aeronautics, parts like turbine blades, compressor discs, and nozzle guide vanes face high-speed flow of air containing dust, sand, and other particulates resulting in solid particle erosion. Another phenomenon, liquid droplet erosion (LDE), takes place in aircraft engines as a result of water or fuel droplets impacting at supersonic velocities. To counteract these effects, metals like nickel-based superalloys (Inconel 718, Hastelloy, Rene 80) and titanium alloys (Ti-6Al-4V) are extensively employed because of their hardness, superior thermal stability, and resistance to oxidation. Newer coatings like WC-Co, Cr_3C_2-NiCr, and TiN are deposited through thermal spraying or physical vapor deposition (PVD) methods to protect against erosion and increase component longevity [59].

Marine environments create their unique erosion problems, especially in propellers, ship hulls, and offshore structures, where materials suffer from cavitation erosion and slurry erosion because of turbulent flow of water, sand, and salt particles. Stainless steels (SS 316L, SS 254), nickel-aluminum bronze, and duplex stainless steels are frequently employed due to their resistance to corrosion and cavitation. Furthermore, oxide dispersion-strengthened alloys and surface processes such as nitriding and laser hardening enhance durability further by increasing surface hardness and inhibiting crack initiation [60]. Power plants, thermal, hydro, and nuclear power plants, use materials that have to withstand intense high-temperature erosion, cavitation, and droplet impingement. Boiler tubes, steam turbine blades, and heat exchangers are subjected to extreme conditions where erosion resistance is essential. Nickel superalloys, cobalt-based alloys (Stellite), and high-chromium steels (9Cr-1Mo, 12Cr steel) are favored for their ability to resist both erosive wear and high temperatures. Protective coatings like thermal barrier coatings (TBCs), HVOF-sprayed WC-Co, and Cr_3C_2-NiCr are often used to prevent material loss and maximize operational effectiveness [61]. Pump impellers, pipelines, and crushers in mining equipment are under extreme slurry erosion and abrasive wear because they are constantly exposed to hard particles in slurries. AISI 440C, AR400, and AR500 high-hardness steels, tungsten carbide composites, and chromium-rich white cast irons are some of the most common materials that resist the harsh conditions of extreme wear. Cold spray and thermal spray coatings also serve as a supplementary protection mechanism, greatly extending the life of components in extreme wear environments [62].

7.7 FUTURE PROSPECTS AND CHALLENGES

Although major breakthroughs have been achieved in the creation of metallic materials with erosion resistance, there are some challenges that remain to limit their large-scale application, especially in cost-efficient processing, scalability, and sustained performance under severe operational environments. These are especially critical in industries like aerospace, power generation, and mining, where high-performance material demand is continually on the rise. Overcoming these challenges demands ongoing innovation in material design, processing technologies, and test methods. Research directions for the future should emphasize the development of coatings, hybrid materials, and computational modeling to improve material performance, sustainability, and cost-effectiveness [63].

Cost-effective processing is still among the principal barriers to the universal application of high-performance erosion-resistant materials. Although some of the newer alloys and coatings such as WC-Co, Cr_3C_2-NiCr, and TiN have demonstrated superior erosion resistance, their manufacturing processes such as HVOF spraying, plasma spraying, and laser cladding are expensive because of the high-quality raw materials used, the processing steps involved, and the energy used. Moreover, certain materials, such as nickel-based superalloys and titanium alloys, are costly to produce and process and hence are restricted to upper-end applications. Thus, one of the most important areas of research in the years to come should be to evolve cost-efficient production methods like additive manufacturing or cold spraying that can offer high-performance materials at lower material loss and lower processing expenditure. Enhancing the scalability of these processes for large-scale manufacturing will be crucial for rendering erosion-resistant materials affordable for wider industrial applications [64].

Scale-up of processing routes is another major challenge. Whereas techniques such as selective laser melting and electron beam melting in additive manufacturing have been touted to create functionally graded materials and complex geometries with potential, these are usually bottlenecked by the processing time involved, material availability, and scaling up to component sizes. More importantly, in some cases, these techniques can be incompatible with the substrate material needed in certain industrial applications. To counter this, future work must concentrate on the optimization of AM processes for large-scale production, minimization of processing time, and the creation of multi-material processes that can support different alloy compositions for a variety of erosion applications. Additionally, the reproducibility and homogeneity of microstructures and properties at scale is important for the reliability of these materials in production environments [65].

Long-term performance under actual use conditions is yet another area where development is important. Resistance to erosion is not merely a matter of the intrinsic characteristics of the materials per se but also of the interaction of the materials with their operating environments over time. Temperature changes, corrosive environments, and high-pressure environments can cause deterioration through time, which classic erosion tests cannot well account for. Thus, long-term performance tests, such as accelerated wear testing and life-cycle analysis, will be critical to determine how materials perform over time when subjected to erosive forces. This information will inform the design of more erosion- and corrosion-resistant coatings and improve the overall wear life of components [66].

In the future, advanced coatings and hybrid materials will be critical to the creation of next-generation erosion-resistant materials. The use of nanotechnology to develop coatings with nano-reinforcements like carbon nanotubes or graphene is promising since they provide added strength and wear resistance while being light in weight. In addition, the blending of ceramic and metal materials in hybrid coatings can provide both toughness and hardness, providing more balanced erosion protection against different types of erosion. Research into functionally graded coatings that offer gradual transitions in hardness or material properties across the coating will also be important for enhancing performance under dynamic and aggressive erosive conditions [67]. Finally, computational modeling will increasingly be used as a more effective tool for the creation of erosion-resistant materials. Sophisticated finite-element analysis and multiscale modeling methods can allow for the characterization of the micromechanical response of the material under erosive conditions so that more efficient design and optimization can be realized. These models can model actual situations and simulate material response to various stress situations, making possible the design of bespoke materials based on specific functions [68].

7.8 CONCLUSION

The production of metallic materials with high erosion resistance is essential for the prolongation of the life and consistency of parts subject to severe erosive conditions in the aerospace, marine, power generation, and mining industries. Progress in alloying, thermal spraying, surface hardening, and additive manufacturing has considerably enhanced the erosion resistance of metallic materials. Alloying with chromium, nickel, and tungsten increases hardness and toughness, thus making materials more resistant to erosive wear. Thermal spraying processes, including HVOF and plasma spraying, enable the deposition of wear-resistant coatings such as WC-Co and Cr_3C_2-NiCr, offering strong protection against solid particle and cavitation erosion. Surface hardening techniques such as carburizing, nitriding, and laser hardening harden the surface of materials, reducing abrasion and erosive forces to which they can be subjected without compromising the ductility of the core material. Additive manufacturing in the form of selective laser melting and electron beam melting allows complex geometries and functionally graded materials that improve performance under harsh erosive conditions. These developments enable the accurate manipulation of material properties, resulting in improved performance in targeted applications. As materials science and engineering advance, future research will continue to refine these methods, which will result in next-generation erosion-proof materials that are more durable, efficient, and cost effective. Such innovations will be the key to enhanced performance and life in key industrial components operating in erosive environments, providing increased reliability and lower costs of maintenance.

REFERENCES

1. Hegde S, Gupta M. Erosion resistance of coatings on stainless steel substrates. J Mater Process Technol. 2008; 202(1–3): 23–30.
2. Yilmaz O, Devrim S. Wear and erosion behavior of NiCrBSi coatings deposited by high-velocity oxy-fuel spraying. Wear. 2006; 261(3–4): 435–442.
3. Patil C, Jadhav M, Karve S, et al. Influence of thermal spraying on wear and erosion resistance of coatings. J Therm Spray Technol. 2013; 22(1): 85–95.
4. Saha P, Das S, Choudhury D, et al. Wear and erosion performance of high-velocity oxy-fuel (HVOF) sprayed coatings: A review. Surf Coat Technol. 2011; 205(3–4): 948–964.
5. Saleh T, Lee Y, Lee H. Investigation of erosion–corrosion behavior of steel in coal slurry. Wear. 2009; 267(5–6): 1079–1085.
6. Liu F, Guo H, Zhang L. Effects of alloying elements on the erosion resistance of materials. Wear. 2008; 264(5–6): 426–432.
7. Siddiquee A, Razzaq H, Ali Z. Erosion resistance of coatings in mineral processing applications. J Mater Sci. 2014; 49(10): 3580–3589.
8. Liu X, Zhang L, Qu J, et al. Role of surface coatings in enhancing erosion resistance of titanium alloys. J Mater Sci. 2012; 47(22): 7841–7850.
9. Deng T. Erosive wear mechanisms of materials—a review of understanding and progresses. Materials. 2025 Apr 2; 18(7): 1615.
10. Liu X, Zhang L, Qu J, et al. Erosion resistance of metallic and composite coatings: A review. Wear. 2010; 268(11–12): 1267–1277.
11. Wang L, Ding Z, Zhang Y. Erosion behavior of high-chromium cast iron. Wear. 2003; 255(1–6): 101–106.
12. Jiang Z, Zhang S, Wei Z, et al. Erosion performance of cold-sprayed coatings: A review. J Therm Spray Technol. 2014; 23(5): 669–686.
13. Zhang Z, Zhang L, Zeng X, et al. Erosion–corrosion behavior of protective coatings in coal-fired power plants. Wear. 2013; 302(1–2): 1159–1168.
14. Xie X, Xu Y, Zhang X, et al. High-velocity oxy-fuel (HVOF) spraying of WC-Co coatings for erosion and wear applications. J Therm Spray Technol. 2011; 20(5): 1050–1059.
15. Liao H, Wei S, Yu M, et al. Mechanical and wear properties of WC-Co coatings for erosion and wear resistance. Wear. 2007; 262(5–6): 594–602.
16. Zhang L, Xie X, Guo H. Erosion resistance of high chromium cast iron with different heat treatments. Wear. 2005; 259(1–6): 466–471.
17. Cizek J, Vopat M, Beranek M, et al. Erosion behavior of thermal spray coatings for improving component life in high-temperature environments. Surf Coat Technol. 2013; 225(1): 91–98.
18. Hu C, Jiang Z, Yu W, et al. Erosion performance of NiCrBSi coatings deposited by HVOF. Surf Coat Technol. 2006; 200(5–6): 1717–1724.
19. Manohar D, Atre S, Mahajan M, et al. Erosion resistance of WC-Co coatings deposited by high-velocity oxy-fuel (HVOF) spraying. Wear. 2010; 268(5–6): 748–753.
20. Kim J, Kim Y, Kim H, et al. Wear and erosion resistance of Ni-based alloy coatings. J Therm Spray Technol. 2009; 18(2): 212–220.
21. Liu C, Wang Y, Liu H. Wear behavior of plasma-sprayed WC-Co coatings for erosive conditions. J Mater Sci. 2009; 44(9): 2372–2378.
22. Rykov D, Il'yushin A, Matveev I. Effect of alloying elements on erosion resistance of Ni-based superalloys. Mater Sci Eng A. 2015; 628: 214–220.
23. Büyükyıldız M, Karaoğlu H, Çolak İ, et al. Erosion behavior of WC-Co-based coatings: The effect of particle size. Surf Coat Technol. 2007; 201(20): 8669–8676.
24. Ramezani M, Mohd Ripin Z, Pasang T, et al. Surface engineering of metals: Techniques, characterizations and applications. Metals. 2023 Jul 20; 13(7): 1299.
25. Mao Z, Li H, Zhang M. Erosion behavior of composite coatings. Surf Coat Technol. 2009; 203(18–19): 2640–2645.
26. Li X, Zhang S, Xue Q, et al. Effect of coating thickness on erosion resistance of NiCr-based coatings. Wear. 2010; 268(5–6): 468–474.
27. Aghaei A, Moghaddam S, Shamanian M. Erosion performance of WC-Co coatings in oil and gas industry applications. Wear. 2015; 332–333: 306–314.

28. Gunturk A, Ince G, Demir A, et al. Erosion resistance of coatings on stainless steel substrates: Effect of process parameters. J Therm Spray Technol. 2007; 16(4): 556–563.
29. Dan M, Dumitru D, Vişan M. Erosion behavior of materials under high-velocity solid particle impingement. Wear. 2004; 256(5–6): 512–518.
30. Dai Y, Kang L, Han S, et al. Surface hardening behavior of advanced gear steel C61 by a novel solid-solution carburizing process. Metals. 2022 Feb 23; 12(3): 379.
31. Kubacki A, Kuczyński J. The erosion behavior of thermally sprayed coatings. J Therm Spray Technol. 2009; 18(3): 480–488.
32. Zhou YL, Xia F, Xie AJ, et al. A review—effect of accelerating methods on gas nitriding: Accelerating mechanism, nitriding behavior, and techno-economic analysis. Coatings. 2023 Oct 27; 13(11): 1846.
33. Kharbuli M, Galvão R, de Souza M. The influence of coating microstructure on erosion performance. Surf Coat Technol. 2011; 205(6): 1713–1719.
34. Maharjan N, Wu N, Zhou W. Hardening efficiency and microstructural changes during laser surface hardening of 50CrMo4 steel. Metals. 2021 Dec 13; 11(12): 2015.
35. Harle R, Sondhi A. A review of erosion in high-velocity impact conditions. Wear. 2004; 257(5–6): 668–677.
36. Ahmadi S, Khalafi S. A study on the erosion resistance of thermally sprayed coatings in different environments. Surf Coat Technol. 2012; 207(1): 57–64.
37. Rahimabadi S, Karimzadeh R. Performance of WC-Co coatings under different erosion conditions. Wear. 2010; 268(3–4): 431–438.
38. Kim J, Lee J, Lee S. Erosion behavior of Al-based and NiCr-based coatings. Wear. 2008; 265(11–12): 1737–1744.
39. Zhang J, Yan Y, Li B. Selective Laser Melting (SLM) additively manufactured CoCrFeNiMn high-entropy alloy: Process optimization, microscale mechanical mechanism, and high-cycle fatigue behavior. Materials. 2022 Dec 1; 15(23): 8560.
40. Dohnal M, Kopecký J, Vopat M, et al. Erosion–corrosion of thermally sprayed coatings in industrial applications. Wear. 2008; 265(11–12): 1745–1752.
41. Neikter M, Olsson M, Olofsson L. Wear and erosion performance of advanced metallic coatings. Wear. 2014; 316(1–2): 28–34.
42. Ye J, Semjatov N, Bidola P, et al. Revealing the mechanisms of smoke during electron beam-powder bed fusion by high-speed synchrotron radiography. J Manuf Mater Process. 2024 May 17; 8(3): 103.
43. Devaraju G, Srinivasan M. Erosion performance of HVOF-sprayed coatings: A review of various alloys. Surf Coat Technol. 2013; 218(1): 37–44.
44. Yang S, Zhuang J. Erosion behavior of thermal spray coatings in acidic environments. Wear. 2009; 267(1–4): 819–824.
45. Tomaz N, Kharab S, Singh M, et al. Erosion resistance of coatings on industrial materials: A review. Surf Coat Technol. 2012; 206(1–2): 252–258.
46. Lascano R, Martinez H, Silva R. Erosion resistance of composite coatings for industrial applications. Wear. 2011; 271(1–2): 99–104.
47. Gupta M, Yilmaz O. Effect of spray parameters on the erosion resistance of coatings. J Therm Spray Technol. 2007; 16(1): 67–75.
48. Wang L, Li Y, Zhang Q. Investigation on the erosion behavior of sprayed coatings in the aerospace industry. Wear. 2011; 271(5–6): 706–712.
49. Liang W, Xie Y, Zhang Z. Effects of spraying techniques on the erosion resistance of metal coatings. J Therm Spray Technol. 2010; 19(4): 719–725.
50. Lee K, Lee H, Park H. Characterization of erosion behavior of HVOF coatings. Wear. 2006; 260(5–6): 510–516.
51. Ganapathy S, Munoz M, Reddy P. Enhanced erosion resistance of coatings on stainless steel substrates. Surf Coat Technol. 2010; 204(12): 1913–1918.
52. Kumar S, Ratol JS. Effects of tribological parameters on slurry erosion behaviour of uncoated and coated materials: A review. Mater Eng. 2013 Jul 1; 20(3): 119–131.
53. Verma PK, Minhas AS, Singh P, et al. Slurry erosion behaviour of thermal sprayed Al2O3 and Cr2O3 coatings for turbine steels. In AIP Conference Proceedings 2024 Feb 20 (Vol. 2986, No. 1). AIP Publishing.
54. Singh SP, Verma PK, Singh S, et al. Microstructural characterization and electrochemical behaviour of HVOF sprayed WC-Co-Cr coatings on 316L boiler steel stainless steel. Results in Surfaces and Interfaces. 2024 Aug 1; 16: 100248.

55. Soltis J, Palacios J, Eden T, et al. Evaluation of ice-adhesion strength on erosion-resistant materials. AIAA J. 2015 Jul; 53(7): 1825–1835.
56. Hansen JS. Relative erosion resistance of several materials. ASTM Int. 1979 Jan 1.
57. Jindal C, Sidhu BS, Kumar P, et al. Performance of hardfaced/heat treated materials under solid particle erosion: A systematic literature review. Mater Today Proc. 2022 Jan 1; 50: 629–639.
58. Krella AK. Degradation and protection of materials from cavitation erosion: A review. Materials. 2023 Mar 2; 16(5): 2058.
59. Wang D, Lin S, Liu L, et al. Micro-nano multilayer structure design and solid particle erosion resistance performance of CrAlNx/CrAlN coating. Vacuum. 2020 Feb 1; 172: 109064.
60. Feltrin AC, Xing Q, Akinwekomi AD, et al. Review of novel high-entropy protective materials: Wear, irradiation, and erosion resistance properties. Entropy. 2022 Dec 30; 25(1): 73.
61. Yang Q, Dolatabadi A, Golovin K. Erosion-resistant materials demonstrate low interfacial toughness with ice and superior durability. Mater Horiz. 2023; 10(10): 4541–4550.
62. Kübarsepp J, Juhani K, Tarraste M. Abrasion and erosion resistance of cermets: A review. Materials. 2021 Dec 22; 15(1): 69.
63. Zhang Z, Kovacevic R. Laser cladding of iron-based erosion resistant metal matrix composites. J Manuf Process. 2019 Feb 1; 38: 63–75.
64. Wright IG, Shetty DK, Clauer AH. Erosion-resistant materials for critical areas of coal liquefaction and coal gasification systems. J Mater Energy Syst. 1984 Dec; 6(3): 172–183.
65. Haugen K, Kvernvold O, Ronold A, et al. Sand erosion of wear-resistant materials: Erosion in choke valves. Wear. 1995 Jul 1; 186: 179–188.
66. Leithead SG, Allan WD, Zhao L. Performance metrics and experimental testing of erosion-resistant compressor blade coatings. InTurbo Expo: Power for Land, Sea, and Air 2014 Jun 16 (Vol. 45578, p. V01AT01A003). American Society of Mechanical Engineers.
67. Tavana SM, Hojjati M, Liberati AC, et al. Erosion resistance enhancement of polymeric composites with air plasma sprayed coatings. Surf Coat Tech. 2023 Feb 25; 455: 129211.
68. Hassani S, Klemberg-Sapieha JE, Bielawski M, Beres W, Martinu L, Balazinski M. Design of hard coating architecture for the optimization of erosion resistance. Wear. 2008 Aug 25; 265(5–6): 879–887.

8 Surface Engineering for Protection against Slurry Erosion

Praveen Kumar Verma, Hitesh Vasudev, and Sule Emmanuel Oshioriameh

8.1 INTRODUCTION

Slurry erosion is a multifaceted material degradation phenomenon that results from the continuous or intermittent impingement of solid particles carried within a liquid medium upon a material surface. The mechanical interaction between the suspended abrasive particles and the target surface induces progressive material loss through mechanisms such as cutting, plowing, micro-fracture, and fatigue. This form of erosive wear is of critical concern in industries that handle solid–liquid mixtures, including but not limited to hydroelectric power generation (turbine runners, guide vanes), oil and gas extraction (pipelines, valves), dredging and marine applications (e.g., ship hulls, pump impellers), chemical processing, and mineral beneficiation. The economic and operational ramifications of slurry erosion are significant, manifesting in reduced equipment efficiency, increased maintenance frequency, unscheduled downtime, and elevated life-cycle costs [1–3]. Figure 8.1 shows factors like coatings and surface treatments in sectors like mining, oil, and power that enhance equipment durability and lifespan and reduce maintenance costs.

The complexity of slurry erosion stems from the synergistic combination of various factors: hydrodynamic flow conditions, particle and fluid properties, dynamics of impingement, and the material's inherent mechanical and microstructural properties. The erosive capacity of a slurry system relies not only on the particle load or hardness but also on parameters like particle size distribution, shape morphology (angularity or roundness), impact velocity, concentration, and the viscosity and density of the slurry. These parameters are responsible for cumulatively defining the erosive energy imparted to the material surface when particles hit. In addition, the angle of impact has a significant effect on the prevailing wear mechanism: ductile materials tend to suffer maximum erosion at low to intermediate angles through cutting and plowing, while brittle materials suffer maximum erosion at near-normal impacts due to chipping and cracking [4–6].

In most engineering environments, slurry erosion is compounded further by superimposed effects like cavitation, corrosion, and thermal fluctuations. Cavitation, which includes the collapse of vapor bubbles in the fluid at solid boundaries, can produce microjets of high local pressure that significantly compromise surface integrity. Corrosive components of the slurry, especially in chemically severe environments, can modify surface chemistry and enhance erosion–corrosion synergy, usually much more erosive than individual influences. High-temperature slurry atmospheres, for instance, in geothermal or thermal power systems, could enhance degradation by decreasing material hardness or speeding up phase transformations. From a materials science point of view, the behavior of a substrate towards slurry erosion is controlled by its hardness; fracture toughness; yield strength; strain-hardening ability; and microstructural properties like grain size, phase distribution, and presence of inclusions or precipitates [7, 8]. Conventional monolithic materials, especially metallic alloys such as carbon steels or stainless steels, tend to suffer from lack of adequate erosion

 DOI: 10.1201/9781003561156-10

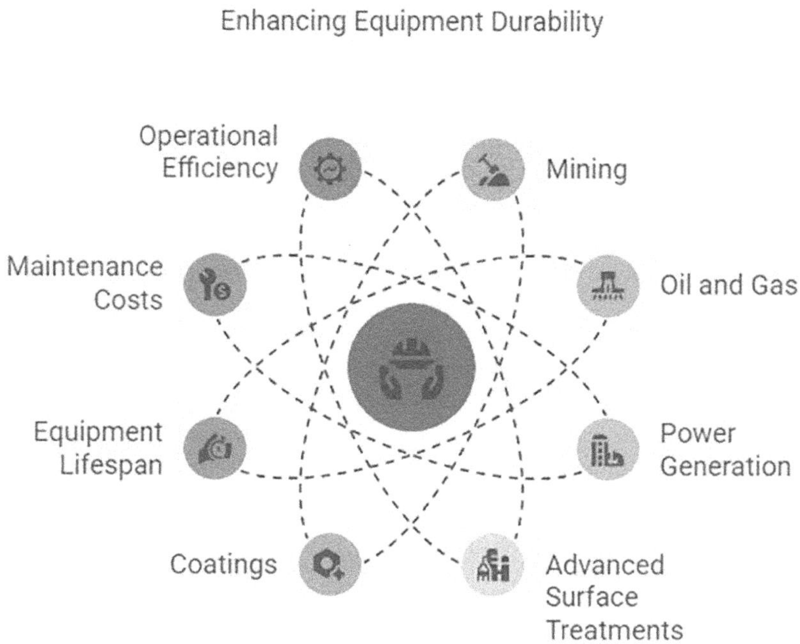

FIGURE 8.1 Key factors and applications enhancing equipment durability.

resistance resulting from either low hardness or poor toughness. Thus, the development of surface-specific properties without losing the bulk mechanical integrity is a potential avenue to prevent erosion damage. Surface engineering has become a key discipline in the development of wear-resistant material design within this regard. Surface engineering involves a wide range of techniques designed to alter the surface characteristics of a material to enhance its performance under particular operating conditions. These processes can include the use of surface coatings (thermal spray, physical vapor deposition, chemical vapor deposition, etc.), surface treatments (laser hardening, carburizing, nitriding, boriding), or surface modification technologies (laser alloying, ion implantation, surface texturing) [9–11]. The aim is to improve surface hardness, toughness, chemical stability, and mechanical impact resistance with the preservation of the favorable bulk properties like toughness and machinability.

Among these methods, newer coatings especially ones based on carbides, ceramics, and metal-matrix composites have proven to be quite effective in slurry erosion protection. Processes like high-velocity oxy-fuel (HVOF) spraying and plasma spraying are capable of depositing dense, adherent films made of tungsten carbide-cobalt (WC-Co), chromium carbide-nickel chromium (Cr_3C_2-NiCr), or aluminum oxide (Al_2O_3). These coatings have higher hardness and erosion resistance, particularly if designed to possess maximum phase composition, porosity, and interface integrity. Next-generation multilayer and functionally graded coatings provide additional benefits by individually adjusting compositional gradients to withstand impact and thermal stresses. In addition, recent advances in laser surface engineering have made it possible to fabricate metallurgically bonded, wear-resistant coatings with finely controlled microstructures and custom-designed hardness profiles. Methods like laser cladding and laser alloying can deposit high-density, defect-free surfaces with better performance than conventional overlays in erosion conditions. Furthermore, micro/nano-textured surfaces fabricated through laser or electrochemical processes are able to affect the behavior of slurry near the surface, which might decrease the number of particle impacts as well as their intensity [12–14].

Notwithstanding these developments, the design of surface-engineered systems to resist slurry erosion is still a difficult endeavor because there are no standardized testing methods; it is hard to reproduce real-world conditions, and erosion rates are multi-parameter dependent. Thus, a good understanding of the fundamental erosion mechanisms, the response behavior of the material, and the effects of microstructural characteristics is crucial. In addition, numerical modeling and data-driven design strategies such as finite-element modeling (FEM) and machine learning (ML) are being combined more and more to forecast erosion behavior and ensure optimal surface engineering measures [15].

This chapter critically reviews the state of the art in surface engineering technologies evolved to address slurry erosion. We first discuss the mechanistic behavior of slurry erosion and conditions that control the severity of erosion. Next, we describe a detailed overview of different surface modification technologies with a focus on their erosion performance, material compatibility, and feasibility of implementation. Last, we outline contemporary research trends, performance assessment methodologies, and future work with the aim to realizing cost-efficient, high-performance, and sustainable erosion-resistant surfaces [16].

8.2 MECHANISM OF SLURRY EROSION

Slurry erosion is a multifaceted tribological process that involves the gradual loss of material from a solid surface by repeated collision of solid particles carried in suspension in a fluid medium. The mechanism of erosion is controlled by a plethora of closely interacting parameters, such as particle kinetic energy, impact angle, particle shape, flow regime, and the inherent properties of the target material. A mechanistic grasp of slurry erosion is fundamental to the rational design of erosion-resistant materials and surface engineering measures [17]. Figure 8.2 shows slurry erosion cutting, plowing, micro-fracture, and fatigue stages where particles damage material through slicing, displacement, cracking, and long-term stress.

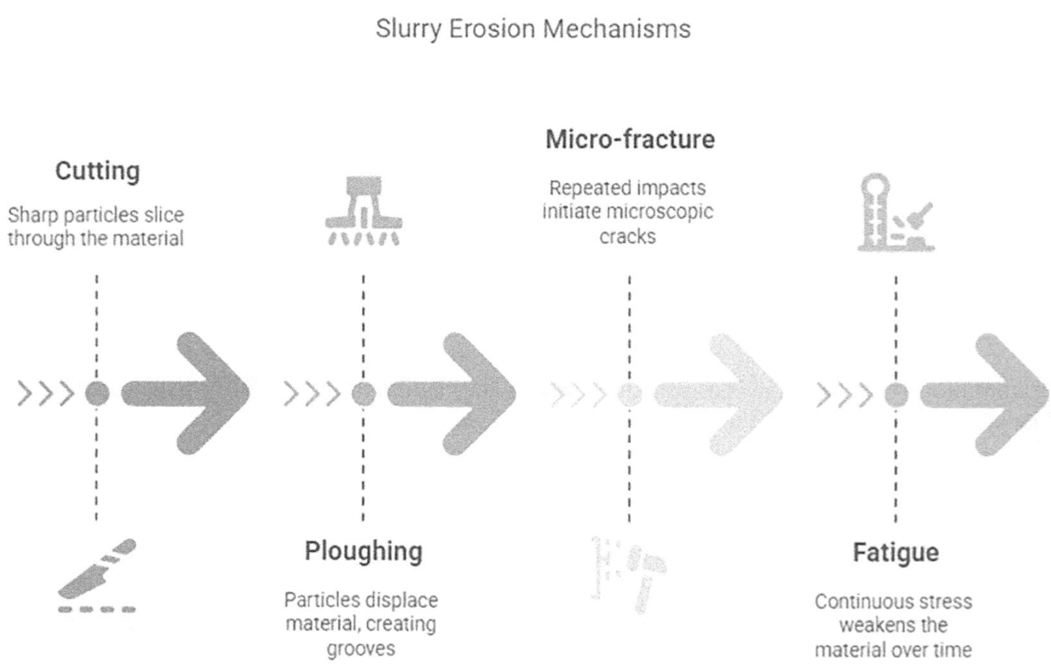

FIGURE 8.2 Mechanisms involved in slurry erosion of materials.

8.2.1 Impact Wear and Particle Kinetics

The dominant mechanism of material removal in slurry erosion is impact wear, wherein erosive particles entrained in the fluid stream impinge on the target surface at varying velocities and angles. The kinetic energy ($E_k = \frac{1}{2}12mv^2$) of these particles, where m is mass and v is velocity, governs the severity of the impact and the resultant mechanical deformation or fracture. At shallow impact angles of 15 to 30 degrees, erosive wear is generally dominated by micro-cutting or plowing especially in ductile materials. In contrast, more brittle materials like ceramics and hardened coatings tend to spall and fracture at higher angles from 60 to 90 degrees. These impacts can also trigger subsurface crack growth and a fatigue-like response which ultimately results in delamination and spallation [18–19]. Figure 8.3 presents the comparison of particle impacts: (a) spherical vs. angular particles causing different penetration depths; (b) ductile materials deform plastically while brittle materials fracture under oblique impacts by solid particles.

8.2.2 Role of Particle Characteristics and Fluid Properties

The rate of material removal in slurry erosion is significantly affected by the nature of entrained abrasive particles and the physical as well as chemical properties of the carrier fluid. The erosive potential of particles depends on a combination of size, shape, hardness, and concentration in the slurry, each of which controls the intensity and nature of impact on the target surface [21].

Particle size has a twofold and frequently multifaceted role in erosion processes. Particles of greater size generally have greater momentum because of their higher mass, which is manifested as greater kinetic energy when striking a surface. This can cause intense plastic deformation or micro-fracturing of the surface, particularly in ductile and semi-brittle targets. Yet very large particles will also be less sensitive to the flow dynamics of the carrier fluid and therefore have fewer direct hits. Smaller particles, though containing less energy per particle, will impart a more homogeneous and steady erosion process. When there is a high concentration of fine particles in dense slurries, cumulative damage takes place due to high-frequency, low-energy hits. The particle shape or morphology also enhances erosion behavior. Angular particles are much more severe than their rounded counterparts because their pointed corners and cutting edges are like tiny cutting tools that can cause shear deformation and plowing effects on the material surface. Round particles, however, tend to deform through blunt impact, resulting mainly in surface fatigue or crater formation but not material cutting. Particle hardness compared to the substrate is also a key parameter in deciding

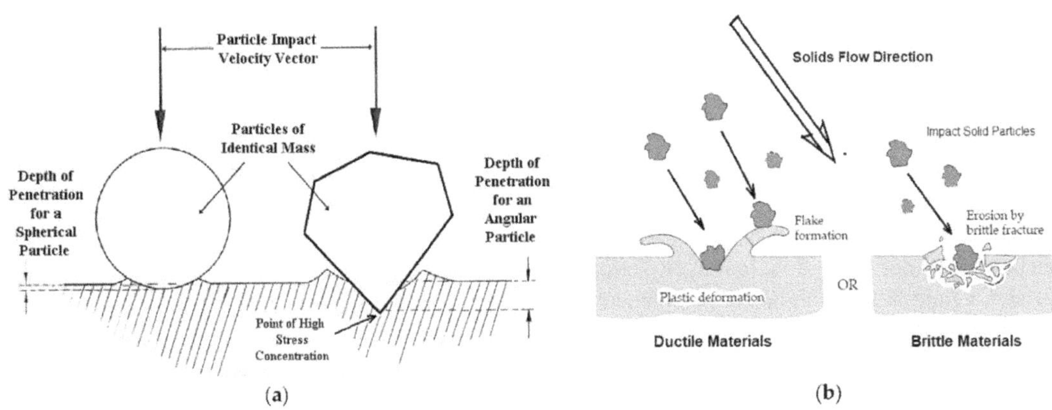

FIGURE 8.3 Surface material under impacts: (a) normal impact; (b) oblique impact [20].

erosion severity. Hard particles like quartz, alumina, or silicon carbide can penetrate and remove material from softer metallic or polymeric surfaces with ease. Softer particles, on the other hand, may be poor wear producers but will cause damage under conditions of high-speed impact through accumulation of repetitive plastic strain [22–24].

The density and viscosity of the carrier fluid directly influence particle motion and momentum transfer. A fluid of low viscosity enables particles to achieve greater velocities, hence higher impact energy. On the other hand, a high-viscosity medium can have a damping effect, slowing impact velocity and potentially lower erosion rates. Instead, it will encourage longer particle suspension time and surface exposure. The fluid density influences the drag and buoyancy forces on the particles and thus their transport velocity and impact dynamics. In addition, chemical and thermal fluid properties like pH, temperature, and concentration of corrosive ions can be responsible for erosion–corrosion synergy. A high temperature can weaken the material matrix, particularly in polymers and heat-affected metals, thus increasing the rate of erosion. Corroding species (e.g., chlorides, sulfates) in the slurry can weaken passive oxide films and cause localized corrosion, which, with the interaction of mechanical wear, results in extremely accelerated material degradation [25, 26].

8.2.3 Flow Dynamics and Erosion Intensification

Hydrodynamic properties of the slurry are central to the regulation of erosion intensities and patterns. Dynamics of flow, especially in the turbulent flow regime, affect the velocity vector, angular momentum, and trajectory of solid particles in the slurry. The flow is turbulent in the majority of practical applications because of high Reynolds numbers and complicated geometries, as in pump housings, turbine blades, elbows, and bends of pipes. Turbulent flow encourages random and high-frequency particle and surface collisions, tending to produce a non-uniform and unpredictable erosion profile [27].

The velocity of the slurry flow dictates the kinetic energy of the particles, which is a principal driver of erosion. Even minor increases in flow velocity can result in exponential increases in erosion rates due to the quadratic dependence of kinetic energy on particle speed. Areas of flow acceleration, abrupt expansion or contraction, and recirculation zones (e.g., downstream from a weld seam or near valves) produce areas of accelerated erosion. Within these regions, turbulent eddies pick up particles and repeatedly and sharply impact the surface, which results in premature failure of the material [28]. Cavitation also makes erosion mechanisms more difficult in high-velocity slurry systems. Cavitation occurs when local pressure falls below the vapor pressure of the fluid, creating vapor cavities or bubbles. The subsequent bubble collapse near a solid surface creates strong microjets and shock waves, producing localized regions of high pressure which induce pitting, fatigue, and microcracking of the surface. The interaction between cavitation and erosion of slurry is very harsh in hydroturbines, impellers of pumps, and ship propellers, where it causes sped-up deterioration that cannot be avoided with normal materials [29].

In addition, flow regime can interact with particle distribution and size to define the level of wall impact. For example, larger particles with high turbulent environments tend to travel in ballistic trajectories and strike walls at sharp angles, leading to heightened wear. Smaller particles, however, can be entrained in boundary layers and cause more diffused but more continual erosion. Smooth surface transitions and flow homogenization are hence design considerations crucial in countering erosion where flow dynamics is heightened [30].

8.2.4 Material Response: Ductile vs. Brittle Behavior

The behavior of a material under slurry erosion is essentially controlled by its mechanical response and microstructural properties, specifically its status as either ductile or brittle. The mechanisms of erosive wear found in these two types of materials are quite different because of

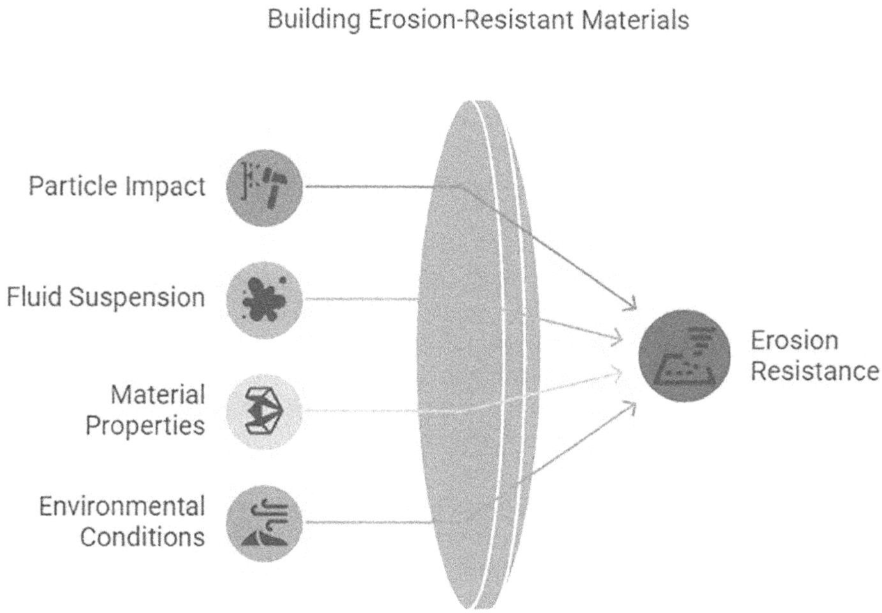

FIGURE 8.4 Influencing factors of erosion resistance in materials.

their dissimilarities in energy absorption and deformation when subjected to high-velocity particle impact [31]. Ductile materials like low-carbon steels, copper alloys, and some engineering polymers deform plastically under impact energy absorption. Upon particle impact, such materials display characteristics like lip formation, crater formation, and smearing of the surface. The deformation is repetitive and localized, resulting in fatigue cracks after some elapsed time. Ductile materials normally exhibit maximum erosion rates at oblique impact angles (20–45 degrees), where dominant cutting and plowing mechanisms occur. The progressive material removal at these angles is similar to abrasive machining, where multiple shear stresses cause surface degradation [32]. Figure 8.4 illustrate four factors, particle impact, fluid suspension, material properties, and environmental conditions, that influence erosion resistance, highlighting their combined role in developing erosion-resistant materials.

Conversely, brittle materials such as ceramics (e.g., SiC, Al_2O_3), hard coatings (e.g., TiN, CrN), and some glass-reinforced composites fracture mainly by crack initiation, extension, and chipping of the surface. These materials possess minimal plastic deformation capability, and at regular or close-to-regular impact angles (60–90 degrees), they experience sudden material loss through tensile and shear cracking. Brittle material erosion is more sporadic, typically loss of large particles because of sub-surface coalescence of cracks. Brittle materials thus show maximum erosion rates at high angle impacts, where the normal particle force component is a maximum [33]. Microstructure elements like grain size, phase morphology, and second-phase particles or porosity also affect erosion resistance. For instance, fine-grained microstructures enhance fracture toughness and decrease crack propagation paths, thus enhancing erosion resistance to brittle erosion. The existence of residual compressive stresses generated through surface treatments such as shot peening or laser shock processing can also enhance erosion resistance by inhibiting crack nucleation. In multi-phase materials, the phase distribution and hardness mismatch may cause delamination or interfacial failure upon repeated impacts, emphasizing the significance of microstructural design in slurry erosion protection [34].

8.3 SURFACE ENGINEERING TECHNIQUES

8.3.1 Thermal Spray Coatings

Thermal spray coatings represent one of the most extensively applied surface engineering techniques for enhancing the erosion and corrosion resistance of components exposed to slurry environments. These processes involve the deposition of feedstock materials—typically in powder or wire form—onto a substrate by heating them to a molten or semi-molten state and accelerating them toward the target surface using a high-energy carrier gas or arc plasma. Upon impact, the particles flatten and solidify rapidly, forming overlapping lamellae or splats that collectively build the coating layer. Among the various thermal spray processes, plasma spraying, high-velocity oxy-fuel spraying, and arc spraying are particularly relevant for slurry erosion applications due to their ability to produce dense, wear-resistant coatings with good adhesion and metallurgical compatibility [35, 36]. Figure 8.5 shows a comparison of plasma spraying, HVOF spraying, and arc spraying techniques, showing their influence on coating properties such as durability, reliability, and longevity for enhanced performance of coated components.

Plasma spray coatings utilize a high-temperature plasma jet (typically exceeding 10,000 K) to melt refractory feedstock materials, such as ceramics and hard metals. This process allows for the deposition of materials with extremely high melting points, including alumina (Al_2O_3), titania, and zirconia, which are well known for their hardness and chemical stability. However, plasma spraying typically produces coatings with higher porosity compared to other methods, which may compromise resistance to slurry penetration and sub-surface cracking. In contrast, HVOF spraying utilizes a combustion-driven high-speed jet (often exceeding 1000 m/s particle velocity) to deposit coatings

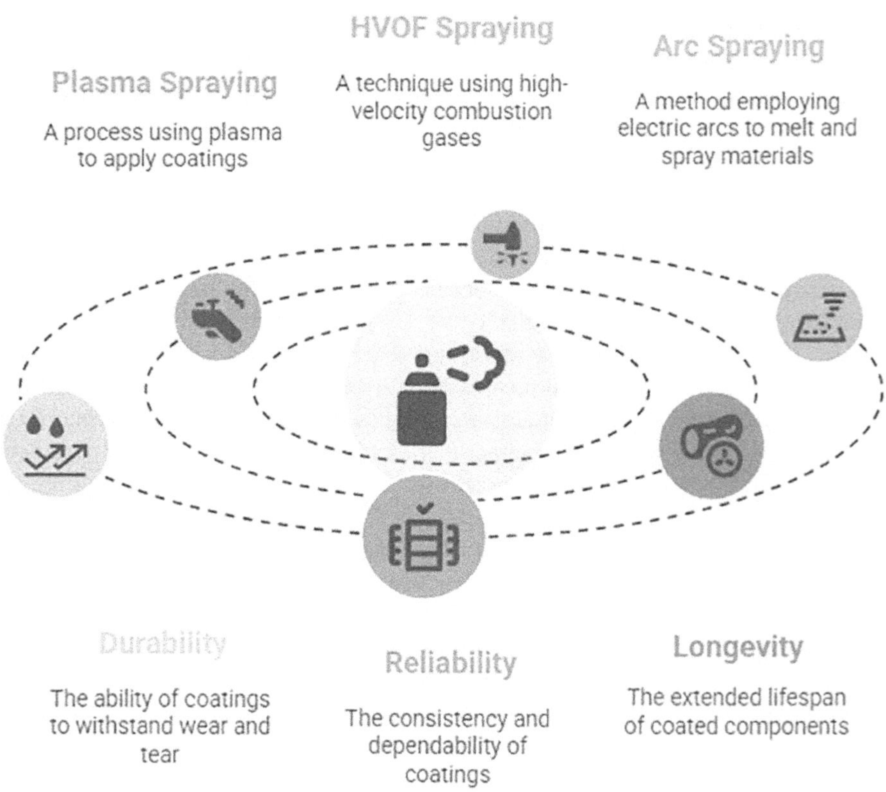

FIGURE 8.5 Thermal spray techniques and their impact on coating performance.

with much lower porosity and superior mechanical interlocking. This technique is particularly effective for depositing tungsten carbide-cobalt (WC-Co) and chromium carbide-nickel chromium (Cr_3C_2-NiCr) coatings, which are widely employed in aggressive slurry environments due to their exceptional combination of hardness, toughness, and corrosion resistance.

Arc spraying, on the other hand, employs an electric arc to melt metallic wires that are then atomized by a compressed gas stream. This method is relatively cost effective and suitable for applying coatings over large areas. Common arc spray materials include Ni-based alloys, stainless steel, and zinc-aluminum composites, which provide moderate erosion resistance with added benefits of sacrificial corrosion protection in marine or saline slurries [37–39].

The effectiveness of thermal spray coatings in resisting slurry erosion can be attributed to several factors. First, the high surface hardness of carbide and ceramic coatings significantly improves their resistance to micro-cutting, indentation, and plowing by abrasive particles. Second, these coatings often exhibit good bond strength, both mechanical and, in some cases, metallurgical, which ensures coating integrity under cyclic impact and vibration. Additionally, corrosion-resistant matrix phases, such as NiCr or NiCo, help prevent the onset of erosion–corrosion synergistic degradation, particularly in chemically active or high-temperature slurries [40].

However, thermal spray coatings are not without limitations. A critical issue is the inherent porosity and the presence of inter-splat boundaries that result from the rapid solidification and layering of molten droplets. These microstructural features can act as preferential sites for slurry ingress, leading to subsurface crack initiation and propagation under repeated particle impact. Furthermore, improper substrate preparation or coating process control may result in weak adhesion or residual stresses that exacerbate spallation or delamination under high erosive loads. Despite these drawbacks, with optimized process parameters and post-treatments such as sealing or laser remelting, thermal spray coatings remain a vital strategy for extending the operational life of components in slurry erosion-dominated environments [41].

8.3.2 Hardfacing

Hardfacing is a surface modification technique that involves the deposition of wear-resistant materials onto the surface of a substrate to enhance its resistance to mechanical degradation, particularly in high-wear environments such as those affected by slurry erosion. This process typically employs welding-based techniques such as shielded metal arc welding (SMAW), gas tungsten arc welding (GTAW), or plasma transferred arc (PTA) welding to apply a layer of durable alloys onto critical areas of components prone to localized damage. Common hardfacing materials include Stellite (Co-based alloys); Inconel (Ni-based superalloys); and various hardfacing steels enriched with chromium, molybdenum, or vanadium carbides. These materials offer a combination of high hardness, thermal stability, and corrosion resistance, making them well suited for combating the complex wear mechanisms associated with slurry impingement [42, 43].

The key advantage of hardfacing lies in its ability to provide localized protection in regions experiencing concentrated erosive or abrasive forces, such as pump impellers, slurry valves, or pipe elbows. By selectively reinforcing these areas, hardfacing significantly extends component life while reducing the need for full material replacement. Additionally, hardfacing alloys are typically engineered to exhibit superior metallurgical bonding with the base material, which enhances coating durability under cyclic loading and impact conditions. However, hardfacing is not without limitations. The deposition process often results in significant heat input, which may lead to thermal distortion, residual stresses, or undesirable microstructural changes in the substrate. Moreover, the additional material layer alters the component geometry, which may require subsequent machining to restore dimensional tolerances. Despite these challenges, hardfacing remains a robust and cost-effective strategy for enhancing slurry erosion resistance in high-stress industrial applications where durability and localized reinforcement are critical [44].

8.3.3 Laser Surface Modification

Laser surface modification has emerged as a highly precise and effective surface engineering technique for enhancing resistance to slurry erosion, particularly in components exposed to aggressive and high-performance environments. Among the various laser-based processes, laser cladding and laser alloying are the most relevant to erosion protection. These techniques utilize a high-energy laser beam to locally melt either pre-placed or simultaneously delivered feedstock materials (in the form of powder or wire), allowing for controlled deposition and integration with the substrate. The laser cladding process creates a metallurgical bond between the coating and the base metal, offering superior adhesion compared to mechanical bonding methods such as thermal spraying. Moreover, due to the rapid solidification rates inherent in laser processing, the resulting microstructure is typically refined, dense, and homogeneous, which contributes to improved hardness and reduced defect density [45, 46].

Laser alloying, in contrast, involves melting the substrate surface along with externally supplied alloying elements to modify the composition and microstructure of the near-surface region. This approach enables in-situ formation of hard phases such as carbides or borides, thereby significantly improving surface hardness, wear resistance, and corrosion performance. Both techniques can be tailored to produce thin, uniform, and crack-free layers with minimal dilution and heat-affected zones, preserving the bulk properties of the substrate [47]. Figure 8.6 present the pros and cons of laser surface modification, highlighting precision and effectiveness versus cost, complexity, and material limitations.

The benefits of laser surface modification in slurry erosion applications are substantial. The resulting surfaces exhibit high hardness, strong adhesion, and superior resistance to impact and abrasion, making them suitable for demanding environments such as slurry pumps, hydrotransport pipelines, and mining tools. Additionally, the process offers excellent control over coating thickness, composition, and geometry, making it ideal for complex components. While equipment costs and process complexity are relatively high, the long-term performance benefits justify the application in critical service conditions [48].

8.3.4 Surface Texturing and Patterning

Surface texturing and patterning have gained significant attention as advanced surface engineering strategies aimed at modifying the tribological interaction between erosive slurry particles and

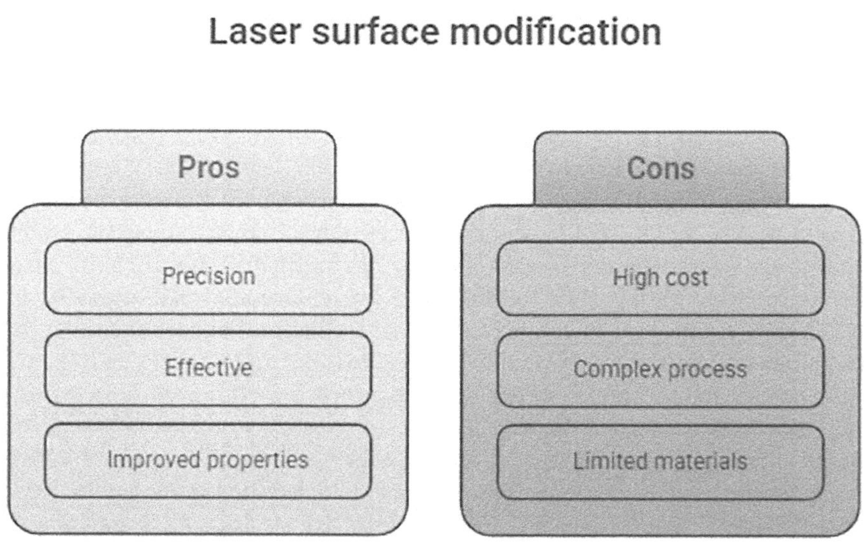

FIGURE 8.6 Advantages and limitations of laser surface modification.

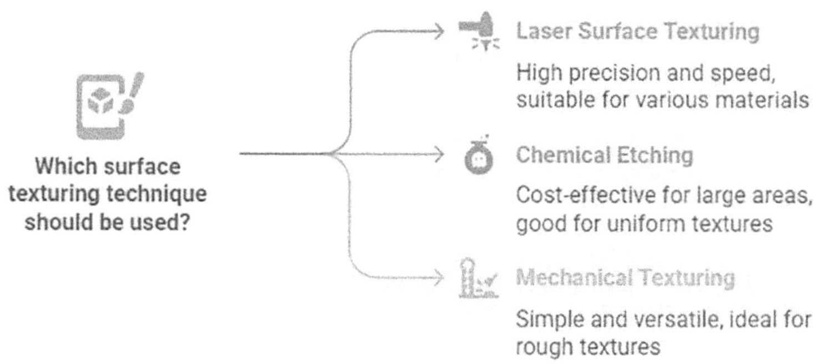

FIGURE 8.7 Comparison of surface texturing techniques.

the material surface. Unlike conventional coatings that provide a barrier layer, surface texturing involves the deliberate creation of micro- or nano-scale features such as dimples, grooves, ridges, or hierarchical structures on the material surface using techniques such as laser surface texturing (LST), electrical discharge machining (EDM), and photolithography followed by etching. These engineered textures influence the dynamics of particle-surface interaction by altering the local flow field and impact conditions [49]. Figure 8.7 comparing laser texturing, chemical etching, and mechanical texturing based on precision, cost-effectiveness, material suitability, and surface finish.

One of the primary benefits of surface texturing is its ability to manipulate particle impingement angles, thereby reducing the effective kinetic energy transferred during impact. For instance, groove-like patterns can deflect particles, leading to glancing impacts rather than normal impingement, which significantly lowers the erosion rate, especially for brittle materials. Additionally, surface textures can act as particle traps, temporarily lodging abrasive particles and preventing them from repeatedly rebounding across the surface an effect that is particularly beneficial in turbulent or recirculating flow conditions [50].

Beyond mechanical interactions, surface patterning can also enhance the hydrodynamic behavior of the surface by reducing fluid–solid contact area or promoting a lubricating film layer, especially under laminar or boundary layer flow regimes. Certain textures may induce micro-vortices or act as flow stabilizers, leading to dampened turbulence and reduced erosive contact frequency. In specific cases, bio-inspired textures (e.g., sharkskin or lotus leaf mimics) have demonstrated improved resistance to both erosion and fouling [51]. However, the performance of surface textures is highly dependent on feature geometry, scale, and uniformity, necessitating precise fabrication control. Despite these challenges, when carefully designed and optimized, surface texturing offers a non-coating-based approach to erosion mitigation with the potential for long-term performance improvements and reduced maintenance in slurry-handling systems [52].

8.3.5 Surface Heat Treatments

Surface heat treatment processes such as nitriding, carburizing, and boriding are thermochemical techniques employed to improve the surface hardness and wear resistance of engineering components without altering their core mechanical properties. These processes involve the diffusion of interstitial elements—namely nitrogen, carbon, or boron—into the surface of a metal substrate at elevated temperatures, leading to the formation of hard, wear-resistant surface layers [53]. Nitriding introduces nitrogen into the surface of steel or other ferrous alloys, forming hard nitrides (such as Fe_4N or CrN) that significantly enhance surface hardness, fatigue strength, and wear resistance. It is particularly advantageous due to its low processing temperature, which minimizes thermal distortion. Carburizing, on the other hand, involves the diffusion of carbon to form a high-carbon martensitic surface layer upon quenching, providing a balance between surface hardness and core toughness. Boriding results in the formation of hard boride phases (e.g., Fe_2B), which offer exceptional hardness and resistance to abrasive and erosive wear, although they may introduce brittleness if not properly controlled [54].

While the depth of hardening achieved through these methods is generally limited—typically in the range of 0.1 to 1 mm—they are cost effective and suitable for components experiencing moderate slurry erosion, such as valves, pump sleeves, and shafts. Additionally, these treatments are compatible with a wide range of steels and alloys and can be integrated into existing manufacturing lines. Despite their limited applicability in extremely aggressive environments, surface heat treatments remain a valuable tool for enhancing erosion resistance in targeted applications with controlled wear conditions [55].

8.4 MATERIAL CONSIDERATIONS

The selection of base materials plays a pivotal role in determining the intrinsic resistance of components to slurry erosion. Different classes of materials—metals, composites, ceramics, and polymers—exhibit varied erosion responses based on their mechanical properties, microstructural features, and failure mechanisms under repeated particle impingement [56]. Metals, particularly stainless steels, duplex steels, and nickel-based alloys, are commonly used in slurry environments due to their balanced combination of toughness, corrosion resistance, and ease of fabrication. Austenitic stainless steels (e.g., 316L) offer good corrosion resistance but moderate erosion resistance due to their relatively low hardness. Duplex stainless steels, with a mixed ferritic-austenitic structure, provide enhanced strength and erosion–corrosion resistance, especially in marine or chemically aggressive slurries. Ni-based superalloys, such as Inconel and Hastelloy, exhibit superior erosion and thermal stability, making them ideal for high-temperature slurry systems, though at a higher material cost [57].

Metal matrix composites (MMCs), reinforced with ceramic particles such as SiC, Al_2O_3, or TiC, provide a synergistic improvement in erosion resistance by combining the toughness of a metal matrix with the hardness of ceramic inclusions. These materials are especially effective in high-energy erosion environments, such as those encountered in mining or dredging, where mechanical and thermal stresses are severe [58]. Ceramics, such as alumina, silicon carbide, and zirconia, offer extremely high hardness and wear resistance, making them suitable for low-impact, high-abrasion conditions. However, their brittle nature limits their use in dynamic or impact-prone environments, where crack initiation and propagation can lead to catastrophic failure [59]. Polymers and elastomers, including polyurethane and rubber linings, are effective in low-impact but chemically aggressive slurry systems. Their energy-absorbing capability and corrosion resistance make them suitable for applications like slurry pipelines, hydrocyclones, and pump linings, although they offer limited resistance to sharp or hard particle impact [60].

In practice, the choice of material must consider not only erosion severity but also temperature, chemical environment, mechanical loading, and cost constraints, often leading to a hybrid solution involving material selection combined with surface engineering treatments [61].

8.5 PERFORMANCE EVALUATION METHODS

Accurate and reproducible evaluation of slurry erosion resistance is essential for understanding material performance under simulated service conditions and for guiding the development of erosion-resistant surfaces. A range of laboratory-scale erosion test rigs and characterization techniques are employed to assess erosion behavior in a controlled and quantifiable manner [62].

Among the most widely used testing systems are the jet-type erosion test rigs, where a slurry mixture is accelerated through a nozzle to impact the specimen at a controlled velocity and angle. This setup enables precise control over parameters such as particle size, concentration, impact

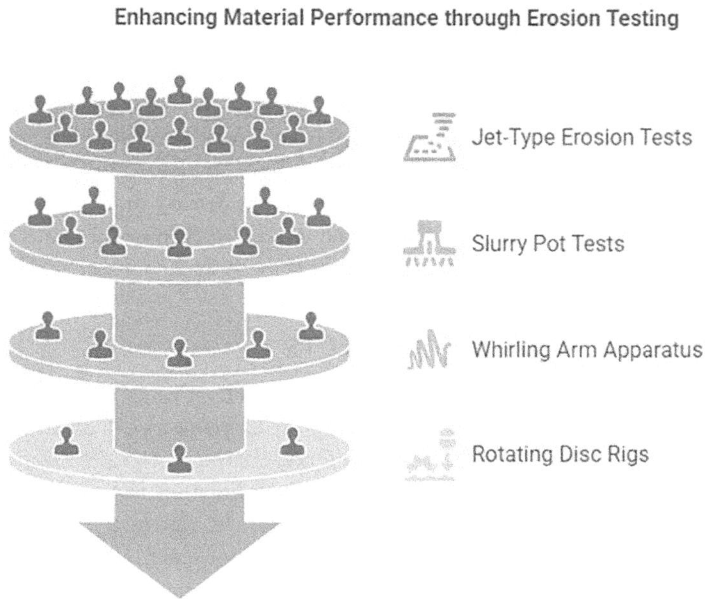

FIGURE 8.8 Erosion testing methods for enhancing material performance.

angle, and temperature. Another common apparatus is the slurry pot test, where specimens are immersed in a rotating slurry mixture, simulating erosive wear under dynamic fluid conditions. The whirling arm apparatus involves a rotating arm that holds specimens exposed to a stationary slurry, often used to simulate centrifugal erosion effects. The rotating disc rig offers continuous particle impact on a surface with adjustable rotational speed and slurry composition, allowing for realistic simulation of components like impellers and pump vanes [63]. Figure 8.8 shows a visual comparison of erosion testing methods: jet-type, slurry pot, whirling arm apparatus, and rotating disc rigs for material evaluation.

Standardized testing protocols have been developed to ensure consistency and comparability. Notably, ASTM G76 describes a method for evaluating erosion using an air-jet system with abrasive particles, particularly useful for high-velocity applications. ASTM G75, on the other hand, defines a slurry abrasion test method using a rubber wheel and slurry media, more suitable for moderate energy erosion–abrasion conditions [64]. Post-test material characterization plays a critical role in understanding erosion mechanisms. Scanning electron microscopy (SEM) is employed to examine the erosion-induced surface morphology, including crater formation, micro-cracking, and lip formation. Energy dispersive spectroscopy (EDS) enables elemental mapping and compositional analysis, particularly for detecting material transfer or oxide formation. Three-dimensional (3D) profilometry allows for non-contact surface topography analysis, providing quantitative data on erosion depth, volume loss, and wear track patterns. Microhardness testing, often performed before and after erosion, helps correlate hardness profiles with erosion resistance and surface degradation patterns [65].

8.6 RECENT ADVANCES AND TRENDS

Progress in slurry erosion protection has been notable with advent of new materials, new surface designs, and computational design approaches. Perhaps one of the most impactful advancements is the introduction of functionally graded coatings (FGCs), which are characterized by changes in composition and microstructure with coating depth. This gradient enables an engineered profile of hardness and toughness, where a hard exterior to withstand impacts from particles and a tougher core to resist stress and minimize delamination. FGCs mitigate differential thermal expansion and mismatched mechanical properties of a coating and substrate, improving dynamic loading durability of the coating [66]. Another critical advancement is nanostructured coatings, where the mechanical properties such as hardness, toughness, and adhesion are enhanced by shrinking the grain size to nanometers. These coatings' high microstructural feature density makes them more resistant to sub-micron cutting and fatigue induced erosion. The nano-crystalline and amorphous constituents of these coatings outperform in slurry erosion in high-velocity environments [67].

8.7 CHALLENGES AND FUTURE PROSPECTS

The area of surface coatings has major challenges that need to be confronted in order to maximize their potential in industry use. Foremost among these challenges is delamination of coatings when subjected to cyclic loading. Fatigue develops at the coating-substrate interface from repeated mechanical strains, resulting in peeling or cracking. Delamination negates the protective role of coatings, particularly in dynamic conditions like automotive, aerospace, and heavy machinery usage, where cyclic loading is rife. An understanding of the intricate relationship among mechanical stresses, temperature effects, and material properties continues to be essential to enhancing coating adhesion and wear resistance [68]. Another significant obstacle is scaling up laboratory-made coatings to industrial dimensions. Numerous coatings prove highly effective in controlled lab environments but struggle when coated on large, intricate surfaces in actual use. Uniformity, reproducibility, and application rate become key issues. Additionally, applying coating methods such as physical vapor deposition, thermal spraying, or chemical vapor deposition to industrial production without compromising quality is of the utmost importance [69].

Cost-effectiveness and sustainability are other challenges. Most advanced coatings use costly raw materials, sophisticated processing, or toxic chemicals, which makes them less desirable for large-scale application. Developments in the future need to be focused on environmentally friendly materials, low-energy processing, and recycling techniques to meet global sustainability priorities. The sector requires solutions with a balance of performance and affordability as well as environmental sustainability [70]. Future directions include designing smart surfaces that have the ability to change dynamically in conditions of wear or environmental conditions. Such surfaces might be capable of self-healing, adjust hardness levels, or change frictional properties according to conditions of use, with resulting longer service life and lower maintenance costs. Embedding responsive materials such as polymers or shape-memory alloys in coatings will be essential in such breakthroughs [71]. Last, the introduction of data-informed material selection through artificial intelligence (AI) and machine learning is transforming coating design. AI algorithms can identify ideal coating compositions and processing parameters based on the analysis of large datasets of material properties, environmental conditions, and performance results, all at a quicker and more accurate rate compared to conventional trial-and-error processes. This method speeds up innovation, lowers development expenditures, and allows for tailored solutions for dedicated applications [72].

8.8 CONCLUSION

Slurry erosion presents a significant challenge to the longevity and reliability of materials in various engineering applications, including pipelines, turbines, and pumps. The continuous impact of abrasive particles suspended in fluid can rapidly degrade surfaces, leading to performance loss and costly maintenance. Surface engineering, through advanced coatings and surface modifications, provides an effective strategy to enhance erosion resistance and extend service life. Selecting the most suitable coating or treatment depends on factors such as the operating environment, erosion severity, cost constraints, and required durability. Techniques must balance performance with feasibility for industrial-scale implementation. Ongoing research is crucial to innovate coatings that offer improved toughness, adhesion, and self-healing capabilities. Additionally, stronger collaboration between academia and industry will accelerate the development of erosion-resistant surfaces that are not only efficient but also sustainable and adaptable to evolving operational demands. This integrated approach is essential to overcoming slurry erosion challenges in future engineering systems.

REFERENCES

1. Tiwari A, Seman S, Singh G, et al. Nanocrystalline cermet coatings for erosion–corrosion protection. Coatings. 2019;9(6):400.
2. Kaya O, Gabatel L, Bellani S, et al. Advances and challenges of hexagonal boron nitride-based anticorrosion coatings. arXiv. 2024 Dec 2.
3. Lelevic A, Walsh FC. Electrodeposition of Ni-P composite coatings: A review. arXiv. 2018 Jun 24.
4. De Nicola F, Castrucci P, Scarselli M, et al. Super-hydrophobic multi-walled carbon nanotube coatings for stainless steel. arXiv. 2015 Mar 18.
5. Kumar R, Goyal K, Bhandari D. Slurry erosion behavior of thermally sprayed ceramic nanocomposite coatings on turbine steel. Int J Appl Ceram Technol. 2022;19(6):3049–3061.
6. Gaurav V, Sankaranarayanan S, Kumareshbabu SP, et al. Slurry erosion behavior of cast 25 Cr austenitic-ferritic steel with niobium addition for subsea pipelines. J Mater Eng Perform. 2021;30:6224–6234.
7. Keshavamurthy R, Naveena BE, Ramesh CS, et al. Evaluation of slurry erosive wear performance of plasma-sprayed flyash-TiO_2 composite coatings. J Bio Tribo Corros. 2021;7:92.
8. Naveena BE, Keshavamurthy R, Sekhar N. Slurry erosive wear behaviour of plasma-sprayed flyash-Al_2O_3 coatings. Surf Eng. 2017;33(7):541–548.
9. Demirsöz R, Singla AK, Uğur A, et al. Understanding the slurry erosion behaviour of iron-based chromium carbide coating material during deposition via arc spray coating method. Proc Inst Mech Eng Part J. 2024;238(1):3–15.
10. Agarwal P, Kumar M, Sharma A, Choudhary M, Shekhawat D, Patnaik A. Experimental and numerical investigation on slurry erosion performance of hybrid glass/steel fiber reinforced polymer composites for marine applications. Polym Compos. 2022;43(8):5592–610.
11. Arabnejad H, Mansouri A, Shirazi SA, McLaury BS. Abrasion erosion modeling in particulate flow. Wear. 2017;376–377:1194–1199.
12. Hall N, Elenany M, Zhu DZ, et al. Experimental study of sand and slurry jets in water. J Hydraul Eng. 2010;136(10):727–738.
13. Naz MY, Sulaiman SA, Shukrullah S, et al. Development of erosion–corrosion mechanisms for the study of steel surface behavior in a sand slurry. Measurement. 2017;106:203–210.
14. Ukpai JI, Barker R, Hu X, et al. Exploring the erosive wear of X65 carbon steel by acoustic emission method. Wear. 2013;301(1–2):370–382.
15. Mansouri A, Arabnejad H, Shirazi SA, et al. A combined CFD/experimental methodology for erosion prediction. Wear. 2015;332–333:1090–1097.
16. Saleh B, Ahmed SM. Slurry erosion–corrosion of carburized AISI 5117 steel. Tribol Lett. 2013;51(2):135–142.
17. Sedrez TA, Rajkumar YR, Shirazi SA, et al. CFD predictions and experiments of erosion of elbows in series in liquid dominated flows. ASME J Fluids Eng. 2018;140(7):071301.
18. Desale GR, Gandhi BK, Jain SC. Development of correlations for predicting the slurry erosion of ductile materials. J Tribol. 2011;133(1):011601.
19. Ndumia JN, Kang M, Gbenontin BV, et al. A review on the wear, corrosion and high-temperature resistant properties of wire arc-sprayed Fe-based coatings. Nanomaterials. 2021;11(10):2527.
20. Deng T. Erosive wear mechanisms of materials—a review of understanding and progresses. Materials. 2025 Apr 2;18(7):1615.
21. Singh G, Ash B. A review on erosion wear of different types of slurry pump impeller materials. Mater Today Proc. 2021;37:2298–2301.
21. Chmielewski T, Siwek P, Chmielewski M, et al. Structure and selected properties of arc sprayed coatings containing in-situ fabricated Fe-Al intermetallic phases. Metals. 2018;8(12):1059.

22. Chailad W, Yang L, Coveney V, et al. Development of slurry-jet erosion test for elastomeric materials. Wear. 2022;488–489:204125.
23. Kumar G, More SR, Bhatt DV, et al. Study of slurry erosion wear behavior of SS-304 in a slurry pot test rig under different operating parameters. SKIT Res J. 2017;7:58–63.
24. Kondul B, Cetin MH. Increasing the wear resistance of railway switches with boron coating and analysis of tribological performance by ANOVA method. Wear. 2022;488–489:204132.
25. Rasse C, Mary N, Abe H, et al. Role of the jet angle, particle size, and particle concentration in the degradation behavior of carbon steel under slow slurry erosion–corrosion conditions. Metals. 2021;11(7):1152.
26. Manisekaran T, Kamaraj M, Sharrif SM, et al. Slurry erosion studies on surface modified 13Cr-4Ni steels: Effect of angle of impingement and particle size. J Mater Eng Perform. 2007;16(5):567–572.
27. Wood RJK. The sand erosion performance of coatings. Mater Des. 1999;20(3):179–191.
28. Lo KH, Cheng FT, Kwok CT, et al. Effects of laser treatments on cavitation erosion and corrosion of AISI 440C martensitic stainless steel. Mater Lett. 2003;58(1):88–93.
29. Kwok CT, Lo KH, Cheng FT, et al. Effect of processing conditions on the corrosion performance of laser surface-melted AISI 440C martensitic stainless steel. Surf Coat Technol. 2003;166(2–3):221–230.
30. Oka YI, Okamura K, Yoshida T. Practical estimation of erosion damage caused by solid particle impact Part 1: Effects of impact parameters on a predictive equation. Wear. 2005;259(1–6):95–101.
31. Clark HM, Wong KK. Impact angle, particle energy and mass loss in erosion by dilute slurries. Wear. 1995;186–187:454–464.
32. Stachowiak GW. Particle angularity and its relationship to abrasive and erosive wear. Wear. 2000;241(2):214–219.
33. Wheeler DW, Wood RJK. Erosion of hard surface coatings for use in offshore gate valves. Wear. 2005;258(1–4):526–536.
34. Kwok CT, Man HC, Cheng FT. Cavitation erosion and pitting corrosion behaviour of laser surface-melted martensitic stainless steel UNS S42000. Surf Coat Technol. 2000;126(2–3):238–255.
35. Clark HM, Hartwich RB. A re-examination of the ‘particle size effect’ in slurry erosion. Wear. 2001;248(1–2):147–161.
36. Verma PK, Reddy BG, Gairola SU. Mechanical behaviour of ceramic coatings for enhancing the erosion resistance of turbine steels. In AIP Conference Proceedings 2024 Feb 20 (Vol. 2986, No. 1). AIP Publishing.
37. O’Flynn DJ, Bingley MS, Bradley MSA, Burnett AJ. A model to predict the solid particle erosion rate of metals and its assessment using heat-treated steels. Wear. 2001;248(1–2):162–177.
38. Verma PK, Reddy BG, Gairola SU. Prediction of slurry erosion behaviour by optimizing the process parameter in ceramic coatings. In AIP Conference Proceedings 2024 Feb 20 (Vol. 2986, No. 1). AIP Publishing.
39. Rajakumar S, Balasubramanian V, Balakrishnan M. Friction surfacing for enhanced surface protection of marine engineering components: Erosion–corrosion study. J Mech Behav Mater. 2016 Aug 28;25(3–4):111–119.
40. Zhao HX, Yabuki A, Matsumura M, et al. Slurry erosion properties of ceramic coatings. Wear. 1999 Dec 1;233:608–614.
41. Kumar S, Ratol JS. Effects of tribological parameters on slurry erosion behaviour of uncoated and coated materials: A review. Mater Eng. 2013 Jul 1;20(3):119–131.
42. Buszko M, Krella A. An influence of factors of flow condition, particle and material properties on slurry erosion resistance. Adv Mater Sci. 2019;19:28–53.
43. Goyal DK, Singh H, Kumar H. An overview of slurry erosion control by the application of high velocity oxy fuel sprayed coatings. Proc IME J J Eng Tribol. 2011 Nov;225(11):1092–1105.
44. Kennedy ZE, Natarajan S. Slurry erosion properties of detonation sprayed and plasma sprayed coatings for materials used in mining environments. Surf Eng. 2009 Aug;25(6):476–481.
45. Singh J, Nath SK. Surface and bulk modification techniques to mitigate silt erosion in hydro turbines: A review of techniques and parameters. Surf Eng. 2022 Mar;38(3):288–302.
46. Matsumura M, Oka Y, Hiura H, et al. The role of passivating film in preventing slurry erosion–corrosion of austenitic stainless steel. ISIJ Int. 1991 Feb 15;31(2):168–176.
47. Grewal HS, Arora HS, Agrawal A, et al. Slurry erosion of thermal spray coatings: Effect of sand concentration. Procedia Eng. 2013 Jan 1;68:484–490.
48. Davis JR, editor. Surface engineering for corrosion and wear resistance. ASM International; 2001.

49. Gupta A, Pandey A, Goyal R, et al. Al2O3-based hybrid composite coatings for mitigating the slurry erosion of an advanced steel in saline and non-saline environments. Surf Coat Tech. 2024 Jan 30;476:130279.
50. Naveena BE, Keshavamurthy R, Sekhar N. Slurry erosive wear behaviour of plasma-sprayed flyash-Al2O3 coatings. Surf Eng. 2017 Dec;33(12):925–935.
51. Xie Y, Jiang JJ, Islam MA. Elastomers and plastics for resisting erosion attack of abrasive/erosive slurries. Wear. 2019 Apr 30;426:612–619.
52. Buszko M, Krella A. An influence of factors of flow condition, particle and material properties on slurry erosion resistance. Adv Mater Sci. 2019;19:28–53.
53. Xie Y, Jiang JJ, Tufa KY, et al. Wear resistance of materials used for slurry transport. Wear. 2015 May 1;332:1104–1110.
54. Lu BT, Luo JL. Correlation between surface-hardness degradation and erosion resistance of carbon steel—effects of slurry chemistry. Tribol Int. 2015 Mar 1;83:146–155.
55. Santa JF, Espitia LA, Blanco JA, et al. Slurry and cavitation erosion resistance of thermal spray coatings. Wear. 2009 Jun 15;267(1–4):160–167.
56. Wood RJ. The sand erosion performance of coatings. Mater Des. 1999 Aug 1;20(4):179–191.
57. Roa CV, Valdes JA, Larrahondo F, et al. Comparison of the resistance to cavitation erosion and slurry erosion of four kinds of surface modification on 13–4 Ca6NM hydro-machinery steel. J Mater Eng Perform. 2021 Oct;30(10):7195–7212.
58. Sharma V, Kaur M, Bhandari S. Micro and nano ceramic-metal composite coatings by thermal spray process to control slurry erosion in hydroturbine steel: An overview. Eng Res Express. 2019 Sep 5;1(1):012001.
59. Yang J, Swisher JH. Erosion–corrosion behavior and cathodic protection of alloys in seawater-sand slurries. J Mater Eng Perform. 1993 Dec;2:843–850.
60. Jones LC. Low angle scouring erosion behaviour of elastomeric materials. Wear. 2011 Jul 29;271(9–10):1411–1417.
61. Chung RJ, Jiang J, Pang C, et al. Erosion–corrosion behaviour of steels used in slurry pipelines. Wear. 2021 Jul 18;477:203771.
62. Vipra R, Singh V, Singh I, et al. Slurry erosion behavior of Ni-Al and Fe-Cr coatings deposited by arc wire spray on hydro machinery SS316 steel. J Mater Eng Perform. 2024 Nov;33(21):11670–11681.
63. Kumar A, Sapra PK, Bhandari S. A review paper on slurry erosion of plasma and flame thermal sprayed coatings. Wear. 2011;1:2.
64. Saini PK, Bansal A, Singh V, et al. Analyzing the mechanisms of Al2O3-TiO2 coating for enhanced slurry erosion resistance on AISI410 stainless steel. Int J Adv Manuf Tech. 2024 Feb;130(11):5837–5851.
65. Kumar R, Bhandari S, Goyal A. Synergistic effect of Al2O3/TiO2 reinforcements on slurry erosion performance of nickel-based composite coatings. Proc IME J J Eng Tribol. 2018 Aug;232(8):974–986.
66. Chen ZX, Hu HX, Guo XM, et al. Effect of groove depth on the slurry erosion of V-shaped grooved surfaces. Wear. 2022 Jan 15;488:204133.
67. Wang Q, Tang Z, Cha L. Cavitation and sand slurry erosion resistances of WC-10Co-4Cr coatings. J Mater Eng Perform. 2015 Jun;24:2435–2443.
68. Rajahram SS, Harvey TJ, Wood RJ. Electrochemical investigation of erosion–corrosion using a slurry pot erosion tester. Tribol Int. 2011 Mar 1;44(3):232–240.
69. Wood RJ, Jones TF, Ganeshalingam J, et al. Comparison of predicted and experimental erosion estimates in slurry ducts. Wear. 2004 May 1;256(9–10):937–947.
70. Levy AV. Solid particle erosion and erosion–corrosion of materials. ASM International; 1995.
71. Lima CR, Mojena MA, Rovere CA, et al. Slurry erosion and corrosion behavior of some engineering polymers applied by low-pressure flame spray. J Mater Eng Perform. 2016 Nov;25:4911–4918.
72. Singh J. Application of thermal spray coatings for protection against erosion, abrasion, and corrosion in hydropower plants and offshore industry. InThermal Spray Coatings 2021 Nov 3 (pp. 243–283). CRC Press.

9 Coating Methods for Protection against Slurry Erosion

Moniya, Jashanpreet Singh, Anuj Kumar Sehgal, and Mohamed Aboughaly

9.1 INTRODUCTION

Slurry erosion is a wear phenomenon caused by solid particles suspended in a liquid medium repeatedly impacting a material surface. The combination of mechanical erosion with corrosion, and sometimes even cavitation, creates fluidically abrasive environments that are especially corrosive. Erosion severity depends on numerous factors, including the size, shape, and hardness of the particles; slurry concentration; impact speed and angle; and the material's hardness, toughness, microstructure, and surface finish. In industry, slurry erosion presents a significant challenge across several fields. In hydroelectric power plants, abrasive silt-laden water exerts powerful erosive forces on turbine blades and associated components, resulting in diminished operational efficiency and heightened maintenance demands. Similar problems exist in the mining and mineral processing industries, where rapid erosion takes place in pipelines, pumps, and chutes used to transport ore slurries [1]. In the oil and gas industry, the erosion of pipelines, valves, and fittings due to the flow of crude oil or gas containing sand particles poses critical safety concerns and operational interruptions. Other affected fields include agriculture, maritime, and chemical processing. Mixers and pumps are examples of agricultural and food processing equipment that comes into contact with slurry flows that cause wear and tear as well as necessitating frequent maintenance. Likewise, ship propellers, hulls, and dredging tools used in marine applications undergo constant erosion from the interaction with abrasive, sand-laden water. In the chemical and process industries, slurry erosion particularly damages the reactors, pipelines, and mixers where continuous flow of abrasive slurries is handled. The detrimental effects caused by slurry erosion include material wastage and surface and geometric inaccuracies, not to mention chronic erosion resulting in equipment failure far before the expected lifespan. These factors all lead to increased equipment downtime, high maintenance expense, and low productivity [2]. Therefore, there is an ever-increasing need for new surface coatings and materials and redesign of the components that can better withstand harsh slurry erosion and prolong equipment service life. Erosion by slurry impacts various components subjected to solid–liquid mixtures, such as hydraulic turbines in hydroelectric power plants and silt-laden water. Components like Kaplan, Francis, and Pelton turbine blades are hydraulically worn. Pumps, valves, and pipeline interiors in mining and oil and gas sectors are particularly vulnerable, as they often convey solid–liquid mixtures under high pressure and high speed. Ship propellers, rudders, and dredging parts in marine industry endure constant wear by sand-laden seawater [3]. Moreover, nozzles, stirrers, and reactor linings in chemical processing plants and agricultural slurry handling units frequently face erosion degradation. The economic and operational impacts are considerable. Material loss decreases efficiency and increases the repair and replacement of worn parts, leading to unplanned stoppages and production breaks, which raises maintenance expenses. In large facilities, such as hydropower stations or mining plants, these cost thousands to millions yearly [4].

DOI: 10.1201/9781003561156-11

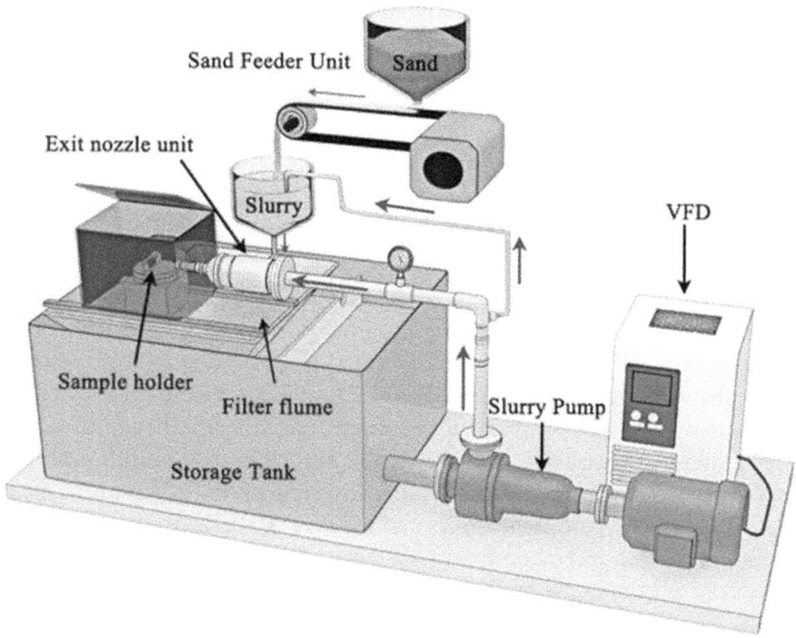

FIGURE 9.1 Schematic view of slurry erosion test rig [6].

Additionally, erosion failures may lead to safety concerns, environmental problems, energy waste, and increased operational burden [5]. As a result, controlling slurry erosion is essential for enduring performance, economically in the long run, and for material savings in maintaining reliability of the system. Significant material erosion occurs due to continuous bombardment of particles from the surface, which leads to aggressive surface destruction, pitting, and component failure. Further, this gear erosion increases equipment downtime maintenance expenses and decreases the operational lifespan of the equipment. The various factors involved in slurry erosion can be tested by using the slurry erosion test rig, as illustrated in Figure 9.1 [6].

In addition, controlled erosion of key components like pipes and pump impellers can be extremely dangerous, as they may cause leaks or mechanical failure, unexpectedly leading to critical safety hazards [7]. Frequent replacements and repairs due to slurry erosion increase operational expenditures, further contributing to significant economic losses. As a protective layer, coatings can mitigate these issues by serving as barriers that shield against surface abrasion and particle impact. Coatings that increase hardness and improve toughness by dissipating impact energy while reducing crack formation are able to resist penetration and cutting by abrasive particles [8]. Coatings are especially vital when slurries are involved, as they often contain corrosive fluids and coatings provide essential corrosion resistance, safeguarding against combined erosion–corrosion damage. In addition, reduction of frictional forces due to certain coatings results in a decrease in the velocity of impacting particles, which subsequently lowers the rate of erosion [9].

9.2 MECHANISM OF SLURRY EROSION

Slurry is a solid–liquid system where solid particles are dispersed in a fluid, usually water. Slurries are classified as settling and non-settling slurries based on the size and behavior of the particles [10]. Settling slurry has larger particles, greater than 100 μm, which tend to settle rapidly and behave like

Newtonian fluids [11]. Non-settling slurries contain smaller particles, in the range of 60–100 μm, which are buoyed up and exhibit non-Newtonian fluid behavior. Erosion severity and nature when slurry interacts with the surface of a material are influenced by many parameters. The characteristics of the particles including size, shape, and hardness are vitally important [12]. Weaker, more rounded particles tend to be eroded less aggressively, whereas larger, harder, angular particles inflict harsher erosive wear due to cutting and fracturing. The particles' speed and angle of impact relative to the surface are crucial as well. Higher speeds increase kinetic energy, which in turn exacerbates the erosion. Impact angle determines the mode of erosion; low angles lead to micro-plowing and micro-cutting (ductile erosion), while high angles cause deformation and cracking (brittle erosion). Slurry-related material degradation occurs in a variety of ways. Ductile erosion is where the material experiences a sort of softening, which results in surface grooves that deepen over time due to cutting and plowing action. Brittle erosion involves the generation of fractures and separation, which usually results in an abrupt loss of material [13]. In many cases, these mechanisms may be working together simultaneously, with one dominating, depending on the material's characteristics and external conditions: impact angle, velocity of the particles, and so forth. The process of erosion from slurry impact occurs stepwise, beginning with solid particles that are suspended in slurry striking the material surface at different angles and speeds [14]. In ductile materials, the slurry flow impacts the surface, resulting in kinetic energy that causes local plastic deformation. Various other mechanisms of slurry erosion in material surfaces are shown in Figure 9.2.

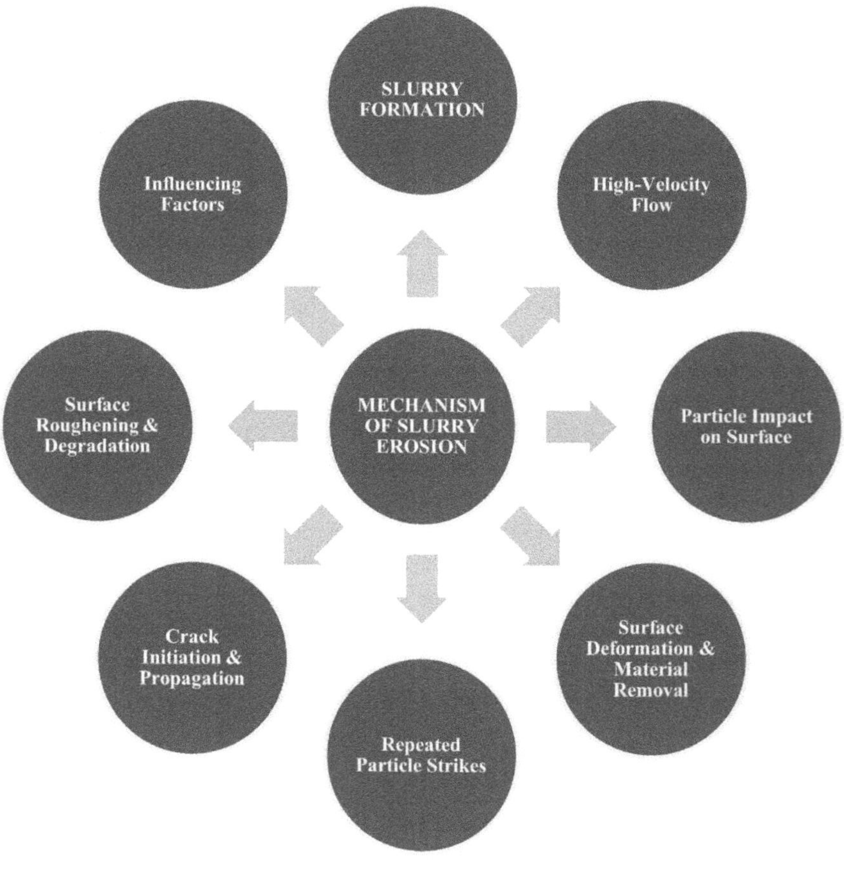

FIGURE 9.2 Various mechanisms of slurry erosion in material surfaces.

Continuation of this impact causes the formation of micro-cracks, which marks the beginning stage of material removal initiation. The rapid formation of these cracks results in the material deterioration. A number of factors influence this material loss, such as the slurry flow rate, hardness of materials, shape and angularity, and sediment concentration. Another important consideration for the erosion process is the impact angle at which the sediments strike. At low striking angles, the primary erosion in ductile materials focuses on cutting and plowing. While striking with higher impact angles, the erosion occurs in terms of a brittle failure mechanism, i.e. cracking. Slurry erosion damage is of a composite nature. Long-term exposure increases surface roughness, which increases both turbulence in the slurry flow and further erosion due to increased interactions between the particles. In understanding the modes and extent of erosion, material characteristics are essential. Ductile materials, like some metals, will mostly undergo plastic deformation and slowly wear as a result of cutting and plowing. Abrupt material loss is more associated with brittle materials, like ceramics, which are susceptible to shattering and fracturing under impact force, resulting in greater localized damage. In slurry-handling systems, understanding these mechanisms enables designers to create erosion-resistant materials while optimizing performance.

9.3 NEED FOR SURFACE COATINGS IN SLURRY EROSION PROTECTION

Base materials are often damaged by slurry impacts because they do not have the mechanical and chemical resistance capabilities needed for protective surfaces. The attack of abrasive particles at high velocity results in severe erosion of the surface in one or more of the following ways: plastic deformation, cutting, micro-cracking, or fatigue [15]. Most structural metals and alloys have adequate fusion strength for stress relief, and while they possess some resistance to mechanical abrasion, they do not epitomize the requisite surface toughness or corrosion resistance to withstand chronic erosion from slurry [16]. To counter this, surface coatings have been developed to combat wear, extend damage tolerance, and improve the reduction of components exposed to extreme strain. Surface coatings are part of an important multifunctional protection system. The hardness of a coating influences ballistic impact erosion resistance; therefore, surface hardness is mandatory. Such a surface reduces the amount of material lost during impact particle erosion. Adhesion with the substrate is important as well; weak bond strength may result in delamination under repetitive loading. Toughness allows a coating to withstand multiple high-speed hypotenuse strikes without crumbling [17]. Simultaneously, erosion of both a ductile and brittle nature may occur. Also, while flowing in a slurry system, aggressive fluids containing corrosive ions can exacerbate material degradation by synergistically enhancing erosion and corrosion damage [18]. There are many different parameters that need to be evaluated with respect to protective coatings. One of the most important basic criteria is the erosion rate, which measures how decay in material occurs over time under controlled slurry impact conditions. Better protective efficiency is indicated by lower erosion rates. The bond strength between the coating and substrate provides reliability of the coating's integrity for mechanical and impact loading. Coating thickness, microstructure, and residual stresses and their thermal stability also matter. Critical components in hydroturbines, pumps, pipelines, and slurry-handling equipment nowadays tend to use advanced coatings like thermal spray deposits, hardfacing alloys, and ceramic composites because of their increased erosion resistance.

9.4 COATING TECHNIQUES FOR SLURRY EROSION RESISTANCE

Components exposed to slurry environments such as those in hydroturbines, pumps, pipelines, and mining equipment must endure extreme mechanical wear caused by high-velocity, abrasive particle impacts. To enhance their durability and operational life, a range of protective coating techniques is employed. These methods are specifically designed to resist material loss due to erosion, minimize maintenance, and improve overall system reliability.

9.4.1 Thermal Spraying: Coatings Formed by Molten Particle Deposition

Thermal spraying is one of the most common techniques for applying erosion-resistant coatings. In this method, coating materials such as hard metals, cermets (WC-Co, Cr_3C_2-NiCr), or ceramics (Al_2O_3) are heated to a molten or semi-molten state and blasted toward the targeted surface at supersonic speeds. The melting energy can be supplied from combustion flames, electric arcs, or plasma gas. When these heated particles collide with the surface, they spread out and cool almost instantly, forming a dense, well-adhered composite coating. Unlike liquid coatings, thermal spray coatings have superior hardness, adhesion, and wear resistance and are impervious to slurry erosion. The raw materials for thermal spray processes are usually metals, alloys, metal matrix composites, or ceramic particles in the form of powder or wire.

9.4.1.1 Plasma Spraying

Schematic diagram of the YSZ coating deposited by the plasma-spraying process is illustrated in Figure 9.3 [19]. Plasma gas employs electric arc technology for spraying, making it the most common choice for creating protective coatings [20], [21]. Electric arcs produce heat, which is capable of melting the required coating materials [22]. Coating materials can also be in the form of argon, helium, and hydrogen ionized gases, with hydrogen yielding higher temperatures. Components such as tungsten carbide, chromium carbide, and aluminum oxide can also be injected into the plasma bob, where they are exposed to extremely high temperatures and "sputtered" to the surface—the end result is dense coatings which enable aluminum hardness [23]. The coatings that are obtained can be seen as the best suited for resisting the impact and cutting forces of slurry erosion [24].

9.4.1.2 High-Velocity Oxygen Fuel Spraying

High-velocity oxygen fuel spraying (HVOF) spraying is a preferred technique for slurry erosion environments because it produces coatings with high hardness, low porosity, and strong bond adhesion [25]. In this case, burning kerosene or hydrogen together with pure oxygen creates a stream of gas at supersonic velocity, accelerating powder particles to the target surface filamentous and supersonic [26]. Because of the short residence time in the flame, oxidation and thermal degradation of the particles are minimal. The HVOF coatings of WC-Co or Cr_3C_2-NiCr provide exceptional erosion resistance in wet abrasive environments and are used to guard turbine blades, pump casings, and pipe bends [27].

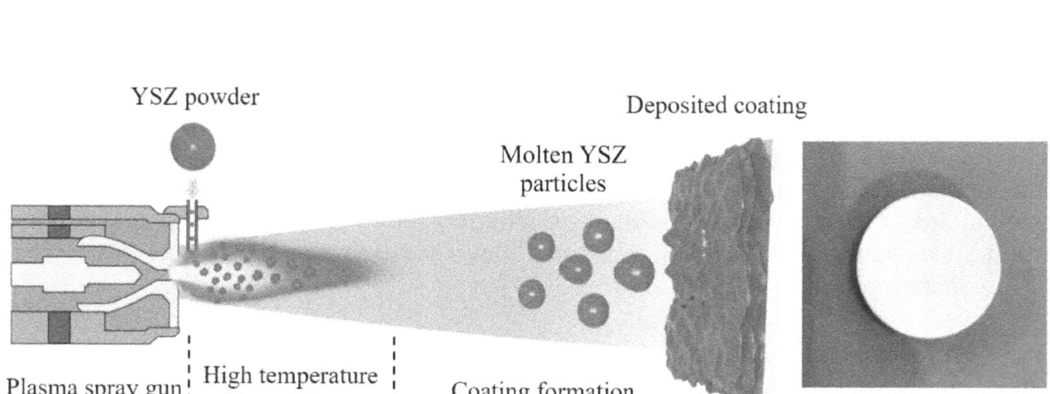

FIGURE 9.3 Schematic diagram of YSZ coating deposited by the plasma-spraying process and images of the prepared coating samples [19].

9.4.1.3 Powder Flame Spraying

An oxy-acetylene flame is employed to heat and project coating powders to a surface using powder flame spraying. This approach is more accessible and cheaper, but HVOF and plasma spraying–produced powder coatings have lower adhesion and high porosity [28]. In many cases, this approach is designated for low to moderate erosion applications or to deposit sacrificial (abradable) coatings that protect pump liners or the surfaces of turbine housings. While not appropriate for hard ceramics or carbides, this method can enable the formation of porous metal coatings for energy-absorbing devices or multi-layered erosion shields [29].

9.4.1.4 Wire Arc Spraying

In wire arc spraying, two consumable wires made from the material to be coated are fed into an electric arc, which melts the tips of the wires [30]. The molten droplets formed are expelled by compressed air [31]. This approach is effective for thick coating of surfaces and is often practiced during maintenance or resurfacing of worn-out parts [32]. It is true that the hardness of the coating is lower than those produced by HVOF or plasma spraying, but it is ideal for the restoration of large, moderately worn areas or for powdered/plasma-sprayed coatings used as a foundation prior to deposition of harder topcoats in multi-layered systems [33].

9.4.1.5 Cold Spray Coatings

Cold spraying is a solid-state technique in which coating particles are heated by pressurized nitrogen or helium gas to high velocities but not to temperatures that would cause melting. Rather the particulate material undergoes plastic deformation and thermally bonds with the substrate upon impact [34]. This method is capable of producing dense and oxidation-free coatings with intricate and elevated structural integrity. Cold spray is exceptionally beneficial in erosion repair and adds protective layers without changing the thermal characteristics of the base material. It serves to repair slurry–erosion damaged parts such as impellers, valve seats, or elbows by resurfacing with coatings that are equal to or greater than the original strength [35].

9.4.2 Physical Vapor Deposition

Physical vapor deposition (PVD) is a method to produce thin films through coating materials which undergo a phase change from solid to vapor, depositing rigid layers that resist abrasion onto the surface of mechanical elements. Schematic diagram of two conventional PVD processes, namely sputtering and evaporating, is shown in Figure 9.4 [36].

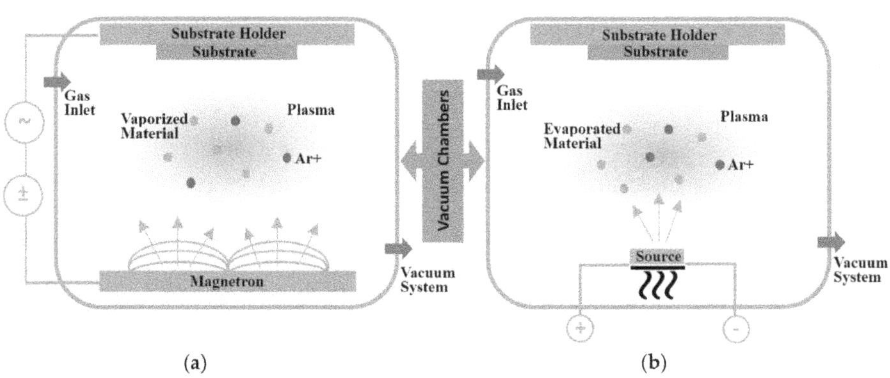

FIGURE 9.4 Schematic diagram of two conventional PVD processes: (a) sputtering and (b) evaporating using ionized Argon (Ar^+) gas [36].

In slurry erosion conditions where high-velocity particle impacts are prevalent, PVD coatings have been found to help improve surface hardness and reduce friction in addition to improving erosion resistance in critical components [37]. The following describes key PVD technologies applicable to slurry erosion mitigation:

9.4.2.1 Arc Technology

In Arc PVD, the coating material is vaporized using a high-energy electric arc with an ionized gas created between two electrodes. The metal atoms are vaporized from the anode, which is normally a solid metal target, and they pass through a vacuum or inert gas medium where they eventually condense into a hard thin film crystalline layer on the part surface. Reaction of metal vapors with nitrogen or carbon within the process chamber forms hard ceramic materials such as titanium nitride (TiN), chromium nitride (CrN), or titanium carbonitride (TiCN). These coatings withstand the micro-cutting and abrasive attacks of slurry particles because they are very hard and chemically stable and provide good adhesion [38]. Arc PVD coatings are highly beneficial for exacting components such as slurry valves, pump impellers, and nozzles which are subjected to fine particle erosion.

9.4.2.2 Sputter Technology

In sputter PVD, a target material is subjected to ion bombardment (usually from argon plasma), which causes the atoms on the surface of the target to vaporize [39]. This vapor is later deposited as a thin film on the substrate. Although the sputter deposition that is plated is usually thinner than that of Arc PVD, sputter coating has one significant advantage: the coatings are extremely smooth and uniform, with microstructure and composition tailored to desired specifications [40]. This makes sputter coatings ideal for high-precision components which must endure erosion due to fine or angular particles, especially in close tolerance assemblies. While sputtered coatings may be thinner, these coatings are hard and possess a dense structure which makes them resistant in low to moderate slurry intensities [41].

9.4.3 Electron Beam Physical Vapor Deposition

In electron beam physical vapor deposition (EBPVD), zirconium oxide (ZrO_2) and other materials with a high melting temperature are deposited through vaporization using a focused electron beam within a vacuum chamber [42, 43]. The vaporization of ZrO_2 results in vapor that condenses into columnar microstructures on the target surface [44]. These microstructures have exceptional resistance to strain and thermal shock, making them useful in slurry erosion applications involving hot erosive flows, such as in geothermal and hydrothermal systems [45]. Though primarily used as thermal barriers, EBPVD porous coatings with resilient structures can be tailored to mitigate the effects of slurry impact fatigue. Nonetheless, due to the pricey, intricate nature of the coating, it is seldom utilized except in cases that demand erosion and thermal resistance [46].

9.4.4 Chemical Vapor Deposition for Slurry Erosion Protection

Schematic diagram of the chemical vapor deposition (CVD) technique is shown in Figure 9.5 [47]. CVD is an extensively used surface engineering procedure for improving the wear resistance of mechanical parts by applying coatings that are hard and adherent [48]. CVD coatings provide protective advantages in slurry erosion environments in which components are subjected to continuous bombardment from fine, high-velocity abrasive particles [49]. The process consists of high-temperature chemical reactions (usually around 1000°C), which lead to strong metallurgical bonding of the coating to the substrate. Such exposure to high temperatures also causes diffusion-driven interface changes, for example, carbon diffusion in cemented carbide substrates (WC) where brittle phases such as WC_{1-x} could be created that may adversely impact erosion behavior [50].

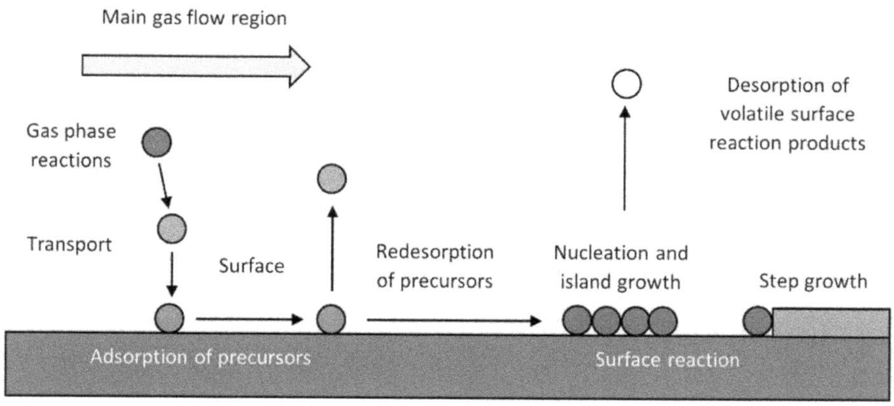

FIGURE 9.5 Schematic diagram of CVD technique (transport and reaction processes) [47]. (CC NY 4.0)

9.4.5 Laser Cladding

Using laser beams, parts that are subjected to wearing and erosion can be treated with enhanced surface coating for protection. The use of high-power lasers with simultaneous deposition of feedstock material (such as powder or wire) upon the workpiece enables clad layers to be formed as the laser moves along the surface [51]. The resultant coating is dense and uniform, adhering at the interface with the base metal due to an intense metallurgical bond that offers much better adhesion than mechanically bonded coatings. This also allows for better load transfer as well as structural integrity [52]. A major advantage of laser cladding is the precise control of thermal input, greatly reducing dilution, distortion, and changes to heat-affected zones. Other commonly used materials for laser cladding are cobalt and nickel alloys, as well as tungsten carbide, chromium carbide, and cermet composites. These are chosen because of their extreme hardness, stability, and corrosion resistance. Industries that have adopted laser cladding technology include power generation, mining, oil, and gas, and it is frequently found in components such as hydro-turbine runners, pump impellers, and drilling tools [53]. Under slurry erosion conditions, laser-cladded surfaces display a remarkable improvement in erosion resistance due to particle impact and material loss, which enhances component lifetime while lowering maintenance expenses. Schematic diagram of laser cladding processing is shown in Figure 9.6.

9.4.6 Electroless Plating

Electroless plating does not require an external electrical current for metal coating deposition, making it a chemical deposition process that forms uniform coatings on segregated metal substrates [55]. An autocatalytic redox reaction drives the process forward. In this reaction, metal ions in solution are reduced onto a surface catalytically active by a chemical reducing agent, which may be sodium hypophosphite or sodium borohydride [56]. This method proves most effective for coating complicated geometric shapes because uniform thickness is achieved even in recesses, blind holes, and complicated shapes. Electroless coatings of nickel-phosphorus (Ni-P) or nickel-boron (Ni-B) alloys are some of the most used. Ni-P coatings provide excellent wear resistance and moderate corrosion protection, while Ni-B coatings have much higher hardness—especially after post-deposition heat treatment—which improves erosion and impact resistance [57]. Usually, these coatings are amorphous or nanocrystalline in structure. Their properties can be tailored by controlling bath chemistry, pH, temperature, and time [58]. Further, heat treatment enhances hardness by inducing

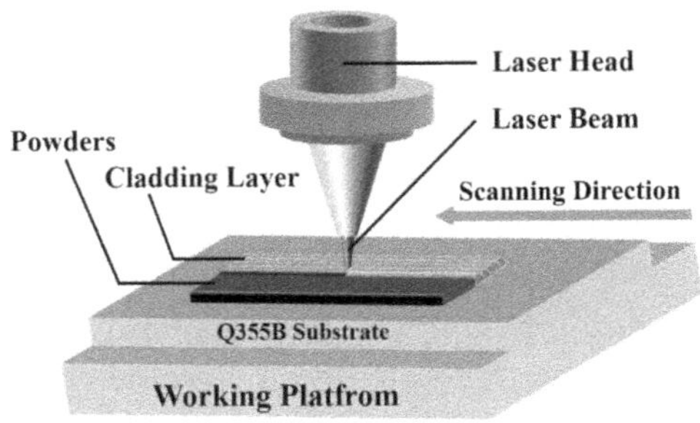

FIGURE 9.6 Schematic diagram of laser cladding processing [54].
(CC NY 4.0)

the precipitation of difficult intermetallic compounds like Ni_3P or Ni_3B. Electroless coatings are commonly employed in components such as valves, pump housings, and impellers used for slurry handling, providing substantial protection against chemical and abrasive wear [59].

9.4.7 Hardfacing Techniques

Hardfacing is a modification method focused on enhancing a component's surface to withstand wear and abrasive impact by applying hard and durable materials to it. This modification can be done using manual or automated overlay techniques, including shielded metal arc welding (SMAW), gas metal arc welding (GMAW), submerged arc welding (SAW), and plasma transferred arc (PTA) welding [60]. This process forms a fusion-bonded overlay which guarantees strong coating to substrate adhesion. The choice of hardfacing materials depends on the specific application [61]. Widely used wrought alloys are cobalt-based Stellite, which has superb toughness and wear resistance at elevated temperatures, while tungsten carbide and chromium carbide possess extreme abrasion resistance. Iron-based alloys and borides are also common in hardfacing where price is a primary consideration. Hardfacing has proven most beneficial in the heavy-duty industries like mining, construction, and agriculture. For example, dredger heads, tiller blades, earth-moving buckets, and slurry pump casings are subjected to intense wear [62]. The coatings may be monolithic or multilayered. Certain applications employ tough buffer layers capped by harder outer coatings to provide durability while retaining impact strength. In slurry-laden conditions, industrial machinery that is equipped with hardfaced surfaces experiences an extended service life, decreased downtime, and increased efficiency in working sustainably [63].

9.5 INDUSTRIAL APPLICATIONS

The use of constraining protective equipment has slurry erosion issues in different industrial areas, which significantly increases the life of the components, decreases the maintenance downtime, and improves the efficiency of the operations. A typical example is Kaplan turbine blades in hydroelectric plants that suffer from material loss because of the cavitation and silted water they deal with. There is a known case from a hydroelectric plant in northern India where the turbine blades were coated with WC-CoCr that was HVOF sprayed [64]. Before coating the blades, there was significantly substantial pitting, and they required servicing every year. With coating, the erosion rate surpassed 60% post-treatment. With routine maintenance performed every two and a half years and significant savings in operational costs against previously unanticipated surge in turbine availability, the results were remarkable [65]. A parallel case can be made from the oil and gas industry, especially when one takes into account offshore drilling, where slurry particles compounded with corrosive fluids result in severe erosion and corrosion. There is a case from a Gulf-based operation which showed application of cermet coatings to mud pump impellers. The use of these coatings achieved a reduction of erosion and corrosion synergy by half, increasing service life of parts from 6 to 18 months. There was significantly reduced downtime, which increased overall productivity [66]. In extractive industry sectors such as dredging and mining, a major South America copper mining company used polyurethane and epoxy-ceramic paints on its pipelines used for transporting slurries. Before coating, pipes would fail every 8 to 10 months. After coating, the service life exceeded 24 months, and the cost of coating was recouped within one maintenance cycle. All these cases illustrate how coatings designed for specific slurries can vastly improve equipment reliability and economy.

9.6 CHALLENGES AND FUTURE DIRECTIONS

Although coating technologies have proved efficient for combating slurry erosion on various industrial applications, there are still many issues that limit their use. Coating materials and their deposition methods, especially for thermal spraying or laser cladding, are expensive and are not economical for SMEs [67]. Efficiency is another issue, as the uniform application of coatings on large, complex geometrical surfaces takes significant time and requires exacting control over many variables. Furthermore, there is increasing scrutiny—from a safety and environmental impact viewpoint—regarding some materials and methods used due to hazardous substances or emissions of harmful pollutants [68]. Several novel approaches are being developed to address these issues. For instance, nanostructured coatings have slurry erosion resistance due to increased hardness, toughness, and compactness. Some nano-WC/CoCr coatings showed over 30–40% greater erosion resistance compared to traditional counterparts. Another promising development is functionally graded coatings (FGCs), which are able to vary composition and mechanical properties with depth. Such a design helps the coating alleviate stress, leading to reduced cracking and improved adhesion to the substrate. Also, smart and self-healing coatings are new areas of research [69]. By means of phase changes or encapsulated healing agents, these materials can detect damage and initiate self-repair processes, which can allow a significant increase of service life and real-time damage control without any manual work. In the area of slurry erosion–resistant coatings, there are multiple avenues for fresh multidisciplinary research that remain unexplored. It is possible to create optimal coating compositions and forecast their performance against operational parameters using artificial intelligence, which makes it possible to provide tailored solutions. There is also an increasing need for coatings that are sustainable, especially those that eliminate hexavalent chromium and other environmentally harmful constituents. The use of coatings and the downtime needed can be optimized with real-time wear tracking through embedded sensors as well as the implementation of predictive maintenance models. Thus, innovative material designs and more accurate predictions can be achieved through multiscale computational simulation of slurry particle interactions and thus accelerate the innovation of next-generation protective coatings.

9.7 CONCLUSION

The problem of slurry erosion is persistent in many industries where the interaction between solids and liquids results in erosion of materials and degradation of performance. This chapter examined protective coatings' effectiveness in reducing slurry erosion by analyzing materials, application techniques, and functionality in real life [70]. The case studies drawn from hydroelectric plants and other oil, gas, and mining industries clearly prove the benefits obtained through the coating technologies, including the HVOF-applied WC-Co-Cr cermet coatings and epoxy-ceramic systems, in terms of enhanced component life and operational efficiency [71]. Although these technologies have limitations, which include high cost and environmental pollution, whereas the self-healing nanocoatings and functionally graded materials pose great promise towards resolving those issues. Selecting the proper coating system is critical, particularly for the designated purpose. Research on AI-based designs, green materials, and real-time monitoring of integrated systems promises to greatly propel advancement in this area [72]. Ultimately, constant modification and introduction of new technologies for advanced coatings is vital in strengthening and increasing the reliability of the components designed for high-endurance flood conditions resulting from slurry erosion [73].

REFERENCES

[1] T. Deng, "Erosive wear mechanisms of materials—A review of understanding and progresses," *Multidisciplinary Digital Publishing Institute (MDPI)*, Apr. 01, 2025, https://doi.org/10.3390/ma18071615.
[2] S. J. Na, Y. S. Kim, and E. S. Jeon, "Analysis of erosion minimization for a slurry pump using discrete phase model simulations," *Applied Sciences (Switzerland)*, vol. 12, no. 3, Feb. 2022, https://doi.org/10.3390/app12031597.
[3] Á. Kalácska *et al.*, "Surface damage analysis on the application of abrasion and slurry erosion in targeted steels using an erosion test rig," *Lubricants*, vol. 10, no. 11, Nov. 2022, https://doi.org/10.3390/lubricants10110316.
[4] Y. Xiao *et al.*, "Slurry flow and erosion prediction in a centrifugal pump after long-term operation," *Energies (Basel)*, vol. 12, no. 8, Apr. 2019, https://doi.org/10.3390/en12081523.
[5] S. Kumar *et al.*, "Neural network prediction of slurry erosion wear of Ni-WC coated stainless steel 420," *Metals (Basel)*, vol. 12, no. 5, May 2022, https://doi.org/10.3390/met12050706.
[6] J. Zhao, A. Ma, X. Ji, J. Jiang, and Y. Bao, "Slurry erosion behavior of AlxCoCrFeNiTi0.5 high-entropy alloy coatings fabricated by laser cladding," *Metals (Basel)*, vol. 8, no. 2, Feb. 2018, https://doi.org/10.3390/met8020126.
[7] Anurag, N. Kumar, and S. Mishra, "Slurry erosion: An overview," in *Materials Today: Proceedings*, Elsevier Ltd, 2019, pp. 659–663, https://doi.org/10.1016/j.matpr.2019.07.717.
[8] K. Kumar, S. Kumar, G. Singh, J. P. Singh, and J. Singh, "Erosion wear investigation of HVOF sprayed WC −10 Co 4 Cr coating on slurry pipeline materials," *Coatings*, vol. 7, no. 4, Apr. 2017, https://doi.org/10.3390/coatings7040054.
[9] S. A. Aldahash, O. Abdelaal, and Y. Abdelrhman, "Slurry erosion–corrosion characteristics of as-built Ti-6Al-4V manufactured by selective laser melting," *Materials*, vol. 13, no. 18, Sep. 2020, https://doi.org/10.3390/ma13183967.
[10] T. Deng, "Erosive wear mechanisms of materials—A review of understanding and progresses," *Multidisciplinary Digital Publishing Institute (MDPI)*, Apr. 01, 2025, https://doi.org/10.3390/ma18071615.
[11] W. Chailad, and L. Yang, "Evaluation of slurry-eroded rubber surface using gloss measurement," *Coatings*, vol. 14, no. 7, Jul. 2024, https://doi.org/10.3390/coatings14070915.
[12] J. Zhao, A. Ma, X. Ji, J. Jiang, and Y. Bao, "Slurry erosion behavior of AlxCoCrFeNiTi0.5 high-entropy alloy coatings fabricated by laser cladding," *Metals (Basel)*, vol. 8, no. 2, Feb. 2018, https://doi.org/10.3390/met8020126.
[13] M. Kerbstadt, E. M. H. White, and M. C. Galetz, "Novel Cr/Si-Slurry diffusion coatings for high temperatures," *Materials*, vol. 16, no. 23, Dec. 2023, https://doi.org/10.3390/ma16237480.
[14] P. Salehikahrizsangi, K. Raeissi, F. Karimzadeh, L. Calabrese, and E. Proverbio, "Effects of surface morphology on erosion–corrosion and corrosion resistance of highly hydrophobic nickel-tungsten electrodeposited film," *Coatings*, vol. 11, no. 9, Sep. 2021, https://doi.org/10.3390/coatings11091084.
[15] P. S. Kevin, A. Tiwari, S. Seman, S. A. B. Mohamed, and R. Jayaganthan, "Erosion–corrosion protection due to cr3c2-nicr cermet coating on stainless steel," *Coatings*, vol. 10, no. 11, pp. 1–17, Nov. 2020, https://doi.org/10.3390/coatings10111042.
[16] B. Fotovvati, N. Namdari, and A. Dehghanghadikolaei, "On coating techniques for surface protection: A review," *MDPI Multidisciplinary Digital Publishing Institute*, Mar. 01, 2019, https://doi.org/10.3390/jmmp3010028.
[17] T. Altamash, J. M. S. S. Esperança, and M. Tariq, "Surface coatings and treatments for controlled hydrate formation: A mini review," *Multidisciplinary Digital Publishing Institute (MDPI)*, Dec. 01, 2021, https://doi.org/10.3390/physchem1030021.
[18] G. Prashar, H. Vasudev, and L. Thakur, "Performance of different coating materials against slurry erosion failure in hydrodynamic turbines: A review," *Elsevier Ltd.*, Sep. 01, 2020, https://doi.org/10.1016/j.engfailanal.2020.104622.

[19] J. Huang, W. Sun, R. Huang, and W. Ma, "Cracking behavior of atmospheric plasma-sprayed 8YSZ thermal barrier coatings during thermal shock test," *Coatings*, vol. 13, no. 2, Feb. 2023, https://doi.org/10.3390/coatings13020243.
[20] C. Guo *et al.*, "Tribological behavior and mechanism of plasma-sprayed high-entropy monoboride coating over a wide temperature range," *Coatings*, vol. 15, no. 2, Feb. 2025, https://doi.org/10.3390/coatings15020159.
[21] A. Kengesbekov, B. Rakhadilov, Z. Sagdoldina, D. Buitkenov, Y. Dosymov, and M. Kylyshkanov, "Improving the efficiency of air plasma spraying of titanium nitride powder," *Coatings*, vol. 12, no. 11, Nov. 2022, https://doi.org/10.3390/coatings12111644.
[22] S. Zhang *et al.*, "Preparation and performance evolution of plasma sprayed abradable CuAl/PHB-NiAl layered seal coatings," *Materials*, vol. 16, no. 1, Jan. 2023, https://doi.org/10.3390/ma16010227.
[23] S. Wang *et al.*, "The effect of plasma spray parameters on the quality of Al-Ni coatings," *Coatings*, vol. 13, no. 12, Dec. 2023, https://doi.org/10.3390/coatings13122063.
[24] R. B. Heimann, "Citation: Heimann, R.B. the nature of plasma spraying," p. 13, 2023, https://doi.org/10.3390/coatings.
[25] M. Oksa, E. Turunen, T. Suhonen, T. Varis, and S. P. Hannula, "Optimization and characterization of high velocity oxy-fuel sprayed coatings: Techniques, materials, and applications," *Coatings*, vol. 1, no. 1, pp. 17–52, Sep. 2011, https://doi.org/10.3390/coatings1010017.
[26] J. Brezinová *et al.*, "High velocity oxygen liquid-fuel (HVOLF) spraying of WC-based coatings for transport industrial applications," *Metals (Basel)*, vol. 10, no. 12, pp. 1–24, Dec. 2020, https://doi.org/10.3390/met10121675.
[27] E. Bakan, G. Mauer, Y. J. Sohn, D. Koch, and R. Vaßen, "Application of high-velocity oxygen-fuel (HVOF) spraying to the fabrication of Yb-silicate environmental barrier coatings," *Coatings*, vol. 7, no. 4, Apr. 2017, https://doi.org/10.3390/coatings7040055.
[28] M. Blum, P. Krieg, A. Killinger, R. Gadow, J. Luth, and F. Trenkle, "High velocity suspension flame spraying (HVSFS) of metal suspensions," *Materials*, vol. 13, no. 3, Feb. 2020, https://doi.org/10.3390/ma13030621.
[29] A. Czupryński, "Flame spraying of aluminum coatings reinforced with particles of carbonaceous materials as an alternative for laser cladding technologies," *Materials*, vol. 12, no. 21, Nov. 2019, https://doi.org/10.3390/ma12213467.
[30] M. Rodriguez Diaz, M. Szafarska, R. Gustus, K. Möhwald, and H. J. Maier, "Oxide free wire arc sprayed coatings—An avenue to enhanced adhesive tensile strength," *MDPI*, Apr. 01, 2022, https://doi.org/10.3390/met12040684.
[31] M. P. Milz *et al.*, "Corrosion fatigue behavior of twin wire arc sprayed and machine hammer peened ZnAl4 coatings on S355 J2C + C substrate," *Corrosion and Materials Degradation*, vol. 3, no. 1, pp. 127–141, Mar. 2022, https://doi.org/10.3390/cmd3010007.
[32] K. Gustke, J. Gebauer, R. Drehmann, A. F. Lasagni, and T. Lampke, "Enhancement of the adhesion of wire arc sprayed coatings on carbon fiber-reinforced plastic by surface laser structuring," *Coatings*, vol. 11, no. 4, Apr. 2021, https://doi.org/10.3390/coatings11040467.
[33] S. Weis, S. Brumm, R. Grunert, J. Morgenschweis, J. Bosler, and T. Grund, "Influence of current modulation on melting behavior during wire arc spraying," *Metals (Basel)*, vol. 12, no. 8, Aug. 2022, https://doi.org/10.3390/met12081347.
[34] D. Klenam *et al.*, "Cold spray coatings of complex concentrated alloys: Critical assessment of milestones, challenges, and opportunities," *MDPI*, Mar. 01, 2023, https://doi.org/10.3390/coatings13030538.
[35] S. Dai, M. Cui, J. Li, and M. Zhang, "Cold spray technology and its application in the manufacturing of metal matrix composite materials with carbon-based reinforcements," *Multidisciplinary Digital Publishing Institute (MDPI)*, Jul. 01, 2024, https://doi.org/10.3390/coatings14070822.
[36] A. Baptista, F. Silva, J. Porteiro, J. Míguez, and G. Pinto, "Sputtering physical vapour deposition (PVD) coatings: A critical review on process improvement andmarket trend demands," *MDPI AG*, 2018, https://doi.org/10.3390/COATINGS8110402.
[37] Y. Zhang, Q. Wang, C. S. Ramachandran, P. Guo, and A. Wang, "Microstructure and performance of high-velocity oxygen-fuel coupled physical vapor deposition (HVOF-PVD) duplex protective coatings: A review," *MDPI*, Oct. 01, 2022, https://doi.org/10.3390/coatings12101395.
[38] C. Mongiello *et al.*, "Plasma-arc-flow technology for sustainable treatment of high-impact fluid waste: A graphene-based material for industrial-wastewater purification," *Processes*, vol. 11, no. 8, Aug. 2023, https://doi.org/10.3390/pr11082307.

[39] P. Panjan, A. Drnovšek, P. Gselman, M. Čekada, and M. Panjan, "Review of growth defects in thin films prepared by PVD techniques," *MDPI AG*, May 01, 2020, https://doi.org/10.3390/COATINGS10050447.
[40] N. F. Ferreira *et al.*, "Tribological comparison of coatings produced by PVD sputtering for application on combustion piston rings," *Coatings*, vol. 14, no. 9, p. 1109, Sep. 2024, https://doi.org/10.3390/coatings14091109.
[41] A. Baptista, F. Silva, J. Porteiro, J. Míguez, and G. Pinto, "Sputtering physical vapour deposition (PVD) coatings: A critical review on process improvement andmarket trend demands," *MDPI AG*, 2018, https://doi.org/10.3390/COATINGS8110402.
[42] N. Ali *et al.*, "Deposition of stainless steel thin films: An electron beam physical vapour deposition approach," *Materials*, vol. 12, no. 4, Feb. 2019, https://doi.org/10.3390/ma12040571.
[43] L. Liu, D. Liu, H. Cai, R. Mu, W. Yang, and L. He, "Failure of electron beam physical vapor deposited thermal barrier coatings system under cyclic thermo-mechanical loading with a thermal gradient," *Coatings*, vol. 14, no. 7, Jul. 2024, https://doi.org/10.3390/coatings14070902.
[44] M. Stange *et al.*, "High-rate epitaxial growth of silicon using electron beam evaporation at high temperatures," *Coatings*, vol. 13, no. 12, Dec. 2023, https://doi.org/10.3390/coatings13122030.
[45] C. L. Tien, C. Y. Chiang, and S. C. Lin, "Optimization of electron-beam evaporation process parameters for ZrN thin films by plasma treatment and taguchi method," *Plasma*, vol. 6, no. 3, pp. 478–491, Sep. 2023, https://doi.org/10.3390/plasma6030033.
[46] B. S. Vasile, A. C. Birca, V. A. Surdu, I. A. Neacsu, and A. I. Nicoară, "Ceramic composite materials obtained by electron-beam physical vapor deposition used as thermal barriers in the aerospace industry," *MDPI AG*, Feb. 01, 2020, https://doi.org/10.3390/nano10020370.
[47] Ł. Łach, and D. Svyetlichnyy, "Recent progress in heat and mass transfer modeling for chemical vapor deposition processes," *Multidisciplinary Digital Publishing Institute (MDPI)*, Jul. 01, 2024, https://doi.org/10.3390/en17133267.
[48] M. Saeed, Y. Alshammari, S. A. Majeed, and E. Al-Nasrallah, "Chemical vapour deposition of graphene—synthesis, characterisation, and applications: A review," *MDPI AG*, Sep. 01, 2020, https://doi.org/10.3390/molecules25173856.
[49] Y. Iwai, T. Miyajima, A. Mizuno, T. Honda, T. Itou, and S. Hogmark, "Micro-slurry-jet erosion (MSE) testing of CVD TiC/TiN and TiC coatings," *Wear*, vol. 267, no. 1–4, pp. 264–269, Jun. 2009, https://doi.org/10.1016/j.wear.2009.02.014.
[50] M. H. Buszko, and A. K. Krella, "Slurry erosion—Design of test devices," *Advances in Materials Science*, vol. 17, no. 2, pp. 5–17, Jun. 2017, https://doi.org/10.1515/adms-2017-0007.
[51] P. Zhao *et al.*, "A review of the laser cladding of metal-based alloys, ceramic-reinforced composites, amorphous alloys, and high-entropy alloys on aluminum alloys," *Multidisciplinary Digital Publishing Institute (MDPI)*, Nov. 01, 2023, https://doi.org/10.3390/lubricants11110482.
[52] M. John, U. B. Kuruveri, and P. L. Menezes, "Laser cladding-based surface modification of carbon steel and high-alloy steel for extreme condition applications," *MDPI*, Oct. 01, 2022, https://doi.org/10.3390/coatings12101444.
[53] K. Wang, Z. Zhang, D. Xiang, and J. Ju, "Research and progress of laser cladding: Process, materials and applications," *MDPI*, Oct. 01, 2022, https://doi.org/10.3390/coatings12101382.
[54] D. Jia *et al.*, "Effects of Y2O3 content on wear resistance and corrosion resistance of 316L/TiC coating fabricated by laser cladding," *Coatings*, vol. 13, no. 8, Aug. 2023, https://doi.org/10.3390/coatings13081348.
[55] B. Vasconcelos, R. Serra, J. Oliveira, and C. Fonseca, "Electroless deposition of Ni-P coatings on HNBR for low friction rubber seals," *Coatings*, vol. 10, no. 12, pp. 1–18, 2020, https://doi.org/10.3390/coatings10121237.
[56] K. Chen, P. Zhang, P. Sun, X. Niu, and C. Hu, "Electroless ni-p plating on mullite powders and study of the mechanical properties of its plasma-sprayed coating," *Coatings*, vol. 12, no. 1, Jan. 2022, https://doi.org/10.3390/coatings12010018.
[57] K. K. Sanap *et al.*, "Development of a simple electroless method for depositing metallic Pt-Pd nanoparticles over wire gauge support for removal of hydrogen in a nuclear reactor," *Materials*, vol. 16, no. 19, Oct. 2023, https://doi.org/10.3390/ma16196541.
[58] B. H. Shin, J. Park, S. Kim, J. W. Ok, D. I. Kim, and J. H. Yoon, "Study of electroless nickel plating on super duplex stainless steel for lithium-ion battery cases: Electrochemical behaviour and effects of plating time," *Metals (Basel)*, vol. 14, no. 3, Mar. 2024, https://doi.org/10.3390/met14030307.

[59] M. Lee, J. Park, K. Son, D. Kim, K. Kim, and M. Kang, "Electroless Ni-P-PTFE composite plating with rapid deposition and high PTFE concentration through a two-step process," *Coatings*, vol. 12, no. 8, Aug. 2022, https://doi.org/10.3390/coatings12081199.

[60] J. Slota, A. Kubit, I. Gajdoš, T. Trzepieciński, and Ľ. Kaščák, "A comparative study of hardfacing deposits using a modified tribological testing strategy," *Lubricants*, vol. 10, no. 8, Aug. 2022, https://doi.org/10.3390/lubricants10080187.

[61] J. Brezinová *et al.*, "Selected properties of hardfacing layers created by pta technology," *Metals (Basel)*, vol. 11, no. 1, pp. 1–20, Jan. 2021, https://doi.org/10.3390/met11010134.

[62] D. Tandon, H. Li, Z. Pan, D. Yu, and W. Pang, "A review on hardfacing, process variables, challenges, and future works," *Multidisciplinary Digital Publishing Institute (MDPI)*, Sep. 01, 2023, https://doi.org/10.3390/met13091512.

[63] D. Arsić *et al.*, "Weldability assessment of various steels by hard-facing," *Materials*, vol. 15, no. 9, May 2022, https://doi.org/10.3390/ma15093082.

[64] C. Mele, F. Lionetto, and B. Bozzini, "An erosion–corrosion investigation of coated steel for applications in the oil and gas field, based on bipolar electrochemistry," *Coatings*, vol. 10, no. 2, Feb. 2020, https://doi.org/10.3390/coatings10020092.

[65] A. Agüero, P. Audigié, M. Gutiérrez, C. Lorente, J. Mora, and S. Rodríguez, "Aluminide coatings by means of slurry application: A low cost, versatile and simple technology," *Coatings*, vol. 14, no. 10, Oct. 2024, https://doi.org/10.3390/coatings14101243.

[66] A. Tiwari, S. Seman, G. Singh, and R. Jayaganthan, "Nanocrystalline cermet coatings for erosion–corrosion protection," *MDPI AG*, Jun. 01, 2019, https://doi.org/10.3390/COATINGS9060400.

[67] M. Kerbstadt, E. M. H. White, and M. C. Galetz, "Novel Cr/Si-slurry diffusion coatings for high temperatures," *Materials*, vol. 16, no. 23, Dec. 2023, https://doi.org/10.3390/ma16237480.

[68] Y. Liu, Q. Li, R. Zhou, and R. Yusufu, "Research progress of superhydrophobic coatings in the protection of earthen sites," *Multidisciplinary Digital Publishing Institute (MDPI)*, Jun. 01, 2024, https://doi.org/10.3390/coatings14060710.

[69] A. Krella, "Resistance of PVD coatings to erosive and wear processes: A review," *MDPI*, Oct. 01, 2020, https://doi.org/10.3390/COATINGS10100921.

[70] B. S. Mann, V. Arya, A. K. Maiti, M. U. B. Rao, and P. Joshi, "Corrosion and erosion performance of HVOF/TiAlN PVD coatings and candidate materials for high pressure gate valve application," *Wear*, vol. 260, no. 1–2, pp. 75–82, Jan. 2006, https://doi.org/10.1016/j.wear.2004.12.045.

[71] N. Chavana, V. Bhajantri F, and S. C. Jambagi, "Improvement in slurry erosion and corrosion resistance of plasma-sprayed fly ash coatings for marine applications," *ACS Omega*, vol. 7, no. 36, pp. 32369–32382, Sep. 2022, https://doi.org/10.1021/acsomega.2c03800.

[72] S. Kumar *et al.*, "Neural network prediction of slurry erosion wear of Ni-WC coated stainless steel 420," *Metals (Basel)*, vol. 12, no. 5, May 2022, https://doi.org/10.3390/met12050706.

[73] S. Kumar *et al.*, "Neural network prediction of slurry erosion wear of Ni-WC coated stainless steel 420," *Metals (Basel)*, vol. 12, no. 5, May 2022, https://doi.org/10.3390/met12050706.

10 Internal Coatings to Prevent Slurry Erosion in Pipelines

Nayana Acharya, Raghavendra Sagar, and Nacer Badi

10.1 INTRODUCTION

The pipes used for transport of oil and gas in industries, ash conveying, chemical industries and mines to transport slurry are subjected to impact of slurry particles of various sizes causing wear of the surface on long time exposure [1]. This problem of surface wear can be addressed by the application of protective coatings for protection against corrosion to reduce friction, improve the flow efficiency, and enhance its service life [2]. Pipeline coating is a procedure involving the application of protective coating to either the inner or outer side, or both, of pipelines. This safeguards the pipe against severe environments and minimizes the occurrence of frequent failures, leading to maintenance and repairs. The different kinds of coatings are epoxy, polyurethane, and ceramic coatings [3]. The choice of type of coating relies on the transported material, working conditions of the pipeline, and environmental circumstances of the location [4]. Internal coating is normally done through the following steps: surface preparation, cleaning, and application of the coating material with the aid of dedicated equipment. These protective coatings can be applied in liquid form or powder on the surface of substrate. Then the process of curing is done through heat treatment or chemical reactions. The protective coating reduces the risk of corrosion and increases the life of the pipe by preventing the metal surface of the pipe from coming into interaction with the fluid or gas that flows through it [5]. Furthermore, the pipeline's flow efficiency is enhanced by the protective coating's smooth, frictionless surface. This helps to decrease energy consumption and increase flow capacity [6]. Also, it prevents leaks and spills caused by pipeline degradation, prevents contamination caused by the working environment, and maintains the quality of the product being transported. Coating can be done internally and externally. Internally coated pipes are used in the oil and gas industry for protection against corrosion caused by transported fluids and harsh environments [7]. Coatings are also employed in water and effluent pipelines, as well as chemical pipelines, to mitigate the deleterious effects of the substances they convey. Additionally, mining and slurry transport pipelines are coated to safeguard piping during mining operations that involve the transportation of abrasive materials. Coatings are utilized to safeguard pipelines that convey steam and water in power facilities [8]. Nevertheless, improvements in deposition techniques, material selection, surface engineering, and post-treatment processes are essential. In this chapter, the various erosion-resistant coatings are elaborated upon, with case studies.

10.2 TYPES OF SLURRY-RESISTANT COATINGS

Erosion-resistant coatings can be broadly classified in to two types based on the area that can be coated during the process. The first is small-scale internal coating, where the area is very small or reported at the research scale, and the other is large-scale internal coating, extensively used in industries [9, 10]. The following subsections explain the types of coatings.

 DOI: 10.1201/9781003561156-12

10.2.1 Small-Scale Internal Coatings

There are different types of materials available for coating the surface of pipes investigated at laboratory scale. The selection of coating material can be influenced by the environmental factors in the area, the operating conditions of the conduit, and the materials being transported in the pipe. The different coating materials used for slurry transport pipelines are as follows:

- **Epoxy coatings:** Epoxy coatings provide excellent corrosion protection and are resistant to chemicals and abrasion. Epoxy coatings are usually preferred in industrial pipelines [11].
- **Polyurethane coatings:** Pipelines that are subject to heavy wear and tear preferably need to be coated with polyurethane coatings, as they offer better corrosion protection compared to epoxy coatings. Polyurethane coatings are preferred in pipelines that transport slurries of large particles, which cause heavy abrasion on impact. Polyurethane provides excellent resistance against abrasions [12].
- **Polyethylene coatings:** Polyethylene coating are used in pipes to protect them from corrosion and abrasion; these are thermoplastic-type coatings. They are primarily used to coat pipelines that transport water and fluids. This coating provides mild resistance to chemicals but not as strong as that of epoxy coats. Polyurethane coatings exhibit excellent UV resistance and are highly durable [13].
- **Ceramic coatings:** Ceramic coatings are done in pipes that are used to transport highly corrosive and abrasive products. Some oxide ceramics (Al_2O_3, ZrO_2, Cr_2O_3) and composite ceramics can be used as coating substances [14]. Since ceramic coats exhibit excellent resistance to corrosion and abrasion, they are popular coating materials. Ceramic coating provides superior insulation, acting as a thermal barrier to high-temperature corrosion. Though ceramic coatings show promising wear resistance, thermal stability, and hardness, they also face some challenges like porosity and defects like pinholes and microcracks formed during deposition, which cause penetration of corrosive agents and degrade the coating [15]. Issues like chemical stability arises as they are degraded in in acidic environments which cause dissolution or flagging of the coating. Adhesion issues like delamination are produced on coating surfaces due to their exposure to corrosion. Mismatch of the thermal expansion coefficient between the coating and substrate can worsen the problem [16].
- **Polymer coatings:** These coatings are capable of offering exceptional corrosion protection and are suitable for a variety of pipeline materials, such as steel, concrete, and plastic [17].

10.2.2 Industry-Scale Internal Coatings

There are many processes for which industry-scale internal coatings are performed. Some of these processes are discussed in following subsection.

10.2.2.1 Fusion-Bonded Epoxy

Fusion-bonded epoxy (FBE) is one of the coating processes used in industry that uses the chemical reaction of cross linking and is seen to provide excellent corrosion protection. FBE coatings have good adhesion to the surface. This process is widely used in the oil and gas industry and can endure prolonged exposure to corrosive environments while preserving properties like thermal insulation, resistance to wearing, and oxidation protection, leading to better performance, dependability, and cost effectiveness [18]. The FBE coating is formed by spraying a thermosetting epoxy powder to the surface of the pipes that is later-melted and fused to the steel pipe surface when heated by means of a high-temperature treatment procedure. FBE forms a continuous, sturdy coating that is extremely resistant to chemical and physical abrasion, hence making it a perfect choice for oil and gas pipelines [19]. The FBE coating process involves applying an electrically charged epoxy powder to a preheated surface, where it melts and fuses to form a durable, protective layer. This method is known as "fusion bonding," and it results in a strong, irreversible coating that cannot be remelted or returned to its original powder form. This means the coating will not melt, even under extreme conditions. Epoxy coatings using the FBE technique are preferred their its exceptional corrosion resistance and strong adhesion. Pipelines used in moderate temperature conditions (generally up to 80°C) widely use FBE coatings. FBE is known to be excellent in adhesion and reliable for long durations and thus provides superior corrosion resistance and mechanical protection compared to alternative materials; thus, pipelines protected by FBE are favored in the transportation of oil, gas, slurry, and water. It has a close connection to the surface of the metal that creates a viable protection against moisture, chemicals, and abrasive materials and extends the lifespan and dependability of the pipeline infrastructure. Also, FBE coating material is environmentally friendly and simple to use since it can be done by electrostatic spray technologies and applied both internally and externally along the pipeline. The dry powder is coated onto the heated surface of a pipe up to a thickness of 400–600 μm [20]. It's easy to apply; environmentally friendly, with no pollution; and exhibits excellent resistance to high temperatures ranging from −40°C to 85°C. FBE-coated steel pipes are buried underground or submerged in water.

The FBE process involves (i) substrate surface preparation by cleaning, degreasing, and blast cleaning, followed by inspection; (ii) preheating; (iii) application of manufactured FBE powder for coating; (iv) post-treatment; and (v) final inspection and quality control [21]. The total time elapsed for pipe surface preparation should be minimized to prevent the formation of oxides on the metal surface. Corrosion due to oxidation and rusting of the steel prior to the coating process in any form is not acceptable. The formation of such oxides requires the surface preparation process to be repeated before the coating procedure is followed. Figure 10.1 shows a block diagram of the FBE coating process.

10.2.2.1.1 Production Process of FBE Powder

The formulation of FBE powder prepared for the pipeline coating process involves accurate measurement and weighing the various components in it to be mixed. These weighed components are first blended in high-speed mixing device. The mixing process ensures that the constituent substances

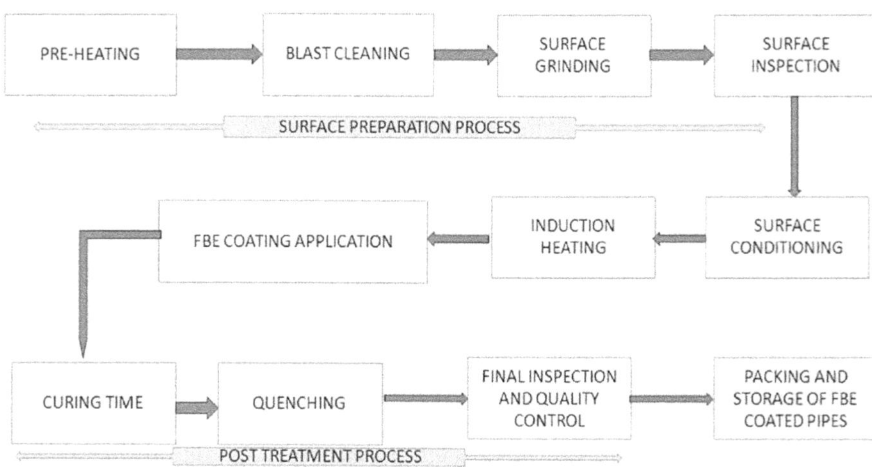

FIGURE 10.1 Block diagram of FBE coating process.

(Created by authors)

are thoroughly mixed before further processing. The mixture is then transported to a high-shear extruder. FBE extruders typically consist of a single- or dual-screw configuration enclosed in a fixed clamshell barrel. The extruder functions in a temperature range of about 50°C to 100°C [22]. Inside the extruder, the FBE mixture is crushed and compressed, followed by heating, and is melted into a molten state. This process causes a complete dispersion of the mixed constituents. The extruder operates rapidly and at a moderately low temperature inside the barrel, preventing the epoxy and hardener components from undergoing significant chemical reactions. The molten extradite is sent between cold rollers, resulting in its solidification to form brittle sheets. Later, it moves to a Kibbler appliance, which cuts the sheet into smaller chips. These chips are further treated in a high-speed crusher to reduce the size of the particles to less than 150 microns [23]. In summary, the production of FBE powders for coatings includes accurate weighing, blending, extrusion, solidifying, sizing, crushing, and packing to produce a superior-quality powder coating product. Appropriate storage conditions are kept to preserve the product's quality.

10.2.2.1.2 Vital Stages in FBE Coating

The FBE coating process consists of three vital stages, irrespective of the type or shape of the steel surface to be coated:

- *Surface Cleaning:* The surface cleaning process involves thorough cleaning of steel surface pipes to eliminate rust, oxide scales, and impurities such as grease or oil. Frequently used methods consist of blast cleaning, solvent cleaning, and burn-off. The surface of steel pipes contains pollutants like salts, soil, grease, oil, and organic residues. These pollutants are to be removed before the first abrasive cleaning step. Failure to eliminate pollutants leads to adhesion problems and film development problems. Later, blast cleaning results in the surface appearing nearly as white.
- *Heating and FBE Powder for Application:* The metal element that has been prepared is heated to the recommended application temperature for FBE powder coatings. Typical application temperatures range from 225°C to 245°C; the components of the powder will melt and convert into a liquid. Heating methods like induction heating, oven heating, and infrared heating are applied in this process. FBE powder is positioned on a fluidization platform, where it behaves like a fluid when suspended in a stream of air. This fluidized powder is then sprayed onto the warm substrate by means of appropriate spray guns. The steel surface to which the molten FBE film is applied is dampened and drifted, and it quickly transforms into a solid coating through the chemical cross-linking process, which is facilitated by heat. Electrostatic spray guns are used to give the powder particles a positive electric charge, guaranteeing even coverage. Spraying lances are used for internal surfaces coating of pipes, moving from one end to the other end while the pipe is made to rotate. Normal coating thickness is in the range 250 to 500 microns. The molten powder bonds with the steel surface and hardens into a coating, undergoing curing in the hot conditions. Comprehensive curing will happen within seconds to minutes, based on the FBE coating system. The coating is irreversibly altered upon the conclusion of the curing procedure. The coating is referred to as a "thermoset" coating because it will not be "melted" by additional heating.
- *Post-Treatment:* To achieve optimum mechanical and protective properties, quenching occurs after the coating is fully cured. The duration of this process is contingent upon the size of the component and the preheat temperature.

There are numerous advantages of the FBE coating method compared to conventional liquid coating methods: ease of coating process, minimal material waste, and fast process. Various types of steel pipes are suitable for coating with FBE, including stainless steel pipes, carbon steel pipes, alloy steel pipes, and nickel alloy pipes [24].

10.2.2.2 Thermal Spraying

Thermal spray coatings are thick coatings with an approximate thickness in the range of 20 μm to several mm dependent on the technique used for thermal coating. Thermal spray coating has a far greater deposition rate than traditional coating processes like electroplating, physical vapor deposition (PVD), and chemical vapor deposition (CVD). Because of its excellent deposition efficiency, thermal spraying is great for big parts or jobs that need thick coatings. Different types of thermal spraying techniques like arc plasma spraying and high-velocity oxy-fuel (HVOF) methods are extensively used for commercial coating applications. For the deposition of nanostructured composites, HVFS is also a good procedure compared with other deposition methods [25].

A variety of materials like metals, alloys, ceramics, plastics, and composites can be coated by the thermal spraying method. Generally, in thermal spraying, the coating precursor in powder or wire form is heated to a molten state by different heating processes and accelerated onto a substrate. Using a gas, the particulates are deposited onto the substrate's surface at a high gas pressure and temperature. Thermal spraying typically employs an electrical arc discharge arrangement as its energy source. The quality of coating is found to increase with the increase in particle velocities evaluated by porosity test, oxide content, macro and micro-hardness, bond strength, and roughness of surface. The material bonds on the surface and forms a corrosion-resistant coating with environmental protection, lubricity, and wear resistance [26]. Thermal spray coating is categorized into various types, namely plasma spray coatings, detonation spray coatings, high-velocity oxy-fuel spray coatings, high-velocity air fuel (HVAF) spray coatings, cold spraying, and wire arc spraying.

Common parts of a thermal spray system are as follows:

- *Spray Gun or Torch:* It is used to melt and accelerate the particles to be deposited on the surface.
- *Feeder:* It is used to supply the coating material in the form of powder or wire to the torch through tubes.
- *Media Supply:* This is the gas or liquid system for the generation of the flame or plasma jet, with gases for carrying the powder.
- *Robot Arm/Manually by Labor:* For positioning and manipulating the spray torch or the substrates to be coated on the surface.

All these types of thermal spray coating processes look similar, but they differ on several points, such as the heating method, the temperatures attained by gas, velocities acquired by the gas stream and coating material, the coating materials used, and the type of bonding that takes place [27]. Different types of thermal spray coating are discussed in the next section.

10.3 DIFFERENT THERMAL SPRAY COATINGS

10.3.1 Plasma Spray Coatings

Plasma spray coating is one of the methods among the various thermal spraying processes that can be used to coat different materials for protection against corrosion. The substrate is cleaned and prepared so that the coating sticks to it properly. The plasma spray technique involves three steps: plasma generation, followed by material injection and deposition of coating. The plasma spray pistol has a copper anode and a tungsten cathode. Both of these parts are cooled by water so they don't become too hot while they work. A high-temperature plasma jet is generated within the gun by ionizing a working gas that is typically a mixture of argon and hydrogen through the establishment of an electric arc between the cathode and the anode. The arc goes around the cathode and through the anode, which is designed like a nozzle that narrows to concentrate the plasma stream. This high-speed plasma jet is injected with a coating material in the form of tiny powder. The powder quickly melts and speeds up towards the substrate [28]. Figure 10.2 shows a schematic drawing of a

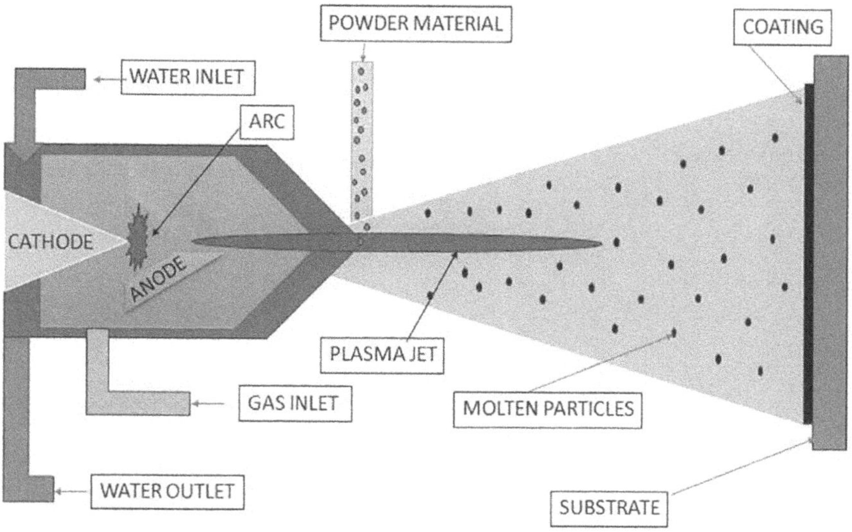

FIGURE 10.2 Schematic representation of plasma spray coating unit.
(Created by authors)

unit for plasma spray coating. The plasma jet may become as hot as 14,000 K, which is hot enough to melt the powder particles. These molten particles are subsequently pushed towards the surface of the substrate at rates of up to 800 m/s. When they hit anything, they quickly flatten and harden, producing a thick, long-lasting covering layer. One of the best things about this method is that the substrate temperature stays low, which lowers the chance of thermal damage, unwanted changes in the metallurgical structure, or deformation of the base material.

The plasma spray coating technique can be used for coating materials like tungsten and ceramics like zirconium oxide which have very high melting points. The coatings formed by this are denser with thickness in micrometer range and stronger and cleaner than other thermal spray processes, with the exception of HVOF, HVAF, and cold spray process. The process can be controlled to achieve specific coating thicknesses and properties [29].

The plasma spray coating method is extensively employed in industries such as aerospace, automotive, and energy sectors that need high-performance coatings. Disadvantages of the plasma spray process are comparatively high cost and complexity of process. The plasma spraying technique is performed at much higher temperature than other thermal spray techniques; hence materials with very high melting points like ceramics can be used for coating the surface. The plasma spray technique forms dense coatings required for better resistance against wear and corrosion. Additionally, plasma sprayed coatings have high bonding strength with the substrate, resulting in a long-wear coating. In industries, the plasma coating process is used for coatings on rollers and machinery parts; in fluid handling industries, components like pumps, valves, engine parts, and propellers can be coated; and in power generation areas, turbines can be coated with protective materials.

10.3.2 Detonation Spray Coatings

The detonation spray coating (DSC) procedure is a kind of thermal spray coating method where a powder feedstock is heated and accelerated at high pressure and high temperature [30]. Figure 10.3 shows a block diagram of the detonation spray coating process. Gaseous products are formed by a controlled explosive combustion of mixture of gases like oxygen and acetylene.

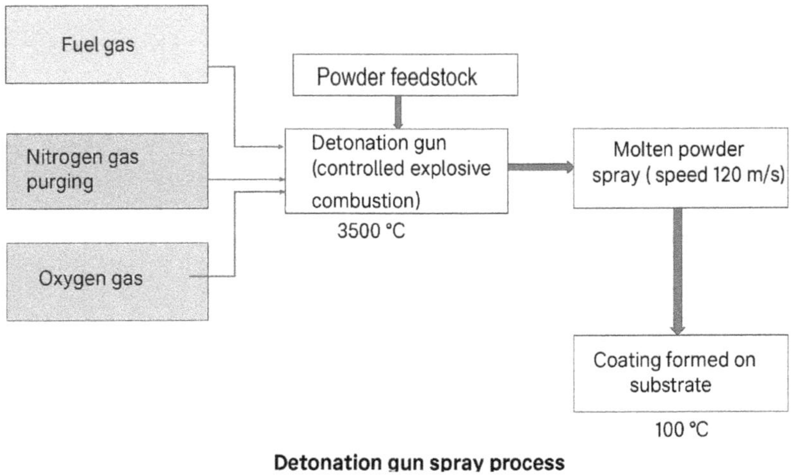

FIGURE 10.3 Block diagram of DSC coating process.

(Created by authors)

A well-adhering coating is generated by the high-speed impact of these accelerated particles on the substrate. The physical condition of the powder particles at the point of effect with the substrate is contingent upon the coating substance. Metals such as Al, Ni, and Cu are in a fully molten state, while ceramics like Al_2O_3 and Cr_2O_3 are in a solid state. Multi-component powders like WC-Co and Cr_3C_2-NiCr are characterized by the partial melting of the low-melting binder phases only [31].

As illustrated in Figure 10.4, the detonation gun system comprises a barrel that is long and water cooled, as well as inlet apertures for gases and coating material (powder). The container is supplied with a charge of powder, as well as fuel such as acetylene and oxygen. In the discharge chamber, a timed explosion is used to detonate measured quantities of acetylene, oxygen, and particulates of a coating material. The gas mixture is ignited by the spark, which causes detonation and increases the powder to supersonic velocity (up to 1200m/s) as it travels down the barrel. After each detonation, the barrel is purified with a nitrogen pulse [32]. This process is executed with a frequency of several times per second. The substrate is struck by the heated powder particles, which travel with high kinetic energy, resulting in a coating that is both dense and durable. The coating applied to the surface exhibits excellent bonding and mechanical strength. The work piece remains below 100°C, thereby eliminating the possibility of microstructure changes in precision parts, despite the fact that temperatures inside the gun reach approximately 3500°C. The substrate's metallurgical properties are unaffected. The coating produced through the detonation gun process exhibits exceptional bond strength, durability, abrasion resistance, and a very low porosity [33].

The coating formed by detonation gun process has a smoother surface finish, dense microstructures, and superior bond strength coating which is the highest among all thermal spray variants, while maintaining a low substrate temperature. Also, the coating exhibits better wear resistance compared to other thermal spray coatings. A comparative study on high-temperature corrosion of NiCoCrAlY coatings deposited by D-gun and APS and exposed to a molten NaSO-KSO-FeO environment illustrated the formation of spinel NiCrO that provided protection against molten salt attack. However, the corrosion resistance of the D-gun coating was superior and credited to the denser microstructure, with 0.9% porosity compared to 2.1% porosity in the APS coating substrate [34]. A variety of powders for coating like pure metals, metal alloys, ceramic oxides, ceramic carbides, and their combinations and mixtures are available for forming protective coatings based on the need and application from the technique.

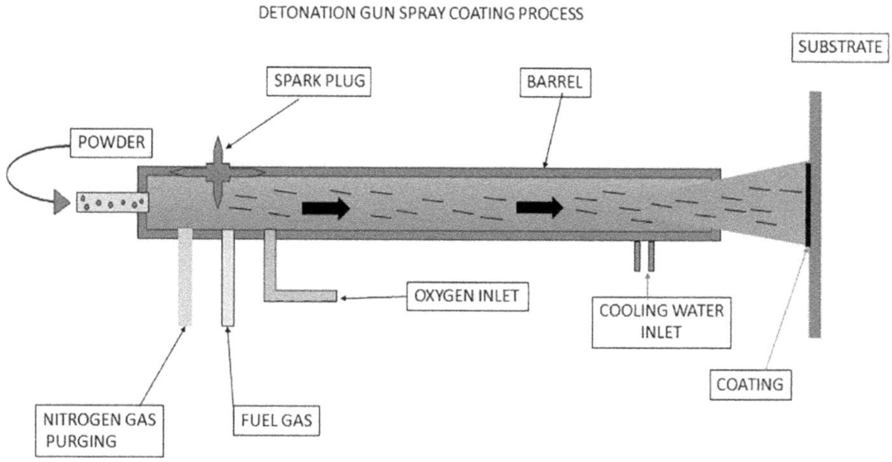

FIGURE 10.4 Diagram of detonation gun system used for deposition of alloys.
(Created by authors)

10.3.3 High-Velocity Oxygen Fuel Spray Coatings

In the 1980s, a supersonic gas jet was used to produce an HVOF dispensing coating through combustion. It has the potential to create coatings that are resistant to erosion and corrosion on ceramic and metallic surfaces [35]. A block diagram of the HVOF coating process is shown in Figure 10.5. In the HVOF technique, a combustion chamber is filled with a mixture of gas or liquid fuel and oxygen, which is ignited and burnt continuously. The heated gas that results is maintained at a pressure of 1 MPa. The gas exits a converging–diverging nozzle and proceeds through a straight section. The fuels may be liquids (e.g., kerosene) or gases (e.g., propane, propylene, acetylene, natural gas, etc.). At the departure of the barrel, the gas jet's velocity exceeds 1000 m/s, surpassing the speed of sound. The temperature of the gas mixture is about 3000°C. A powder feed stock is injected into the high-velocity gas stream. This stream of hot gas carries the powder coating feedstock towards the substrate to be coated, which is placed in its path. The powder melts partly in the stream due to the high temperature and then deposits onto the substrate on impact with high speed. The resulting coating has low porosity and high bond strength.

Figure 10.6 shows the schematic representation conventional HVOF system. Here coatings may be as thick as half inch. Coatings with exceptional density, adhesion, and hardness are produced as a result of the high kinetic energy. The powder feedstocks of tungsten carbide cermet materials (WC-Co, etc.), chromium carbide, MCrAlY, and alumina are used along with chromium carbide and stainless steel [36]. Coating done by the HVOF method is thin but strong and adheres well. HVOF-coated material exhibits better resistance to corrosion compared to coatings done by using the plasma spraying technique, but the HVOF process is not well suited to high temperature ranges. The oil and gas and power generation industries prefer coating by HVOF thermal spray techniques. HVOF techniques create long-lasting and dense coatings that improve wear and corrosion resistance; hence they are used for coating turbine blades, engine components, and industrial valves.

In HVOF coating, high-speed nozzles are employed, due to which the flame velocity exceeds the speed of sound. This results in spray particles of high kinetic energy and impact force, which is responsible for the density and bond strength of the coating. The temperature of the HVOF coating process is lower compared to other thermal spray technologies. This reduces thermal deformation of the substrate and prevents change in material properties; hence HVOF is suitable for temperature-sensitive applications. Since the particles are sprayed with high kinetic energy from the high-speed

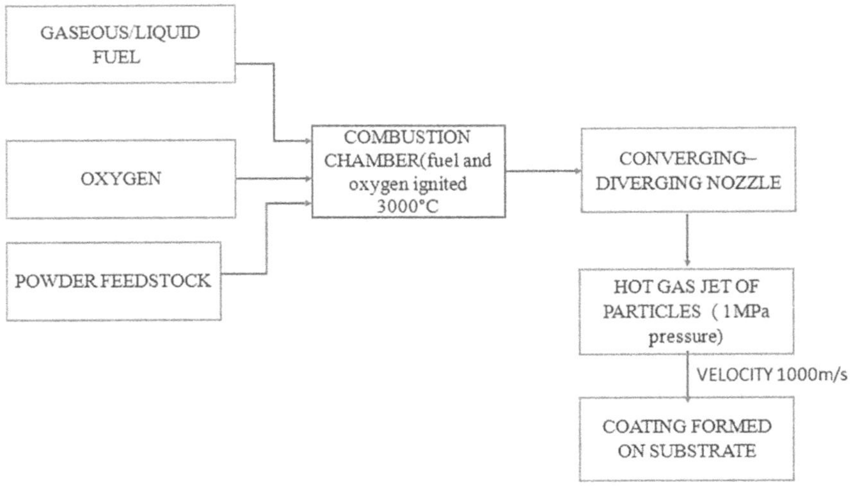

FIGURE 10.5 Block diagram of deposition process using HVOF system.

(Created by authors)

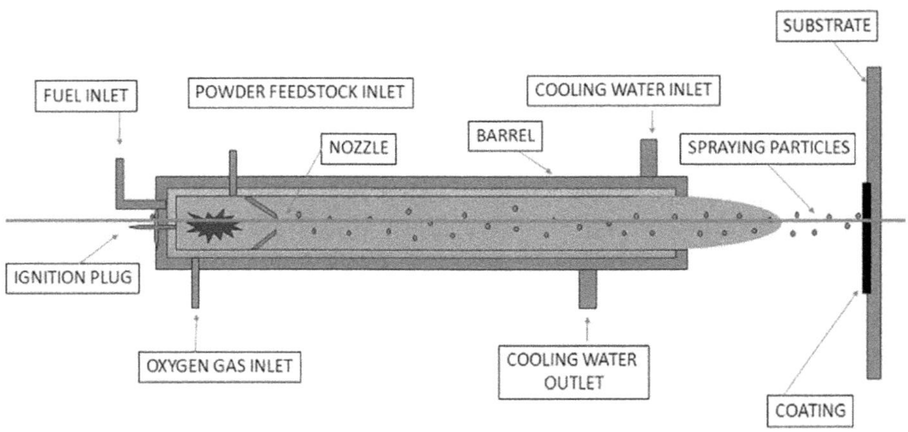

FIGURE 10.6 Schematic representation of a conventional HVOF system.

(Created by authors)

flame jet of HVOF, it results in strong mechanical bonding on the substrate, which enhances the bond strength and durability of the coating, as it provides a robust protective layer which has better wear, corrosion, and high-temperature resistance. Since the thickness of the coating can be precisely controlled by adjusting spraying time and powder flow rate, one can complete coatings in a shorter time than traditional methods, improving production efficiency and saving time.

Despite the many advantages of HVOF spray coating, there are some disadvantages, like high equipment cost, complex operation, and high energy consumption during the high-speed spraying process [37]. Some of the important components in a HVOF spray coating system include:

- *Spray Gun:* The spray gun is a crucial component that propels the coating material onto the substrate. The HVOF spray gun design provides high flame velocity and accurate and precise control over the spraying procedure.

- *Powder Feeder:* The powder feeder is responsible for delivering the coating powder to the spray gun. It must guarantee a steady and precise flow of feedstock powder to attain uniform coatings on surface.
- *Control System:* The control system monitors and adjusts various parameters of the HVOF spraying process, including pressure, temperature, and spraying distance. This ensures the desired coating thickness and quality.
- *Cooling System:* A cooling system is essential to prevent overheating of both the substrate and the HVOF coating equipment. It may involve water or other cooling methods.
- *Gas Supply System:* The gas supply system provides the necessary gases, such as oxygen and fuel, for the combustion process that generates the high-velocity flame used in the HVOF coating process.
- *Automation and Robotics:* In advanced HVOF coating systems, automation and robotics may be integrated to enhance precision and repeatability, especially in large-scale production environments.

The HVOF coating process is done in the following seven stages:

- *Stage 1:* The surface of the substrate material to be coated is pre-treated to eliminate contaminants to enhance adhesion of coating on the substrate. Surface cleaning is carried out by grit blasting, bead blasting, and chemical treatment.
- *Stage 2:* The powder to be coated is ground well to obtain uniform particle size and then mixed with a solvent in mixer for proper blending.
- *Stage 3:* The HVOF machine parameters spraying pressure, spraying distance, and spraying speed are adjusted according to the coating material properties to ensure uniform and stable spraying of coating materials onto the substrate surface.
- *Stage 4:* The coating material is poured into the spray gun of the HVOF machine and then ejected at a very high speed onto the substrate surface through the nozzle. To guarantee continuous and uniform spraying on the substrate, steady spraying distance and speed should be maintained during the entire spraying procedure.
- *Stage 5:* The sprayed coating materials evaporate instantly and cool down on contact with the substrate, leading to formation of a uniform and dense coating. The high spray velocity in HVOF thermal spraying gives sufficient kinetic energy to the particles, resulting in improved adhesion and dense coating.
- *Stage 6:* On completion of the HVOF spraying process, post-treatment work is done on the coating to further improve their performance. The common post-treatment approaches are sintering and heat treatment, which makes the coating denser and more robust, leading to improved resistance to wear and corrosion.
- *Stage 7:* Testing and evaluation are done to check the quality of coating. This involves the process of checking hardness, adhesion, thickness, and porosity detection.

The HVOF technique uses coating different materials like metal alloys such as cobalt alloys, cobalt-chrome alloys, and cobalt-chrome-tungsten alloys. Nickel alloys have good corrosion and high temperature resistance and are suitable for aerospace and energy industries.

10.3.4 High-Velocity Air Fuel Spray Coatings

HVAF is a thermal spray process that puts coatings on parts, vessels, and structures to protect them from wear, erosion, and corrosion. It can also restore surfaces and spray the inside of pipes in industries like aerospace, automotive, and marine where high-quality coatings are very important [38]. Tungsten carbide is the predominant material applied by HVAF for abrasion wear resistance. HVAF uses high-velocity air and liquid fuel combustion to create a high-speed gas jet. The combustion process can be precisely controlled using this technique, producing coatings with exceptional microstructural characteristics, low porosity, a high level of corrosion resistance, and mechanical strength. HVAF coatings are similar to coatings made using HVOF and cold spray methods. HVAF is a kind of "warm spray" method that is colder than HVOF but hotter than the cold spray process. HVAF process guns use an axial powder injection technology to shoot powder into a jet of air fuel that is between 1900°C and 1950°C, so this approach can effectively apply carbide-based materials. However, the air-fuel jet in HVAF creates a lot fewer oxides than high-temperature oxy-fuel jets. This means that the HVAF method can coat metal powders with nearly no oxidation, like cold spray. HVAF may be used to apply all kinds of thermal spray powders, except for ceramics.

As HVAF spray guns are of very small size, on the order 55 mm, they can be utilized to coat the internal surfaces of pipes and portions as narrow as 80 mm [39]. The length of the pipe to be coated is not a constraint; pipelines that are 50 feet or longer can be internally coated. The combustion of fuels such as propane occurs in the compressed air stream, and this coating technology generates a uniform high-velocity flow, similar to the HVOF technique. The thermal discharge mechanisms are further stabilized by the inclusion of a heat baffle in HVAF. This method involves the injection of material into the air–fuel stream, which propels coating particles towards the component. HVAF has a maximum flame temperature ranging from 3560°C to 3650°C. Because the highest combustion temperature is close to the melting point of most spray materials, HVAF creates a coating that is more even and flexible. High-speed air fuel sprinkling adds powder to a gas jet made of compressed air and fuel, which is commonly propane, propylene, or natural gas. HVAF may spray any kind of particle material except ceramics as long as the temperature is about 2000°C, which is higher than the melting point of the coating material but lower than its evaporation temperature [40]. The particulates that are discharged have a velocity of 1200 m/s in this location. In contrast to HVOF, the metals that are applied using the HVAF procedure exhibit minimal oxide formation.

10.3.5 Cold Spraying Technique

Particulates are accelerated to high velocities in the cold spraying technique by a carrier gas that is propelled over a converging-diverging de Laval–type nozzle. Solid particulates with an adequate amount of kinetic energy undergo plastic deformation and mechanically adhere to the substrate, resulting in the formation of a coating upon impact [41]. Figure 10.7 illustrates a cold spraying unit used in research laboratories and industries. The critical velocity required to establish bonds varies with the material's characteristics, size of the powder, and temperature.

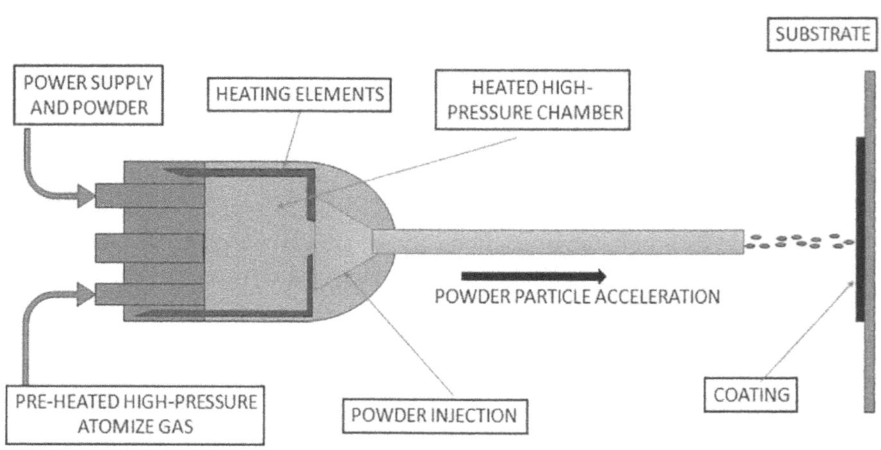

FIGURE 10.7 Conventional cold spraying unit used in research laboratories and industries. **(Created by authors)**

One can use the cold spraying approach to put down a variety of materials, including metals, ceramics, polymers, composites, and nanocrystalline powders. Cold spraying works well with light metals like Cu and Al, but it has also been used to coat harder materials like W, Ta, Ti, MCrAlY, and WC-Co [42–45]. However, the efficiency of deposition is typically low for alloy powder materials, and the powder sizes required are limited. In order to impart high velocity to powders, finer particles of less than 20 µm are used. These powder particles can be accelerated to much higher velocities by employing helium gas as processing gas instead of nitrogen to achieve speeds greater than the speed of sound. Nevertheless, the cost of helium is high, and its discharge rate, which results in a higher consumption, is also higher. Nevertheless, in order to enhance the acceleration capability, the deposition efficacy and tensile strength of the deposits are enhanced by heating the nitrogen gas to approximately 900°C [46].

In cold spraying, the materials that are sprayed don't melt, which sets it apart from other thermal spray methods. An inert gas, such nitrogen or helium, is heated up electrically to a temperature between 0°C and 900°C. After that, a high-pressure gas stream with supersonic speeds of up to 1200 m/s speeds up the coating powder [47]. This technique relies on the kinetic energy of the spray particles, which makes them stick to the surface when they hit it. There are no heat-affected zones or oxidation of sprayed materials in this method, which is different from conventional thermal spray procedures. Since the cold gas spray technique does not depend on high-temperature combustion like conventional thermal spray techniques, it instead propels powder particles onto the substrate by imparting high kinetic energy using carrier gas streams. This produces a highly cohesive and dense coating, and, as there is no scope for thermal degradation, this method can be used to coat the surfaces of heat-sensitive materials [48]. Cold spray coating is extensively used in the aerospace, automotive, and electronics industries, where delicate components need protective coatings without the risk of thermal distortion.

10.3.6 Wire Arc Spraying

Figure 10.8 shows a diagram of a wire arc spraying system used for the coating of various substrates. Wire arc spraying, also referred to as twin wire arc spraying, generates an electrical discharge that arcs between two electrically conductive wires, one of which is positively charged and the other negatively charged [49]. The temperature of the electric discharge can reach approximately 5000°C when the two conductors connect in the spray cannon. The arc facilitates heating, which in turn induces deposition and melting that is a process comparable to the use of a torch for combustion flame spraying. This high heat developed causes melting of the wires, and a stream of compressed air or gas is used to spray the molten metal in the form of droplets that interlock one on top of another on the substrate surface. Here the sprayed particles move with low speeds, about 60–300 m/s, as compared to other methods [50].

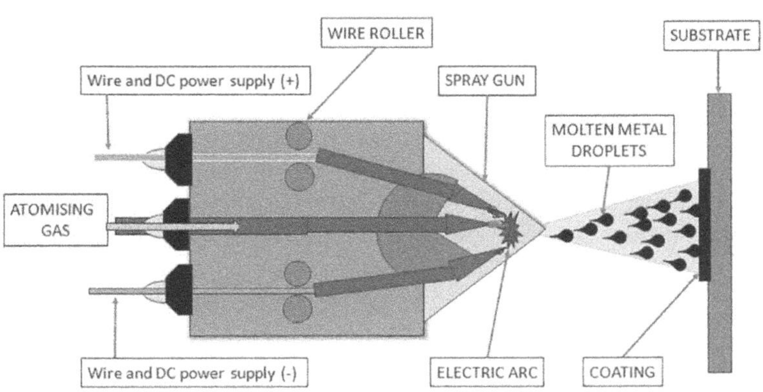

FIGURE 10.8 Wire arc spray system used for coatings on various substrates.

(Created by authors)

10.4 CASE STUDIES ON VARIOUS COATING MATERIALS AND COATING TECHNIQUES

There are various reports on different classes of materials that are used as protective coatings in a variety of substrates to prevent erosion [51–55]. In Rajeev et al. [56], accelerated examination of the slurry erosion properties of flame-sprayed Ni-20Al_2O_3 and Ni-10Al_2O_3-10TiO_2 coating showed that the former did not exhibit erosion resistance as well as the latter. Grewal et al. [57] considered the slurry erosion of Ni-Al_2O_3–based composite coating and demonstrated in a fracture contact model that alumina content had a large effect on the erosion rates. Liu et al. [58] investigated the properties of wear and erosion of WC10Co_4Cr coatings enriched with 2–15 percent nano-WC12Co using the AC-HVAF method. They found that the coatings were harder and less porous, with 15% nano-WC12Co providing the stiffest properties. Thakur et al. [59] investigated WC-10Co4Cr nanostructured coatings incorporated using multi-walled carbon nanotubes (MWCNTs) through HVOF, enhancing fracture toughness and slurry erosion resistance. According to Singh et al. [60], atmospheric plasma spraying deposition of nano-Al_2O_3 on stainless steel resulted in increased sliding and erosive wear resistance compared to conventional alumina coating, as further confirmed by FE-SEM/EDS. Zhang et al. [61] reported the plasma-sprayed nanostructured Al_2O_3-13wt.% TiO_2 coatings on substrates of varying morphology to study the effect of interfacial microstructures on the performance. Sharma et al. [62] sprayed multi-phase Ni-based coatings (Ni-40TiO_2 and Ni-20TiO_2-20 Al_2O_3) via HVFS on CA6NM turbine steel as an intervention for slurry erosion, resulting in low porosity (0.8 to 2.0%), high hardness, and very strong bond strength. Goyal et al. [63] coated CF8M turbine steel with WC10Co4Cr and Al_2O_3+13TiO_2 coatings using HVOF and showed that Al_3TiO_2 experienced both ductile and brittle erosion mechanisms, but the WC in a 10Co4Cr coating had a mixed mechanism and highly increased its erosion resistance. The review of Singh et al. [64] considered cavitation erosion in hydroturbines and highlighting the success of different coatings such as carbides, cermets, intermetallics, and nitrides deposited to the surface by thermal spray, HVOF, arc plasma spray, laser hardening, CVD, and PVD. Finally, Malik et al. [65] determined the efficiency of fusion-bonded epoxy as a corrosion protection coating on cement concrete pipes carrying water; FBE showed good adhesion, mechanical strength, and resistance to corrosion under both static and dynamic water environments, and FBE is a potential candidate for corrosion protection coating to water-transmitting structures.

10.5 CONCLUSION

In this chapter, the techniques used to develop protective coatings on mining and industrial pipes and hydroturbines are discussed. The development of problems like corrosion, erosion, or abrasion normally observed in these pipes that hinder the transport of industrial slurry is discussed. The techniques that help to develop large-scale integral coatings like FBE and thermal spraying are discussed in detail to ensure proper conditions required for coating pipes at the industrial scale. Among the discussed techniques, FBE and cold spray are the simplest and capable of large-scale coating. Nevertheless, the development of coatings using ceramic or alloys is challenging and needs other systems such as plasma spray or HVOF and HVAF techniques.

REFERENCES

1. N.K. Singh, G. Vinay, A.S.M. Ang, D.K. Mahajan, H. Singh, Cavitation erosion mechanisms of HVOF-sprayed Ni-based cermet coatings in 3.5% NaCl environment. Surf. Coat. Technol. 434, 128194 (2022). https://doi.org/10.1016/j. surfcoat.2022.128194.
2. G. Sundararajan, A comprehensive model for the solid particle erosion of ductile materials. Wear. 149, 111–127 (1991). https://doi.org/10.1016/0043-1648(91)90368–5.

3. R. Kumar, D. Bhandari, K. Goyal, Slurry erosion behaviour of HVOF sprayed coatings on hydro turbine steel: A review. IOP Conf. Ser. Mater. Sci. Eng. 1033 (2021). https://doi.org/10.1088/1757-899X/1033/1/012064.
4. S. Saurabh, M.K. Singhal, R.P. Saini, Hydro-abrasive erosion in hydro turbines: A review. Int. J. Green Energy. 15, 232–253 (2018). https://doi.org/10.1080/15435075.2018.1431546.
5. N. Lin, Y. He, C. Wu, S. Liu, X. Xiao, Y. Jiang, Influence of TiC additions on the corrosion behaviour of WC-Co hard metals in alkaline solution. Int. J. Refract. Met. Hard Mater. 46, 52–57 (2014). https://doi.org/10.1016/j.ijrmhm.2014.05.009.
6. V. Kumar, R. Verma, Tribology international effect of GNP and laser-surface texturing on HVOF sprayed WC10Co4Cr coatings for high-wear resistance. Tribol. Int. 178, 108057 (2023). https://doi.org/10.1016/j.triboint.2022.108057.
7. Q. Wang, S. Zhang, Y. Cheng, J. Xiang, X. Zhao, G. Yang, Wear and corrosion performance of WC-10Co_4Cr coatings deposited by different HVOF and HVAF spraying processes. Surf. Coat. Technol. 218, 127–136 (2013). https://doi.org/10.1016/j.surfcoat.2012.12.041.
8. T. Denis, G. Martin, G. Stéphane, The effect of materials properties on the reliability of hydraulic turbine runners. Int. J. Fluid Mach. Syst. 8, 254–263 (2015). https://doi.org/10.5293/IJFMS.2015.8.4.254.
9. H.A. Webster, The costs of corrosion: A perspective. J. Air. Pollut. Control Assoc. 26, 302–303 (2012). https://doi.org/10.1080/00022470.1976.10470252.
10. H.Y. Al-fadhli, J. Stokes, M.S.J. Hashmi, B.S. Yilbas, The erosion—corrosion behaviour of high velocity oxy-fuel (HVOF) thermally sprayed inconel-625 coatings on different metallic surfaces 200, 5782–5788 (2006). https://doi.org/10.1016/j.surfcoat.2005.08.143.
11. R. Verma, J.S. Randhawa, S. Kant, N.M. Suri, Characterization studies of slurry-sprayed mullite-nickel coatings on ASTM 1018 steel. Arab. J. Sci. Eng. (2019). https://doi.org/10.1007/s13369-019-03753-6.
12. E. Xavier, E. Eduard, F. Mohamed, A. François, C. Miguel, Detection of cavitation in hydraulic turbines. Mech. Syst. Signal Pract. 20, 983–1007 (2006). https://doi.org/10.1016/j.ymssp.2004.08.006.
13. J.A. Picas, M. Punset, E. Ruperez, S. Menargues, E. Martin, M.T. Baile, Corrosion mechanism of HVOF thermal sprayed WC-CoCr coatings in acidic chloride media. Surf. Coat. Technol. 371, 378–388 (2019). https://doi.org/10.1016/j. surfcoat.2018.10.025.
14. J. Ramasamy, M. Amanullah, Nanocellulose for oil and gas field drilling and cementing applications. J. Pet. Sci. Eng. 184 (2020). https://doi.org/10.1016/j.petrol.2019.106292.
15. D. Xiang, K. Du, Y. Chengqing, D. Zhangxiong, C. Xudong, Microstructure and cavitation erosion resistance of HVOF deposited WC-Co coatings with different sized WC. Coatings 8, 307 (2018). https://doi.org/10.3390/coatings8090307.
16. A.A.M.M. Singha, P.A. Francob, Enhancement of corrosion resistance on plasma spray coated mild steel substrate exposed to marine environment. Mater. Today Proc. 15, 84–89 (2019). https://doi.org/10.1016/j.matpr.2019.05.028.
17. A.K. Basak, J.P. Celis, P. Ponthiaux, F. Wenger, M. Vardavoulias, P. Matteazzi, Effect of nanostructuring and Al alloying on corrosion behaviour of thermal sprayed WC-co coatings. Mater. Sci. Eng. A 558, 377–385 (2012). https://doi.org/10.1016/j.msea.2012.08.015.
18. S. Somasundaram, B.R. Ramesh Bapu, R. Raj Jawahar, Tribological characteristics of n-(GO/WC-10Co-4Cr) HVOF-coatings under biolubricant conditions. Surf. Eng. 37, 91–100 (2021). https://doi.org/10.1080/02670844.2020.1762467.
19. A. Cassidy, J.A. Miwa, L. Hornekær, Graphene as an anti-corrosion coating layer, 495–509 (2015). https://doi.org/10.1039/C4FD00259H.
20. G. Singh, M. Kaur, High-temperature wear behaviour of HVOF sprayed 65% (NiCrSiFeBC) 35% (WC-Co) coating. Surf. Eng. 36, 1139–1155 (2020). https://doi.org/10.1080/02670844.2019.1639932.
21. W. Dongsheng, T. Zongjun, W. Songlin, S. Lida, H. Yinhui, Solid particle erosion behaviour of plasma-sprayed conventional and nanostructured Al2O3-13 wt% TiO2 ceramic coatings. Trans. Indian Ceram. Soc.74, 90–96 (2015). https://doi.org/10.1080/0371750X.2015.1036169.
22. H. Masoumi, S.M. Safavi, M. Salehi, S.M. Nahvi, Effect of grinding on the residual stress and adhesion strength of HVOF thermally sprayed WC-10Co-4Cr coating. Mater. Manuf. Process. 29, 1139–1151 (2014). https://doi.org/10.1080/10426914.2014.930893.
23. K. Rajeev, B. Sanjeev, G. Atul, Synergistic effect of Al2O3/TiO2 reinforcements on slurry erosion performance of nickel-based composite coatings. Part J J. Eng. Tribol. 232, 974–986 (2018). https://doi.org/10.1177/1350650117736487.

24. R. Kromer, S. Costil, J. Cormier, L. Berthe, P. Peyre, D. Courapied, Laser patterning pretreatment before thermal spraying: A technique to adapt and control the surface topography to thermomechanical loading and materials. J. Therm. Spray Technol. 25, 401–410 (2017). https://doi.org/10.1007/s11666-015-0352-x.
25. V.A.D. Souza, A. Neville, Mechanisms and kinetics of WC-co-cr high velocity oxy-fuel thermal spray coating degradation in corrosive environments. J. Therm. Spray Technol. 15, 106–117 (2006). https://doi.org/10.1361/105996306X92677.
26. S. Vibhu, K. Manpreet, B. Sanjeev, Development and characterization of high-velocity flame sprayed Ni/ TiO_2/Al_2O_3 coatings on hydro turbine steel. J. Therm. Spray Technol. 28, 379–1401 (2019). https://doi.org/10.1007/s11666-019-00918-5.
27. A comparative study on wear and corrosion behaviour of HVOF- and HVAF-sprayed WC-10Co-4Cr coatings. Surf. Eng. 33, 63–71 (2017). https://doi.org/10.1080/02670844.2016.1218194.
28. J.A. Picas, M. Punset, E. Rup´erez, S. Menargues, E. Martin, M.T. Baile, Corrosion mechanism of HVOF thermal sprayed WC-CoCr coatings in acidic chloride media. Surf. Coat. Technol. 371, 378–388 (2019). https://doi.org/10.1016/j. surfcoat.2018.10.025.
29. J.D. Mangadlao, P. Cao, R.C. Advincula, Smart cements and cement additives for oil and gas operations. J. Pet. Sci. Eng. 129, 63–76 (2015). https://doi.org/10.1016/j.petrol.2015.02.009.
30. M.T. Alsaba, M.F. Al Dushaishi, A.K. Abbas, A comprehensive review of nanoparticles applications in the oil and gas industry. J. Pet. Explor. Prod. Technol. 10, 1389–1399 (2020). https://doi.org/10.1007/s13202-019-00825-z.
31. H. Panchal, H. Patel, J. Patel, M. Shah, A systematic review on nanotechnology in enhanced oil recovery. Petrol. Res. 6, 204–212 (2021). https://doi.org/10.1016/j.ptlrs.2021.03.003.
32. H.S. Grewal, A. Agrawal, H. Singh, Slurry erosion performance of Ni-Al2O3 based composite coatings. Tribol. Int. 66, 296–306 (2013). https://doi.org/10.1016/j.triboint.2013.06.010.
33. A.A.H. Abdel, T.S. Jamil, M.S. Shalaby, A.M. Shaban, E.R. Souaya, Application of (polyaniline/zeolite X) composite as anticorrosion coating for energyrecovery devices in RO desalination water plants. Int. J. Ind. Chem. 10, 175–191 (2019). https://doi.org/10.1007/s40090-019-0182-7.
34. F. Walsh, G. Ottewill, D. Barker, Corrosion and protection of metals: II. Types of corrosion and protection methods. Trans. IMF. 71, 117–120 (1993). https://doi.org/10.1080/00202967.1993.11871002.
35. M. Horynová, L. Klakurková, J. Švejcar, M. Juliš, P. Gejdoš, L. Čelko. Failure analysis ofcasing of draft tube of turbine used in hydropower plant. Eng. Fail. Anal. 82, 848–854 (2017). https://doi.org/10.1016/j.engfailanal.2017.08.002.
36. P. Linhardt, Twenty years of experience with corrosion failures caused by manganese oxidizing microorganisms. Mater. Corros. 61, 1034–1039 (2010). https://doi.org/10.1002/maco.201005769.
37. A. Arghavan, R.K. Kashyzadeh, A. Amiri Asfarjani, Investigating effect of industrial coatings on fatigue damage. Appl. Mech. Mater. 87, 230–237 (2011). https://doi.org/10.4028/www.scientific.net/AMM.87.230.
38. H. Abdollahnia, M.H. Alizadeh Elizei, K.R. Kashyzadeh, Fatigue life assessment of integral concrete bridges with H cross-section steel piles mounted in water. J. Fail. Anal. Prev. 20, 1661–1672 (2020). https://doi.org/10.1007/s11668-020-00976-w.
39. M.B. Bishwakarma, H. Støle, Real-time sediment monitoring in hydropower plants. J. Hydraul. Res. 46, 282–288 (2008). https://doi.org/10.1080/00221686.2008.9521862.
40. A.J. Ninham, A.V. Levy, The erosion of carbide-metal composites. Wear. 121, 347 (1988).
41. M.K.A. Ehsan, G.H. Farrahi, M.K.A. Mahdi, A.Z. Amir, P. Siavash, Failure analysis of agas turbine compressor in a thermal power plant. J. Fail. Anal. Prev. 3, 313–319 (2013). https://doi.org/10.1007/s11668-013-9663-8.
42. D. Xiang, K. Du, Y. Chengqing, D. Zhangxiong, C. Xudong, Microstructure andcavitation erosion resistance of HVOF deposited WC-Co coatings with different sized WC. Coatings 8, 307 (2018). https://doi.org/10.3390/coatings8090307.
43. M. Ning, G. Lei, C. Zhenxiong, W. Huantao, Y. Fuxing, Z. Keke, Improvement onmechanical properties and wear resistance of HVOF sprayed WC-12Co coatings byoptimizing feedstock structure. Appl. Surf. Sci. 320, 364–371 (2014). https://doi.org/10.1016/j.apsusc.2014.09.081.
44. J.M. Guilemany, S. Dosta, J.R. Miguel, The enhancement of the properties of WC-CoHVOF coatings through the use of nanostructured and microstructured feedstock powders. Surf. Coat. Technol. 201, 1180–1190 (2006). https://doi.org/10.1016/j.surfcoat.2006.01.041.
45. B.E. Naveena, R. Keshavamurthy, N. Sekhar, Slurry erosive wear behaviour of plasma-sprayed flyash-Al_2O_3 coatings. Surf. Eng. 33, 925–935 (2017). https://doi.org/10.1080/02670844.2017.1288341.

46. A.F. Ahmed, Applications of nanomaterials in corrosion protection coatings andinhibitors. Corros. Rev. 38, 67–86 (2020). https://doi.org/10.1515/corrrev-2019-0011.
47. S. Dawei, A. Jinliang, W. Gang, Y. Jinglei, Double-layered reactive microcapsules with excellent thermal and non-polar solvent resistance for self-healing coatings. J. Mater. Chem. A. 3, 4435–4444 (2015). https://doi.org/10.1039/C4TA05339G.
48. S. Hong, Y.P. Wu, Q. Wang, G.Y. Li, G.B. Ying, B. Wang, W.W. Gao, Cavitation silt erosion behaviour of HVOF sprayed nickel-based alloy coatings. Surf. Eng. 29, 588–593 (2013). https://doi.org/10.1179/1743294413Y.0000000168.
49. S.L. Liu, X.P. Zheng, G.Q. Geng, Influence of nano-WC-12Co powder addition in WC-10Co-4Cr AC-HVAF sprayed coatings on wear and erosion behaviour. Wear. 269, 362–367 (2010). https://doi.org/10.1016/j.wear.2010.04.019.
50. V. Nitesh, R.K. Khatirkar, S.G. Sapate, Tribological behaviour of HVOF sprayed WC-12Co, WC-10Co-4Cr and Cr3C2-25NiCr coatings. Tribol. Int. 105, 55–68 (2017). https://doi.org/10.1016/j.triboint.2016.09.025.
51. L. Yang, L. Ying, C. Jingjing, L. Li, Solid particle erosion behavior of HVOF/HVAF sprayed WC-Co-Cr coatings. Proc. Inst. Mech. Eng. Part J J. Eng. Tribol. 230, 634–643 (2016). https://doi.org/10.1177/1350650115608209.
52. N. Gaurav, K. Satish, Studies on slurry erosion behavior of HVOF applied Y_2O_3-added WC-10Co-4Cr cermet on 4 martensitic stainless steel. Metallogr. Microstruct. Anal. 7, 133–142 (2018). https://doi.org/10.1007/s13632-018-0426-5.
53. R.J.K. Wood, The sand erosion performance of coatings. Mater. Des. 20, 179–191 (1999). https://doi.org/10.1016/S0261-3069(99)00024–2.
54. P.S. Vinay, S. Anjan, R.A. Jayaganthan study on sliding and erosive wear behaviour of atmospheric plasma sprayed conventional and nanostructured alumina coatings. Mater. Des. 32, 584–591 (2011). https://doi.org/10.1016/j.matdes.2010.08.019.
55. Z. Jianxin, H. Jining, D. Yanchun, L. Xiangzhi, Y. Dianran, Microstructure characteristics of Al_2O3–13 wt.% TiO_2 coating plasma spray deposited with nanocrystalline powders. J. Mater. Process. Technol. 197, 31–35 (2008). https://doi.org/10.1016/j.jmatprotec.2007.06.001.
56. K. Rajeev, B. Sanjeev, G. Atul, Synergistic effect of Al2O3/TiO2 reinforcements on slurry erosion performance of nickel-based composite coatings. Part J J. Eng. Tribol. 232, 974–986 (2018). https://doi.org/10.1177/1350650117736487.
57. H.S. Grewal, A. Agrawal, H. Singh, Slurry erosion performance of Ni-Al2O3 based composite coatings. Tribol. Int. 66, 296–306 (2013). https://doi.org/10.1016/j.triboint.2013.06.010.
58. S.L. Liu, X.P. Zheng, G.Q. Geng, Influence of nano-WC-12Co powder addition in WC-10Co-4Cr AC-HVAF sprayed coatings on wear and erosion behaviour. Wear. 269, 362–367 (2010). https://doi.org/10.1016/j.wear.2010.04.019.
59. L. Thakur, N. Arora, A study of processing and slurry erosion behaviour of multi-walled carbon nanotubes modified HVOF sprayed nano-WC-10Co-4Cr coating. Surf. Coat. Technol. 309, 860–871 (2017). https://doi.org/10.1016/j.surfcoat.2016.10.073
60. V.P. Singh, A. Sil, R. Jayaganthan, A study on sliding and erosive wear behaviour of atmospheric plasma sprayed conventional and nanostructured alumina coatings. Mater. Des. 32, 584–591 (2011). https://doi.org/10.1016/j.matdes.2010.08.019
61. J. Zhang, J. He, Microstructure characteristics of Al2O3–13wt.% TiO2 coating plasma spray deposited with nanocrystalline powders. J. Mater. Process. Technol. 197 (1–3), 31–35 (2008). https://doi.org/10.1016/j.jmatprotec.2007.06.001
62. V. Sharma, M. Kaur, S. Bhandari, Performance investigation of high velocity flame sprayed multi-dimensional Ni-TiO2 and Ni-TiO2-Al2O3 coated hydro turbine steel under slurry erosion. Wear. 462–463, 203498 (2020). https://doi.org/10.1016/j.wear.2020.203498
63. K.G. Deepak, S. Harpreet, K. Harmesh, S. Varinder, Slurry erosion behaviour of HVOF sprayed WC-10Co-4Cr and Al2O3 + 13TiO_2 coatings on a turbine steel. Wear. 289, 46–57 (2012). https://doi.org/10.1016/j.wear.2012.04.016.
64. R. Singh, S.K. Tiwari, S.K. Mishra, Cavitation erosion in hydraulic turbine components and mitigation by coatings: Current status and future needs. J. Mater. Eng. Perform. 21, 1539–1551 (2012). https://doi.org/10.1007/s11665-011-0051-9.
65. U. Malik, I. Andijani, S. Ahmed, F. Al-Muaili, Corrosion and mechanical behavior of fusion bonded epoxy (FBE) in aqueous media. Desalination. 150, 247–254 (2002), https://doi.org/10.1016/S0011-9164(02)00980–3.

11 Surface Treatment Methods for Protection against Slurry Erosion in Marine Environments

Praveen Kumar Verma, Hitesh Vasudev, and Bhupinder Singh

11.1 INTRODUCTION

The functional integrity and lifespan of naval and submarine parts are severely jeopardized by intense erosive processes inherent to their in-service environments. Of these, slurry erosion, defined by the mechanical loss of material resulting from the impact of solid particles suspended in liquid media, is a leading mode of deterioration. In submerged and coastal applications, particularly those with high-speed flows and turbulent interactions with seawater laden with entrained sediments, parts like ship hulls, propellers, rudders, sonar domes, and pump housings experience constant erosive wear. Increased severity of slurry erosion is caused by a set of factors that include flow-induced pressure gradients, particle impact velocity, angle of impingement, and hardness of particles, as well as secondary mechanisms like cavitation that increases material loss due to cyclic micro-jet impingement and shock wave propagation. These processes build up into a synergistic degradation mode that is not only mechanically intensive but also chemically reactive as a result of the saline environment of the marine setting [1–4]. Figure 11.1 shows slurry erosion mechanisms, including suspended solids, high-pressure flow, and cavitation affecting submarine components like propellers.

Slurry erosion is especially a cause for concern in submarine environments, where stealth, reliability, and endurance in operational capability are paramount considerations. Submerged components are required to sustain structural and functional integrity over long periods under circumstances that rule out routine maintenance or component replacement. The erosive behavior of suspended particulate matter, from organic detritus to mineral abrasives like silica and alumina, poses an imposing challenge in the design and upkeep of marine-grade materials. Furthermore, in the case of erosion coupled with corrosion, particularly in seawater environments that are full of chlorides, it results in an even more aggressive mode of deterioration referred to as erosion–corrosion that cannot be reduced through basic design enhancements only. To overcome such adversity, the use of surface protective treatments has become a common solution. Among diverse methods, the use of advanced coatings appears to be a specific and efficient method for preventing slurry erosion. Protective coatings function as a sacrificial or hard barrier layer that separates the base material from erosive elements. Their performance is a function of a combination of hardness, toughness, adhesion strength, and corrosive species resistance. In addition, coatings can be designed to have a controlled microstructure and phase composition, which provides multifunctional protection against mechanical and chemical degradation. In submarine systems where the operational conditions are demanding and reliability is not an option, protective coatings are enablers of system long life and mission-readiness [5–7].

DOI: 10.1201/9781003561156-13

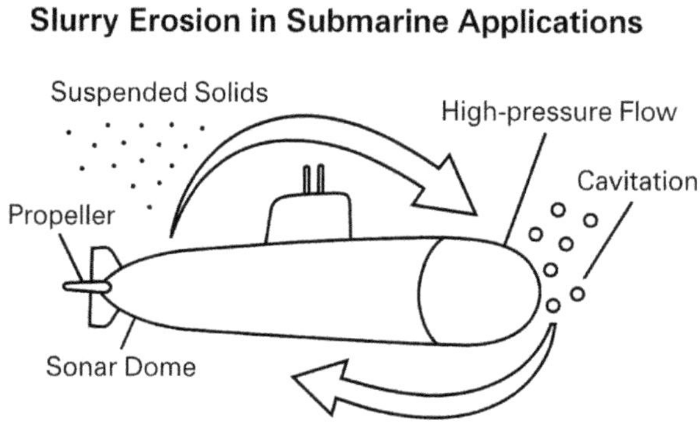

FIGURE 11.1 Schematic representation of slurry erosion in submarine applications.

High-velocity oxy-fuel (HVOF) thermal spray coatings have attracted a great deal of research and industry attention as a potential candidate to improve the erosion resistance of materials employed in marine applications. The HVOF technology allows the deposition of dense, well-bonded coatings with low porosity and high cohesive strength, attributes essential for withstanding the erosive action of slurry particles. By propelling powdered feedstock materials onto target surfaces utilizing supersonic gas streams, HVOF coatings are able to deposit high particle impact velocities and controlled heat input, resulting in fine-grained microstructures and reduced thermal degradation of the substrate. This is especially ideal for depositing hard carbide-based coatings like WC-Co, Cr_3C_2-NiCr, cermets, and even metal matrices reinforced with ceramics, all of which have superior wear and corrosion resistance [8–9].

The purpose of this review is to deliver an extensive evaluation of surface treatment approaches for the prevention of slurry erosion in submarine and marine environments, with a focus on application and performance using HVOF-sprayed coatings. Although numerous surface engineering techniques, ranging from plasma spraying to laser cladding, physical vapor deposition (PVD), and chemical vapor deposition (CVD), have been investigated for such applications, the emphasis here is on elucidating the mechanistic strengths and weaknesses of HVOF coatings in the context of their microstructural development, erosion–corrosion synergy, and long-term exposure in actual or simulated seawater environments. The review also discusses how coating composition, thickness, processing parameters, and post-treatment processes play a part in controlling the erosion resistance. Besides documenting cutting-edge advancements in HVOF technology, the chapter presents recent experimental and modeling strategies that have been utilized to predict and assess coating behavior under slurry erosion situations. There is a focus on case studies and comparative studies that compare the advantages of HVOF coatings to uncoated materials and other treatment methods. The purpose is to provide researchers, naval engineers, and surface technologists with a critical understanding of how to select, optimize, and apply HVOF coatings for marine durability [10–11].

Through this focused review, we seek to bridge the knowledge gap between material science innovation and practical marine engineering demands, thereby supporting the development of next-generation surface protection strategies tailored to the harsh realities of underwater operations.

11.2 MECHANISMS OF SLURRY EROSION IN SUBMARINE APPLICATIONS

Submarine slurry erosion is a multifaceted degradation process controlled by intricate interactions among hydrodynamic forces, particulate impact dynamics, and environmental parameters. Terrestrial erosion contrasts with the marine environment, where influencing variables such as salt corrosion, alternating hydrostatic pressure, and biological activity are introduced and compound to intensify the accelerated wear of structural components. It is necessary to understand the basic mechanisms of slurry erosion in such an environment for the formulation of effective mitigation methods, especially via innovative surface engineering processes like thermal spray coatings [12]. Slurry erosion primarily arises from the cyclic impingement of solid particles suspended in a carrier liquid, typically seawater, on material surfaces. In underwater systems, such particles can be derived from sediment-payload currents, resuspension of the seabed, or human activities like dredging and propulsive forces. When the fluid medium moves over underwater components (like rudders, propellers, and pump casings), suspended abrasive particles strike the surface with different intensities depending on local flow conditions.

The erosive wear mechanism is triggered by a series of micro-mechanical occurrences, such as local plastic deformation, surface microcracking, and final removal of the material fragments. The interaction depends mainly on the transfer of momentum from the particle to the surface, whose magnitude depends on the particle's mass and velocity vector at impact. In marine high-speed propulsion systems, for example, tangential and normal velocity components due to cavitation effects and turbulent flow increase the frequency and intensity of particle collisions, thus enhancing erosion intensity [13]. The slurry erosion rate and severity depend on three highly interactive variables: fluid velocity, particle impact angle, and particle properties (hardness, size, shape). Of these, velocity of the fluid has the greatest effect, being exponentially related to erosive wear rate. Empirical evidence indicates that the erosion rate is proportional to velocity to a power in the range of 2 to 4 depending on the erodent and erodent medium. In submarine thrusters or seawater intake systems, where the velocities can be several meters per second, even marginal increases in speed can result in disproportionately high wear rates [14]. Figure 11.2 shows the particle velocity, impact angle, and material response during slurry erosion, illustrating mechanisms of surface wear and erosion.

The angle with which particles impinge the surface also plays a significant part in identifying the wear mechanism. Ductile materials tend to show peak erosion at low angles (~15°–30°), when cutting and plowing are predominant. Brittle materials, on the other hand, experience more at impact angles of about normal or near-normal (~90°), where fracture and spalling are the dominant modes of wear. This angular dependence requires case-specific surface treatments depending on component material as well as operational orientation. Particle size and morphology also influence erosion dynamics. Larger particles possess higher kinetic energy and are capable of penetrating deeper and inducing more damage per hit. Angular or sharp-edged particles are especially injurious because they have a higher potential to nucleate micro-cracks. Smaller particles, although less damaging per event, are responsible for cumulative fatigue and roughening of the surface because of their high number density and incidence of interaction [16–17].

Material degradation in submarine environments is usually not the result of one specific cause. Instead, it is the combined effect of mechanical abrasion, electrochemical attack, and cyclic loading

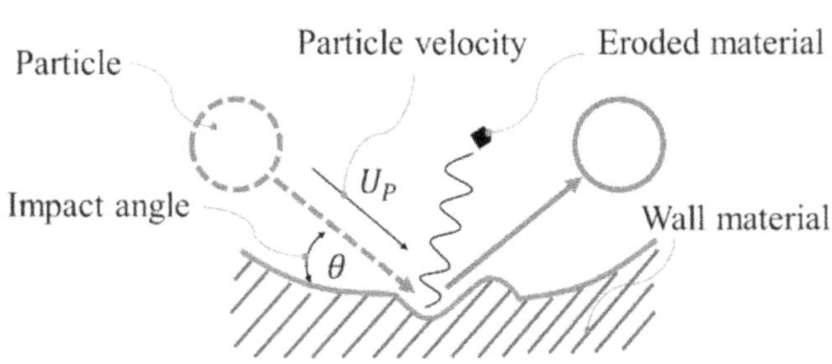

FIGURE 11.2 Schematic illustration of particle impact mechanics in slurry erosion [15].
(CC BY 4.0 attribution)

(fatigue) that results in increased material loss. In salt seawater, the passive oxide coating that normally protects metal surfaces is being constantly broken up by erosive particle impact, exposing fresh substrate to corrosive attack. This passivation–disruption cycle results in pitting and localized corrosion cells, which are sites for crack initiation and also stress concentrators under cyclic loading conditions [18]. Under cyclic loading conditions, repeated erosion–corrosion cycles cause marked reduction in fatigue life of materials in rotating machinery and high-pressure systems. Surface imperfections developed by erosion become micro-notches, increasing local stress and facilitating crack growth. Occasionally, the damage degrades to delamination or full material failure long before the nominal fatigue limit is attained. These synergistic factors require materials and coatings that are not only resistant to mechanical impacts but also retain chemical stability and toughness under dynamic loading conditions [19].

Submarine and marine environments bring with them a number of special environmental parameters that further affect slurry erosion behavior. One important factor that accelerates electrochemical reactions is salinity, which reduces the corrosion resistance of the vast majority of metal materials. The high chloride levels in seawater destabilize passive films on stainless steels and aluminum alloys, enhancing their susceptibility to erosion–corrosion [20]. Biofouling as a result of settlement of marine life like algae, barnacles, and mollusks changes local flow characteristics and can establish differential aeration cells, thereby enhancing corrosion phenomena. In addition, biofilm development can trap sediments and enhance localized erosion as a result of greater particle residence time at the surface [21]. Fluctuations in hydrostatic pressures found in deep-sea environments also affect slurry erosion through changes in cavitation behavior and affecting the properties of protective coatings. Higher pressure can result in more aggressive cavitation collapses, increasing the mechanical damage on exposed surfaces. It further influences the solubility of gases and diffusion kinetics, hence altering corrosion dynamics [22].

11.3 SURFACE TREATMENT METHODS FOR SLURRY EROSION PROTECTION

11.3.1 Conventional Methods

Traditional surface treatment methods have been the front line of protection against erosive wear in sea and submarine conditions for many years. Traditional methods, such as hardfacing, cladding, chemical passivation, and anodizing, are the usual methods used to enhance the service life of metallic parts through increased surface hardness, reactivity reduction, or protective barrier layer formation. But in their performance against slurry erosion in dynamic seawater conditions, their effectiveness is commonly restricted by aspects such as porosity, inadequate adhesion under hydrodynamic stress, and sensitivity to erosion–corrosion synergy [23–24].

Hardfacing and weld cladding processes entail the application of wear-resisting materials usually metallic or metal-ceramic composite onto the base material using welding processes like gas metal arc welding (GMAW), tungsten inert gas (TIG) welding, or shielded metal arc welding (SMAW). These overlays normally include hard phases like chromium carbides, tungsten carbides, or borides in order to improve wear resistance. Effective as they are in raising surface hardness, hardfaced layers tend to show high residual stresses and dilution effects as a result of welding thermal input. Metallurgical discontinuities at the coating-substrate interface occur, which can undermine integrity under cyclic loading or cavitation-induced fatigue conditions frequently experienced in marine propulsion and pump systems [25]. Chemical passivation, particularly in stainless steels, is used to create a thin, protective oxide surface layer, mainly chromium oxide. The passive film is a corrosion inhibitor, shielding against local attack in chlorides from seawater. These passive layers are mechanically vulnerable and are easily compromised by repeated particle impacts typical for slurry erosion. Localized corrosion accelerates once the passive layer is stripped, especially in crevices or geometrically intricate parts. Anodizing is another traditional process specifically applied to aluminum alloys, in which a porous, thick layer of oxide is electrochemically grown. Anodized coatings may provide modest corrosion and surface hardness enhancement. They do not possess the impact resistance and toughness needed for high-velocity slurry erosion conditions. Moreover, their porous nature acts as a chloride ion trap that will further enhance corrosion under submerged service [26].

Although heavily utilized, such traditional methods are severely lacking in extreme marine environments. The main disadvantages are insufficient resistance to erosion–corrosion–fatigue synergism, absence of specially designed microstructure for resisting impact, and poor adhesion under turbulent flow conditions. Further, most of these coatings are unsuitable for geometries of complex shape or for parts subject to high mechanical loads and alternating hydrostatic pressure. As a result of these shortcomings, attention has increasingly turned towards newer surface engineering technologies that provide greater performance, greater control over microstructural properties, and flexibility for application to a broad spectrum of material and shapes. These newer strategies, such as thermal spray coatings and surface alloying methods, offer a viable route towards designing durable surface protection systems that can withstand the hostile conditions in submarine service [27].

11.3.2 Advanced Surface Engineering Approaches

With the growing need for better performance, endurance, and reliability of underwater components in marine and subsea applications, new surface engineering techniques have become highly prominent. Such techniques are intended to overcome the limitations of traditional treatments by providing better mechanical properties, enhanced microstructural control, and improved corrosion-erosion resistance. Methods like laser cladding, plasma transferred arc (PTA) welding, cold spray deposition, sol-gel coatings, and thermal spray processes, especially high-velocity oxy-fuel (HVOF) spraying, are leading this technological advancement [28–29].

Laser cladding is the controlled deposition of metal or composite powders on a substrate by using a high-intensity laser beam as a heat source. Metallurgical bonding between the substrate and coating is made possible through this process, and dilution is avoided while there is control over the fine microstructure. The rapid solidification accompanying laser cladding results in coatings of improved hardness, low porosity, and high wear resistance. Additionally, the possibility to vary the composition of the clad material makes it amenable to adjusting erosion and corrosion resistance in extreme marine environments [30]. Plasma transferred arc welding is yet another sophisticated surface modification method that provides strong metallurgical bonding with high deposition efficiency. By employing a restricted arc plasma for melting and depositing erosion-resistant alloys like Ni-based or Co-based composites, PTA coatings may exhibit superior mechanical erosion and thermal shock resistance. The process flexibility enables gradient coatings, which can be tailored for optimum surface hardness and toughness of the substrate. Yet dilution and high thermal input are still difficulties in the case of applying PTA to thin-walled or heat-sensitive marine structures [31].

Cold spray technology is one of the solid-state deposition processes where powder particles are accelerated to supersonic speeds and plastically deformed during impact to create dense coatings with minimal thermal degradation. Cold spray technology is very promising for marine applications because it can deposit oxidation-sensitive material (such as Al, Cu, and Ti alloys) with little porosity and great corrosion resistance. However, the process is typically restricted to ductile materials and demands high kinetic energy for proper particle bonding [32]. Sol-gel coatings, though found comparatively less in structural applications, provide a cost-effective and environmentally friendly means of generating thin, consistent coatings with designed chemical functionality. The coatings can be structured to offer corrosion inhibition and mild erosion resistance using ceramic nanoparticles and corrosion inhibitors. However, their mechanical adhesion and strength are usually less than that which can be obtained by thermal spray or cladding techniques and are therefore restricted to secondary protective coatings or low-stress applications [33]. Among the thermal spray processes, high-velocity oxy-fuel spraying has proven to be the most promising method for protection against slurry erosion in submarine applications. HVOF coatings are developed by injecting the feedstock powder into a high-pressure combustion chamber where it is heated and accelerated to supersonic speeds before striking the substrate. The outcome is a dense, well-bonded coating having low oxide levels and small porosity. Coatings of WC-Co, Cr_3C_2-NiCr, and Fe-based amorphous alloys are frequently applied with HVOF because they are very hard, very tough, and highly resistant to corrosion [34]. The mechanical strength and chemical stability combined in HVOF coatings make them the perfect fit for components operating in particle-laden seawater flows. Their microstructure may be designed by modifying spraying parameters like flame temperature, spray distance, and powder morphology to fine-tune hardness, residual stress, and adhesion strength. In addition, HVOF coatings are highly performing under combined erosion–corrosion–fatigue conditions and thus form a desirable solution for essential marine systems [35].

11.4 HVOF-SPRAYED COATINGS

High-velocity oxy-fuel spraying is a thermal spray method that has received extensive consideration due to its outstanding performance under conditions experiencing harsh erosion, corrosion, and fatigue. The technology is especially beneficial in submarine use where parts are permanently exposed to particle-laden streams, saltwater environments, and varying mechanical loads. The HVOF process yields high-density, adhesion-strong, and low-oxide-content coatings, making it an indispensable surface engineering technology for the optimization of durability and service life of key submerged parts like propellers, pumps, rudders, and sonar domes [36]. The underlying principle of the HVOF spray is the combustion of a fuel gas (like hydrogen, propane, or kerosene) with oxygen in a high-pressure combustion chamber to produce a high-velocity, high-temperature gas stream. The feedstock powder is axially or radially injected into this stream, where it becomes partially or completely molten and is accelerated to supersonic speeds (up to 1000 m/s). When they strike the substrate surface, the particles' high kinetic energy enables deformation and mechanical interlocking, creating a dense, strongly adherent coating. The most important operating parameters controlling the process are fuel-to-oxygen ratio, combustion pressure, spray distance, and powder feed rate. Accurate control of these factors allows for the adjustment of coating microstructure and performance characteristics to particular environmental and mechanical needs [37]. Figure 11.3

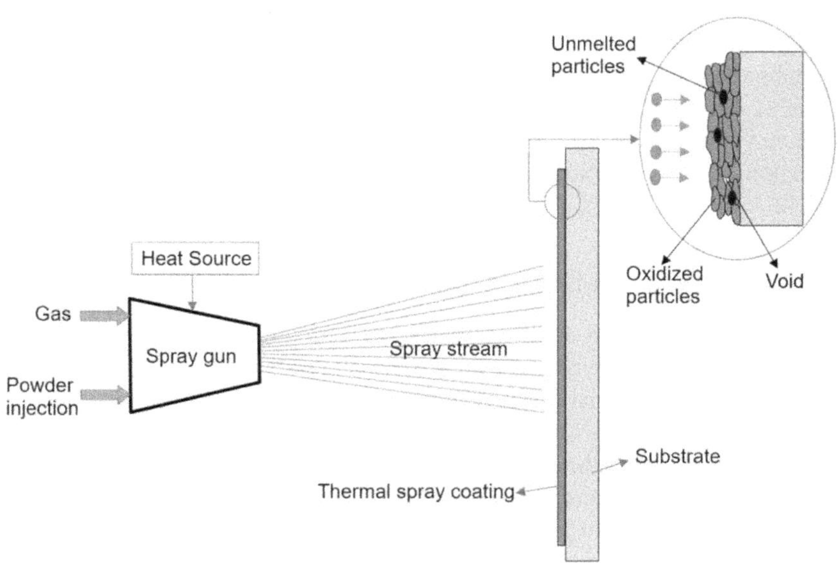

FIGURE 11.3 Schematic illustration of the thermal spray coating [38].

shows the thermal spray coating process showing feedstock injection, heat source, and deposition onto substrate for protective layer formation.

Compared to other thermal spray processes like wire arc spraying and atmospheric plasma spraying (APS), HVOF has some unique advantages. Plasma spraying, as a high-temperature process that is capable of melting high-temperature ceramics, tends to cause extensive oxidation of metallic phases because molten particles come into contact with ambient air, resulting in oxide-rich, porous coatings and poor inter-splat bonding. Wire arc spraying, while inexpensive and ideal for bulk applications, typically forms coatings that are more porous and less scratched because of the larger droplet size and reduced particle velocities involved. In contrast, HVOF coatings are formed with much higher particle velocities and reduced particle temperatures, which reduce oxidation and thermal degradation en route, thus forming coatings of excellent mechanical integrity and corrosion resistance [39]. The microstructural characteristics of HVOF-sprayed coatings are low porosity (characteristically $<1\%$), low oxide inclusions, refined lamellar structure, and high cohesive strength. Feedstock particle rapid solidification and deformation by impact induce a tightly packed dense coating with good interparticle bonding. This high-density microstructure not only improves the mechanical strength of the coating but also acts as a good barrier for the entry of corrosive species, like chloride ions in sea water. Additionally, the low oxide content guarantees the chemical stability of the coating in harsh marine environments, with no tendency for early degradation through galvanic action or pitting [40].

In submarine applications, these microstructural features equate to key functional benefits. HVOF coatings exhibit very high hardness, usually 1000–1500 HV for the material system (WC-Co, Cr_3C_2-NiCr). The very high hardness plays directly into their erosion resistance by minimizing the tendency towards plastic deformation and microcutting under high-velocity particle impact. The high adhesive and cohesive strength of the coating also leads to stability under varied mechanical loads and hydrodynamic forces encountered in submerged systems. Corrosion resistance is a second significant advantage of HVOF coatings in marine environments. The closed, low-porosity structure reduces paths for seawater intrusion, and the application of corrosion-resistant feedstock materials (Ni-based alloys, stainless steels, cermets) ensures long-term electrochemical stability. In service conditions such as seawater pumps and intake valves, in which erosion and corrosion operate synergistically, HVOF coatings have been found to minimize material loss rates considerably and maximize maintenance intervals [41]. In addition, the versatility of the HVOF process towards various metallic and composite materials for submarine applications is further promoted by its adaptability. Complex geometries and also large components are able to be coated with uniform thickness and microstructural consistency. The comparatively low process temperature relative to other thermal spray methods also reduces thermal distortion of the substrate, thus enabling it to be applied on dimensionally sensitive components [42].

11.5 MATERIALS USED IN HVOF COATINGS FOR MARINE USE

The behavior of HVOF-sprayed coatings in marine and submarine environments is intrinsically related to the choice of appropriate feedstock materials. The key desire in these applications is to increase surface resistance against a range of erosive, corrosive, and mechanical degradation processes. Materials employed for HVOF coatings are thus designed to possess high hardness, exceptional toughness, chemical inertness, and microstructural stability under marine operating conditions. Among the most heavily employed materials for sea HVOF coatings are tungsten carbide-based systems, chromium carbide composites, Ni-based superalloys, cermets, corrosion-resistant stainless steel powders, and nanostructured or functionally graded materials [43]. Figure 11.4 flowchart illustrating selection of HVOF coating materials (WC-Co/Cr_3C_2-NiCr) based on hardness, corrosion resistance, and marine operational conditions.

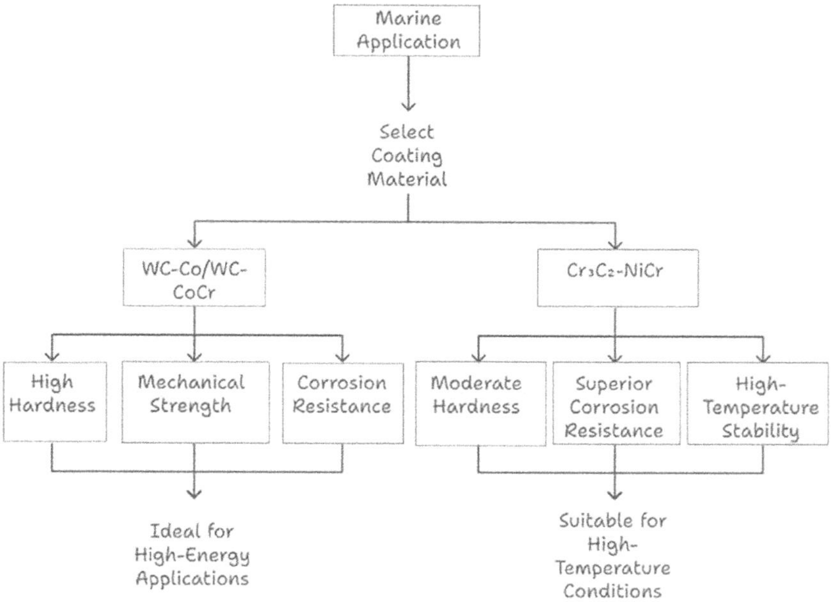

FIGURE 11.4 Decision flowchart for HVOF coating selection in marine applications.

Tungsten carbide-cobalt (WC-Co) and tungsten carbide-cobalt-chromium (WC-CoCr) systems are two of the most widely used cermet materials for protection against erosion in submerged environments. These coatings are described by their very high hardness (normally in excess of 1200 HV) and mechanical strength, both inherited from the hard WC phase and the tough metal binder matrix. A certain level of ductility and impact resistance as well as increased bonding strength between the carbide grains is achieved through the cobalt or cobalt-chromium matrix. The addition of chromium to WC-CoCr further improves corrosion resistance by stabilizing passive oxide films and minimizing galvanic corrosion susceptibility in seawater. The coatings are particularly ideal for high-energy applications like pump impellers, valve seats, and propeller blades, where particle-transporting flows impose heavy erosive stress. Their microcutting, plowing, and cracking resistance guarantee long-term integrity under cyclic mechanical and hydraulic loading [44]. Chromium carbide-nickel chromium (Cr_3C_2-NiCr) coatings are yet another group of high-performance HVOF coatings, widely applied in marine environments. They display a good combination of moderate hardness, superior corrosion resistance, and high-temperature oxidation stability. The wear resistance is conferred by the chromium carbide phase, while the NiCr binder phase improves ductility and achieves outstanding corrosion protection through formation of a protective chromium oxide layer in chloride-rich environments. Cr_3C_2-NiCr coatings are especially beneficial for use in conditions with high temperatures and alternating thermal loads, for instance, for steam turbines or heat exchanger components working within or close to marine environments. Cr_3C_2-NiCr coatings also provide sufficient resistance to erosion–corrosion synergy and contain less brittleness compared to WC-based coatings, thus ideal for components that experience repeated impact and bending stresses [45].

Nickel-based superalloys, including NiCrBSi, are commonly applied for corrosion and wear protection in naval systems as they possess high metallurgical compatibility with steel substrates and can deposit dense, well-bonded HVOF coatings. The incorporation of silicon and boron improves flowability during deposition and enables the development of hard intermetallic phases. Ni-based coatings also show excellent resistance to localized corrosion phenomena like pitting and crevice corrosion in saltwater environments and thus are found to be very effective for components submerged in seawater for prolonged periods of time. In addition, these coatings possess good tribological properties and may be applied to applications that involve both sliding wear and corrosion protection [46].

Ceramic-metallic composites, provide an even-handed solution to obtaining both high hardness and structural toughness. These materials generally consist of hard ceramic phases (like carbides, borides, or oxides) bonded with metallic binders (like Co, Ni, or Cr alloys), giving coatings that are resistant to abrasive and erosive wear but also have sufficient impact resistance. Cermets impart a synergistic benefit in marine use by resisting particle impingement while providing chemical passivity in seawater. The cermets' microstructural architecture enables optimization of particle size, binder chemistry, and phase distribution in order to achieve performance customization for any category of underwater mechanical loading [47].

Powders based on stainless steel, especially with high chromium, molybdenum, and nickel content, are employed in HVOF spraying to impart corrosion protection to structural parts exposed in marine environments. Although less brittle than carbide-based materials, stainless steel coatings provide better ductility, good adhesion, and good corrosion resistance through the establishment of stable oxide layers. Such coatings are frequently used on large or geometrical complex structures where toughness, machinability, and corrosion resistance must be balanced, like submarine hulls [48]. Nanostructured coatings and functionally graded materials (FGMs) are new approaches to improving HVOF coating performance in marine environments. Nanostructured coatings, made of ultra-fine grains or nanoparticle-reinforced matrices, have better hardness, lower porosity, and stronger bonding as a result of the high surface area of the reinforcement phases. Such coatings are best suited to withstand slurry erosion because of their excellent resistance to crack initiation and propagation. Functionally graded coatings, in contrast, offer a continuously changing composition or microstructure from substrate to surface. This gradient reduces thermal and mechanical mismatch, minimizing residual stresses, enhancing coating adhesion, and increasing durability under conditions of dynamic loading [49].

11.6 PERFORMANCE OF HVOF COATINGS UNDER SLURRY EROSION CONDITIONS

Testing of high-velocity oxy-fuel sprayed coatings under conditions of slurry erosion is a must for assessing their applications in marine and submarine environments, where exposure to continuous and aggressive particle-laden fluid flow is inevitable. Regularized test techniques like ASTM G75 and ASTM G76 are commonly used to evaluate coatings' erosion resistance under standardized laboratory conditions that mimic actual conditions of erosion. ASTM G75, the Miller slurry abrasion test, utilizes a steel rubber-lined wheel that turns against a specimen that is immersed in an abrasive slurry to provide testing for erosion under a low-velocity, high-abrasive-load condition. ASTM G76, on the other hand, assesses erosion resistance at high-velocity conditions through the application of a stream of abrasive particles onto the coated surface with a gas jet, thereby modeling impact erosion at different angles and particle velocities. Such standardized test procedures give significant insight into relative performance of coating materials as well as deposition methods [50–51].

Common performance metrics used to measure HVOF coatings' erosion resistance are mass loss, change in surface roughness, and degradation in microhardness. Mass loss in the units of mg or mg/cm^2 is a direct measure of material loss through slurry particle impingement. HVOF coatings like WC-CoCr and Cr_3C_2-NiCr always exhibit much lower mass loss when compared to the substrates or coatings that have been deposited using less energetic methods. Their enhanced erosion resistance is due to their high hardness, dense microstructure, and high interfacial particle–substrate adhesion. Roughness of the surface, usually determined by profilometry prior to and after testing, is an indication of the level of surface deformation and topographical alteration. Coatings with smoother surface conditions after erosion are less susceptible to hydrodynamic turbulence and cavitation phenomena, both of which worsen material loss in immersed parts. In addition, variations in microhardness as a result of cyclic particle impacts or heat effects can signal degradation of the coating structure or phase transformation. HVOF coatings are better at preserving their level of hardness compared to other weld-sprayed or thermally sprayed coatings because they have less oxide content and more stable phase structure [52].

The angle of impact of slurry particles is a very important aspect in controlling the erosion rate of coatings. At shallow angles (usually 15–30°), the erosion mechanism is microplowing and cutting, which prefers ductile materials. At steep angles (60–90°), brittle fracture and microchipping prevail, rendering hard, brittle coatings more prone to damage. HVOF coatings, especially WC-based coatings, show a semi-brittle character with balanced performance at varying impact angles. Optimization of binder level and particle size distribution of these coatings can further control their erosion characteristics to meet particular operating conditions. Hardness of erodent particles is another important factor determining coating performance. Hard particles, like silica or alumina, cause more wear by cutting and microfracture, particularly when they impinge at higher velocities. HVOF coatings with high-volume fractions of hard phases (WC or Cr_3C_2) exhibit excellent resistance to such stressful wear because they have high intrinsic hardness and can withstand localized plastic deformation [53].

Furthermore, temperature has a synergistic function in coating erosion performance. Parts may see higher temperatures in marine environments as a result of fluid friction, mechanical loading, or being near heat sources. High temperatures can soften and reduce the toughness of some coatings, increasing their erosion rate. Nonetheless, HVOF Cr_3C_2-NiCr coatings have been demonstrated to maintain structural integrity and phase stability at moderately high temperatures (up to 800°C), due to the oxidation resistance of the NiCr matrix and thermal stability of the carbide phase [54]. Figure 11.5 presents the

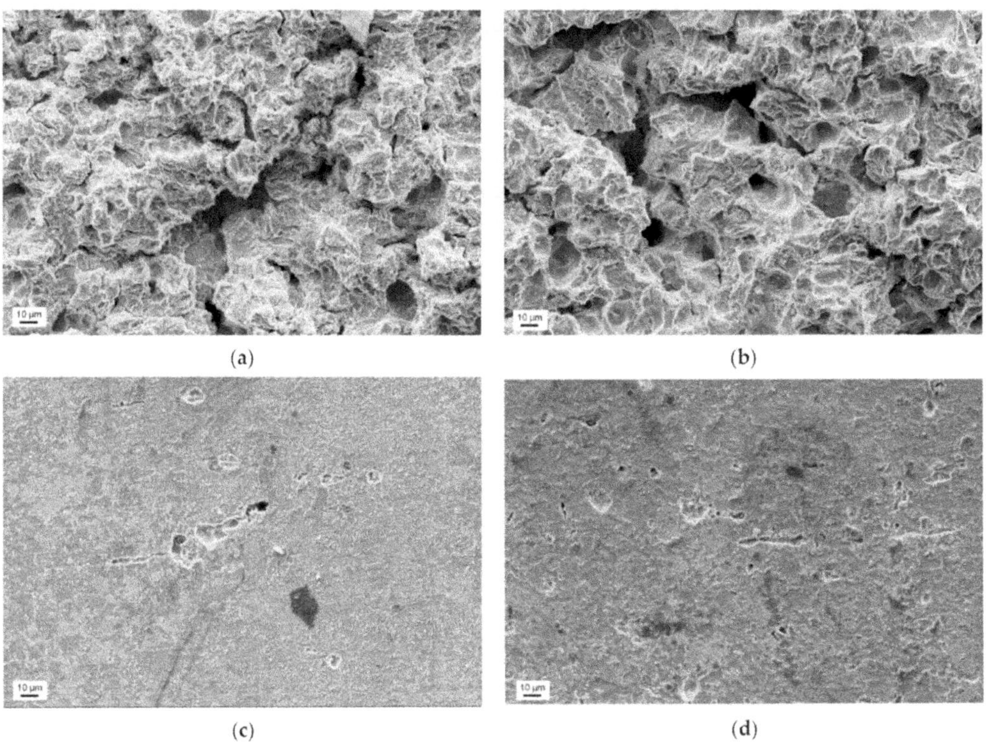

FIGURE 11.5 Cavitation Erosion Damage on S235JR-AR and S235JR-TT Steels [55].

surface morphologies of cavitation erosion damage observed on S235JR-AR and S235JR-TT steels subjected to a flow velocity of 35 m/s. Images (a) and (b) represent the high erosion areas of S235JR-AR and S235JR-TT steels, respectively, with extensive material loss, surface pitting, and plastic deformation. Pictures (c) and (d) show the low erosion regions of the same steels, with relatively less surface damage and fewer erosion pits. The microstructural contrast and heat treatment effect become evident in the different erosion behaviors, which emphasize the contribution of material condition to the resistance against cavitation under high-velocity flow conditions.

11.7 COATING DESIGN CONSIDERATIONS FOR SUBMARINE APPLICATIONS

The development of protective coatings for use in submarines necessitates a thorough knowledge of the environmental and operational issues confronting underwater components. Submerged surfaces are subjected to incessant contact with multifaceted synergistic degradation phenomena such as hydrostatic pressure, slurry erosion, corrosion by seawater, and biological colonization. Hence, coating engineering for such conditions must extend beyond mere selection of the material used and include structural integrity, resistance to corrosion, prevention of biofouling, and long-term maintainability. High-velocity oxy-fuel sprayed coatings, although providing considerable improvement in mechanical performance, need to be optimized and specified further through intelligent design approaches to meet the multi-faceted requirements of the submarine operating environment [56].

Adhesion and structural integrity in hydrostatic pressure is a key design criterion, especially for coatings deposited on parts working at very deep levels. The very high hydrostatic pressures that are found in deep-sea environments have the potential to create tensile stresses at the coating-substrate interface, which results in delamination or spallation if adhesion is poor. HVOF coatings possess inherent superiority in this respect because of their high kinetic energy deposition, thereby creating strong mechanical interlocking and, in certain situations, limited metallurgical bonding. These are, however, usually followed by surface pre-treatments in the form of grit blasting and application of bond coats (such as NiCr or NiAl) to further improve adhesion and minimize residual stresses. The coating thickness needs to be optimized as well to provide mechanical durability while allowing for flexibility in terms of substrate deformation without cracking during pressure cycling [57]. Figure 11.6 shows (a) a 3D schematic of a submarine showing rudder and dihedral coordinate systems affecting flow dynamics and (b) a 2D aerodynamic analysis of rudder angle, flap deflection, and resulting lift and drag forces.

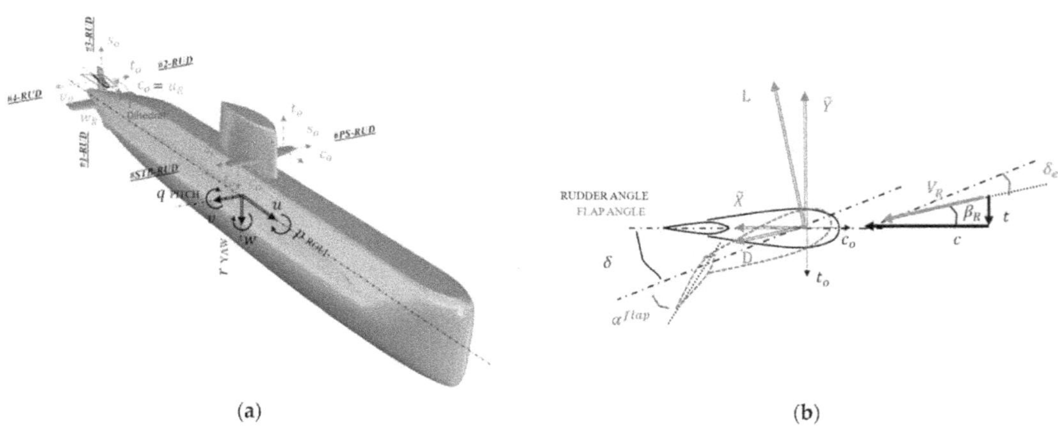

FIGURE 11.6 Hydrodynamic flow and rudder force analysis in submarine navigation [58].

Resistance to corrosion caused by seawater is also of similar importance, as chloride-containing marine environments favorably encourage initiation of localized corrosion mechanisms like pitting, crevice corrosion, and stress corrosion cracking. HVOF coatings based on corrosion-resistant matrices like NiCr, NiCrBSi, or stainless-steel–based powders exhibit superior passivation capability as a result of the development of adherent and stable oxide films. The addition of chromium and molybdenum to the binder phase enhances corrosion resistance even further by increasing the stability and self-healing capacity of these protective films. Porosity in the coating, an important microstructural variable, needs to be reduced to avoid electrolyte penetration and substrate corrosion. Dense microstructures attained with optimized HVOF parameters in this scenario significantly reduce corrosion pathways and increase component longevity [59].

Mitigation of biofouling is another critical element of coating design, particularly for extended-duration submarine and unmanned underwater vehicle (UUV) operations. Biofouling, which consists of the growth of microorganisms, algae, and invertebrates on immersed surfaces, results in added hydrodynamic drag, lower maneuverability, and increased fuel use. HVOF coatings themselves are not antifouling but can be tailored to discourage biological settlement. Strategies involve the use of topcoats filled with biocidal compounds (e.g., nanoparticles of silver or copper), hydrophobic or superhydrophobic treatments of surfaces, and incorporation of textured or patterned surfaces with reduced settlement affinity for fouling organisms. The incorporation of such antifouling features into HVOF coatings or as part of a multi-layer hybrid structure is a developing field of research in marine coatings science [60].

Hybrid coating systems consisting of HVOF-applied base layers with one or more additional sealing or functional layers have emerged as a primary method for optimizing overall performance in marine environments. Sol-gel–based sealers or polymeric topcoats are applied over HVOF coatings to seal residual porosity, diminish surface roughness, and deliver secondary protective functions like corrosion inhibition or antifouling behavior. For example, WC-CoCr or Cr_3C_2-NiCr coatings sprayed by HVOF can be coated with a ceramic sol-gel overlayer which not only reduces permeability to chloride ions but also offers a biofilm-resistant surface. These hybrid concepts allow the surface properties to be engineered while leaving the mechanical strength of the base HVOF layer intact. Additionally, modularity of such systems allows functional updates and targeted repairs to be made [61].

Repairability and life in service are paramount design considerations both from an operational and a financial perspective. Submarine parts are access-tough and expensive, and therefore need coating systems with extended service lifetimes and stable degradation behavior. HVOF coatings have high wear resistance and low erosion rates and thus provide extended service lifetime even under extreme operating conditions. In localized damage or wear cases, restoration methods like localized re-spraying or surface polishing can be used without the need for the complete replacement of components. Additionally, the ability of HVOF coatings to work with many substrate materials, such as stainless steel and nickel alloys, makes their integration into different component geometries and applications possible [62].

11.8 FUTURE DIRECTIONS

The evolution of protective coatings for submarine use stands to gain enormously from the incorporation of novel technologies that move beyond traditional material science paradigms. The most promising evolution is the creation of smart coatings tailored systems with the capability of autonomous response to environmental cues. These coatings can be integrated with self-healing mechanisms, which are often facilitated by microencapsulated healing agents or shape memory alloys and can be used to heal microcracks or localized damage caused by slurry erosion or mechanical fatigue. In addition, sensing functionalities embedded in coatings such as piezoelectric or conductive nanoparticles can provide real-time monitoring of stress, corrosion initiation, or coating deterioration, allowing for proactive maintenance in submerged systems [62].

At the same time, the use of machine learning (ML) and artificial intelligence (AI) for slurry erosion prediction is on the rise. Experimental parameter-driven models learned from data on parameters like impact angle, velocity, erodent size, and coating composition can predict erosion rates and failure modes quickly and decrease dependency on time-consuming physical testing. ML methodologies can even be used to find optimal HVOF process parameters for specific performance parameters like erosion resistance, hardness, and adhesion strength. The incorporation of such predictive models within design processes has the potential to greatly speed up the generation of high-performance coatings for marine conditions. Another area where future attention will be drawn is the introduction of in-situ monitoring and diagnosis within operational submarines. The integration of sensors into essential coated components enables dynamic monitoring of parameters like vibration, temperature, pressure, and erosion-induced mass loss. With a convergence of wireless communication and onboard data analysis, this strategy can revolutionize conventional maintenance practices to predictive and condition-based routines, thus improving operating reliability and minimizing downtime [63].

The ecological impact of coating process fabrication and application is under growing scrutiny. New coating technologies of the future will be required to conform to green manufacture techniques, focusing on the employment of non-toxic precursors, diminished energy usage, and recyclable or biodegradable elements. For this purpose, research into solvent-free sol-gel synthesis, cold spray methods, and environmentally friendly binders is holding out very attractive alternatives to conventional processes dependent on extensive thermal use and toxic chemicals. These innovations not only minimize the environmental footprint of maritime coatings but also enhance safety for manufacturing staff and advance regulatory compliance in vulnerable marine environments [64–65].

11.9 CONCLUSION

High-velocity oxy-fuel coatings represent a significant advancement over conventional surface protection methods in combating slurry erosion and corrosion in submarine environments. Compared to traditional techniques such as weld overlays, anodizing, and basic passivation, HVOF offers superior microstructural control, resulting in dense, low-porosity coatings with strong adhesion and minimal oxidation. These characteristics translate into exceptional resistance against mechanical wear, cavitation erosion, and aggressive chloride-induced corrosion, all of which are prevalent in marine settings. The ability to tailor the composition, such as the use of WC-CoCr, Cr_3C_2-NiCr, and nanostructured cermets, further enhances performance in harsh underwater conditions. However, the complexity of submarine environments demands tailored coating systems that extend beyond general-purpose erosion resistance. Submarine components are subject to unique mechanical stresses, fluctuating hydrostatic pressures, and biofouling phenomena that necessitate multifunctional coatings. Integration of antifouling properties, self-healing functionalities, and corrosion resistance within a single layered or hybrid coating architecture remains a critical challenge and an area of active investigation.

Future research should focus on validating HVOF coatings under simulated submarine conditions, employing standardized erosion and corrosion testing aligned with defense-grade specifications. Additionally, incorporating in-situ diagnostics and leveraging data-driven approaches such as machine learning for predictive modeling can greatly enhance design accuracy and operational reliability. A cross-disciplinary strategy involving material science, ocean engineering, and defense technology will be essential in realizing next-generation coatings that not only protect but also extend the functional longevity of strategic submarine systems.

REFERENCES

1. Neville A, Hodgkiess T, Wood RJK. A review of erosion–corrosion of engineering steels. Wear. 2001;249(10–11):144–153.
2. Stack MM, Jana S. Slurry erosion–corrosion of passive alloys: Synergism and effect of material parameters. Wear. 2002;253(1–2):229–236.

3. Shipway PH, Hutchings IM. Mechanisms of sliding wear in metallic materials. Wear. 1995;181–183:260–271.
4. Misra RD. Erosion–corrosion of engineering materials. J Mater Sci. 1991;26:3479–3488.
5. Matthews A, Leyland A. Engineering coatings for corrosion and erosion protection. Surf Coat Technol. 2003;163–164:1–7.
6. Pawlowski L. The Science and Engineering of Thermal Spray Coatings. 2nd ed. Chichester: Wiley; 2008.
7. Sidhu T, Prakash S, Agrawal RD. Studies on the properties of HVOF sprayed Cr_3C_2-NiCr coatings on Ni- and Fe-based superalloys. J Mater Process Technol. 2006;173(3):331–339.
8. Bolelli G, Lusvarghi L, Barletta M. HVOF-sprayed WC-CoCr coatings: Tribological behavior and surface properties. Surf Coat Technol. 2006;201(5):2998–3006.
9. Davis JR. Surface Engineering for Corrosion and Wear Resistance. ASM International; 2001.
10. Davis JR. Thermal spray technology. ASM Handbook. 2004;5:1–18.
11. Levy AV. The erosion–corrosion behavior of materials. J Mater Sci. 1986;21:1–12.
12. Finnie I. Erosion of surfaces by solid particles. Wear. 1960;3(2):87–103.
13. Meng H, Ludema KC. Wear models and predictive equations: Their form and content. Wear. 1995;181–183:443–457.
14. Hattori S, Ishigaki M. Erosion wear of metallic materials under slurry jet impingement: Influence of impact angle and velocity. Wear. 2001;250:861–867.
15. Na SJ, Kim YS, Jeon ES. Analysis of erosion minimization for a slurry pump using discrete phase model simulations. Applied Sciences. 2022 Feb 2;12(3):1597.
16. Springer GS. Erosion by solid particle impact. J Fluids Eng. 1972;94(4):563–571.
17. Bitter JGA. A study of erosion phenomena Part II. Wear. 1963;6(3):169–190.
18. Hack HP. Corrosion issues in military applications. Mater Perform. 1999;38(2):32–36.
19. Burstein GT, Liu C. Stress corrosion fatigue and synergy with erosion in marine environments. Electrochim Acta. 2007;52:5279–5290.
20. Landolt D. Corrosion and Surface Chemistry of Metals. CRC Press; 2007.
21. Delauney L, Compère C, Durrieu G. Biofouling protection for marine sensors. Ocean Sci. 2010;6(2):503–511.
22. Verma PK, Reddy BG, Gairola SU. Mechanical behaviour of ceramic coatings for enhancing the erosion resistance of turbine steels. In AIP Conference Proceedings 2024 Feb 20 (Vol. 2986, No. 1). AIP Publishing.
23. Saha GC, Khan TI. Characterization of HVOF sprayed WC-Co coatings and slurry erosion–corrosion performance. Surf Coat Technol. 2009;204(3):376–382.
24. Basak A, Prasad BK, Yegneswaran AH. Erosion behaviour of weld hardfacings: Influence of test parameters and base material. Wear. 1995;181–183:130–140.
25. Dey A, Das S, Ghosh A. A review on weld cladding by flux cored arc welding process. Mater Today Proc. 2020;26:952–957.
26. Kwok CT, Cheng FT, Man HC. Synergistic effect of erosion and corrosion of laser surface-modified stainless steel. Surf Coat Technol. 2003;166(2–3):221–229.
27. Brown SA, Lemons JE. An evaluation of passivation for implant alloys. J Biomed Mater Res. 1997;38(3):291–297.
28. Sharma V, Sidhu BS. Slurry erosion and corrosion behavior of conventional and advanced coatings: A review. Mater Res Express. 2021;8:082001.
29. Choi HJ, Lee CS, Kim SJ. Characteristics of laser cladding and HVOF coating for marine application. J Mater Sci Technol. 2005;21(6):803–808.
30. Choi H, Kim S, Shin C. Wear characteristics of PTA and HVOF coatings under erosive conditions. Surf Coat Technol. 2006;200(24):6659–6666.
31. Toyserkani E, Khajepour A, Corbin SF. Laser Cladding. CRC Press; 2004.
32. Davis JR. Welding, brazing, and soldering. ASM Handbook. 1993;6:789–807.
33. Champagne V. The Cold Spray Materials Deposition Process. Woodhead Publishing; 2007.
34. Hench LL, West JK. The sol-gel process. Chem Rev. 1990;90(1):33–72.
35. Sidhu T, Prakash S, Agrawal RD. Hot corrosion and performance of HVOF sprayed coatings. Surf Coat Technol. 2006;200(12–13):3755–3769.
36. Dey A, Das S, Ghosh A. HVOF sprayed cermet coatings for erosion–corrosion protection: Recent developments. Trans Indian Inst Met. 2022;75(7):1755–1765.

37. Pawlowski L. Thermal Spray Fundamentals. Springer; 2008.
38. Ramezani M, Mohd Ripin Z, Pasang T, Jiang CP. Surface engineering of metals: Techniques, characterizations and applications. Metals. 2023 Jul 20;13(7):1299.
39. Fauchais P, Vardelle A, Dussoubs B. Quo vadis thermal spraying? J Therm Spray Technol. 2001;10(1):44–66.
40. Pawlowski L. Suspension and solution thermal spray coatings. Surf Coat Technol. 2009;203(19):2807–2829.
41. Lima RS, Marple BR. Superior performance of HVOF-sprayed carbide coatings compared to plasma-sprayed counterparts. J Therm Spray Technol. 2007;16(1):40–47.
42. Espallargas N. Future directions in erosion-resistant coatings. In: Erosion by Particulates. Woodhead Publishing; 2020. p. 345–372.
43. Bolelli G, Candeli A, Lusvarghi L. Sliding and slurry erosion behavior of HVOF coatings. Wear. 2007;262(3–4):301–312.
44. Espallargas N, Fischer WJ. HVOF coatings for wear resistance in offshore systems. Surf Eng. 2010;26(6):404–412.
45. Matthews A. Coatings for extreme environments. J Phys D Appl Phys. 2003;36(10A):A137–A145.
46. Sahasrabudhe H, Bandyopadhyay A. WC-Co-Cr coatings for erosion resistance. J Mater Sci. 2014;49:6828–6836.
47. Bolelli G, Cannillo V, Lusvarghi L. Corrosion and erosion behavior of Cr3C2-NiCr coatings. Surf Coat Technol. 2006;201(5):2981–2990.
48. He JW, Liang W, Xu BS. Performance of NiCrBSi coatings in marine applications. Surf Eng. 2009;25(6):451–458.
49. Sampath S, Jiang X. Functionally graded thermal barrier coatings. Mater Sci Eng A. 1999;245(2):186–204.
50. Guilemany JM, Benito N, Cano IG. Stainless steel coatings via HVOF. Surf Coat Technol. 2006;200(14–15):4479–4485.
51. Chattopadhyay K. Nanostructured Materials: Processing, Properties and Applications. Elsevier; 2013.
52. ASTM G75–15. Standard Test Method for Determination of Slurry Abrasivity and Resistance of Materials to Abrasive Wear. ASTM International; 2015.
53. ASTM G76–13. Standard Test Method for Conducting Erosion Tests by Solid Particle Impingement Using Gas Jets. ASTM International; 2013.
54. Rao KP, Sundararajan G. Erosion behavior of HVOF coatings at elevated temperatures. Wear. 2001;249(10–11):763–774.
55. Krella AK, Zakrzewska DE, Buszko MH, Marchewicz A. Effect of thermal treatment and erosion aggressiveness on resistance of S235JR steel to cavitation and slurry. Materials. 2021 Mar 16;14(6):1456.
56. Arslan E, Totik Y, Durgutlu A. Effects of impact angle on erosion of thermal sprayed coatings. Mater Design. 2008;29(8):1595–1600.
57. Azarmi F, Coyle T, Mostaghimi J. Mechanical properties of HVOF thermal sprayed coatings. Surf Coat Technol. 2008;202(18):4473–4480.
58. Piaggio B, Vernengo G, Ferrando M, Mazzarello G, Viviani M. Submarine Manoeuvrability design: Traditional cross-plane vs. x-plane configurations in intact and degraded conditions. J Mar Sci Eng. 2022 Dec;10(12):2014.
59. Guilemany JM, Espallargas N. Thermal and corrosion performance of carbide coatings. Surf Coat Technol. 2006;201(5):265–272.
60. Dorfman M, Rybicki E. Thermal spray coating design for marine use. Naval Eng J. 2000;112(4):73–82.
61. Sun C, Wang Y, Zhang D. Adhesion of thermal sprayed coatings under pressure. Surf Eng. 2008;24(5):356–361.
62. Aruna ST, Mukasyan AS. Corrosion-resistant coatings in marine environment. Surf Coat Technol. 2009;203(24):3650–3660.
63. Magin CM, Cooper SP, Brennan AB. Non-toxic antifouling strategies. Biofouling. 2010;26(6):623–634.
64. Chen H, Li CJ, Ohmori A. Sealing effect on HVOF coatings. Surf Coat Technol. 2004;179(2–3):267–275.
65. Kumar P, Bansal A, Sidhu BS. Repair of worn turbine blades using thermal spraying. Mater Today Proc. 2020;26:2261–2268.

12 Post-Processing of Surfaces for Improving Their Performance against Slurry Erosion

Praveen Kumar Verma, Hitesh Vasudev, and Jürgen Markus Lackner

12.1 INTRODUCTION

Slurry erosion is a critical and persistent issue that affects a broad range of industries such as hydroelectric power generation, mineral processing, oil sands extraction, chemical slurry transport, and agriculture. It is mostly described by material removal from a solid surface caused by continuous impingement of high-speed abrasive particles contained within a fluid medium. It is a multifaceted tribological phenomenon that frequently causes enhanced deterioration; loss of efficiency; more frequent maintenance; and, in the worst-case scenario, failure of vital components such as pipelines, pump impellers, turbine blades, and valves. In contrast to traditional wear processes, slurry erosion constitutes concurrent interactions between solid particles, fluid flow, and surface properties of the material, thus calling for a multidisciplinary solution to alleviating its impact [1–3].

Slurry erosion is governed by a broad range of process parameters such as, but not limited to, particle size and concentration, fluid velocity, impact angle, temperature, and chemical makeup of the erodent and medium. On the material side, factors like hardness, toughness, ductility, microstructure, surface finish, and residual stresses significantly affect the amount and mode of erosion. Typically, high hardness and high interfacial bonding strengths result in resistance to micro-cutting and deformation behavior, and materials with good fracture toughness are more effective at withstanding the stresses generated by impact and arresting further propagation of the crack. Despite this knowledge, the intrinsic properties of bulk materials often fall short of meeting the demands of modern engineering systems operating in harsh erosive environments. To overcome this challenge, researchers and engineers have increasingly turned to surface engineering and post-processing techniques aimed at enhancing the surface properties of materials without compromising their bulk characteristics. Surface post-processing encompasses a range of physical, chemical, and mechanical treatments applied after initial manufacturing or forming processes. These treatments modify the surface layer in terms of hardness, phase composition, grain refinement, texture, residual stress, and morphology, which collectively influence the erosion resistance of the component [4, 5]. Figure 12.1 shows various parameters affecting slurry erosion, including impact angle, particle velocity, concentration, particle size, and fluid properties. Arrows and symbols depict their interactions with a surface, emphasizing how each parameter contributes to erosion damage and surface wear.

Among the most commonly employed post-processing methods are thermal spray coatings, laser surface modification, shot peening, surface alloying, and heat treatment. Each of these techniques offers unique advantages and can be tailored to specific operational conditions. For example, thermal spray coatings such as high-velocity oxy-fuel (HVOF) or plasma spray can deposit dense, wear-resistant layers of ceramics or cermets like WC-Co, Cr_3C_2-NiCr, and Al_2O_3-ZrO_2 composites. These coatings act as sacrificial barriers that absorb erosive impacts and prevent damage to the substrate. Similarly, laser surface treatment methods, including laser cladding and laser melting, can locally

DOI: 10.1201/9781003561156-14

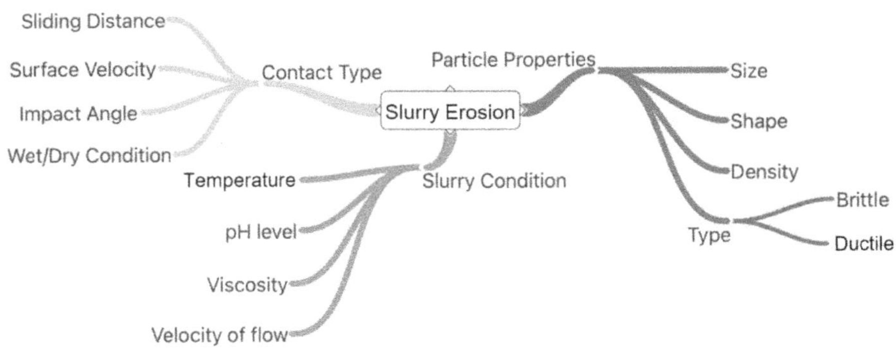

FIGURE 12.1 Different parameters involved in slurry erosion [6].
(CC BY 4.0 attribution)

modify the microstructure of metallic surfaces, resulting in grain refinement, phase transformation, and the introduction of compressive residual stresses that improve resistance to particle impingement [7, 8]. Shot peening, another widely used technique, improves surface toughness and delays the initiation of microcracks by inducing compressive stresses and surface hardening. On the other hand, surface alloying methods such as friction stir processing or tungsten inert gas (TIG) alloying allow for the incorporation of erosion-resistant elements like chromium, molybdenum, and boron into the surface matrix, forming hard intermetallic or carbide phases [9, 10]. Additionally, traditional heat treatment processes like carburizing, nitriding, and induction hardening, though historically used for enhancing wear resistance, are also being repurposed to improve slurry erosion performance by increasing surface hardness and modifying the microstructural stability under dynamic loading [11]. It can be mentioned here that the effectiveness of such post-treatments not only depends on material but also significantly depends on process parameters. Misuse can lead to issues such as coating delamination, development of residual tensile stresses, and undesirable phase formation, which can enhance erosion instead of preventing it. Therefore, accurate understanding of the interdependence of the surface treatment parameters, the developed microstructural properties, and erosion behavior is necessary for the effective application of such treatments in industrial processes [12].

In addition, advances in characterization methods such as scanning electron microscopy (SEM), atomic force microscopy (AFM), X-ray diffraction (XRD), and 3D profilometry have enabled thorough examination of post-treatment surface topography and erosion-induced damage. The methods provide information on wear mechanisms such as cutting, plowing, fatigue spallation, and delamination, each of which reacts differently to treated surfaces. This information is critical in the optimization of post-processing parameters to avoid certain erosion mechanisms based on the application. In recent years, there has been a move towards the integration of hybrid and smart surface modification technologies to meet the ever-increasing demands of performance. Hybrid processes combine two or more post-processing operations, for example, laser cladding with shot peening or thermal spraying with surface alloying in a bid to reap the synergistic benefit of two or more mechanisms. Likewise, with the availability of digital manufacturing and smart materials, researchers are studying self-healing coatings and surfaces with adaptive response to dynamic erosive conditions [13–15].

This chapter aims to provide a critical overview of surface post-treatments with a special focus on the improvement of slurry erosion resistance. It will outline the scientific basis, selection criteria of the material, process parameters, and performance testing of such treatments. Case studies and experimental findings from the literature will be summarized to illustrate the practical utility and limitations of each method. To conclude, future directions in the area, such as the integration of additive manufacturing, functionally graded materials (FGMs), and smart coatings, will be discussed to provide an overview to next-generation surface engineering techniques against slurry erosion.

12.2 FUNDAMENTALS OF SLURRY EROSION

12.2.1 Mechanism of Slurry Erosion

Slurry erosion is a multifaceted phenomenon that results when an aggregate of liquid and solid particles, referred to as a slurry, strikes a surface and causes material deterioration. The phenomenon is common in most industrial processes like mining, oil and gas, and chemical processing where process equipment and pipelines are regularly subjected to abrasive slurries. The slurry erosion mechanism consists of a chain of physical and chemical reactions that lead to gradual removal of material from the surface [16]. The main slurry erosion mechanism is the repetitive impact of abrasive particles on the surface. The impact can lead to various modes of wear, such as plastic deformation, micro-cutting, pitting, and material removal. Each of these mechanisms of wear is affected by a number of factors, which are important in the understanding and prevention of slurry erosion. The angle and velocity at which particles strike the surface are of major importance in the erosion process. More energetic impact velocities tend to cause greater erosion through the higher kinetic energies of the particles. This energy is imparted to the surface on striking, producing greater plastic deformation and material removal. Angle of impact is just as critical. Normal impacts (90°) produce more material removal and plastic deformation, while oblique impacts (less than 90°) may produce more cutting and scratching. The best angle for the highest erosion is generally between 30° and 45°, where cutting and deformation are combined to the best effect [17–19]. Figure 12.2 shows how impact angle and velocity affect slurry erosion. Particles are shown impacting on a surface at different angles and speeds, and variations in these factors influence material loss, severity of erosion, and patterns of surface deterioration.

The angularity and hardness of the abrasive particles also play important roles in the erosion process. Angular and sharp particles are more efficient in scratching and cutting the surface and, in turn, causing higher erosion rates. Round or spherical particles cause more plastic deformation but less material loss. The relative hardness of the particles to the surface material is also an important consideration. Harder particles can cause more severe damage, especially if they are harder than the surface material. For example, quartz particles are much harder than most metals and can cause extensive erosion at low velocities [20]. The surface roughness and

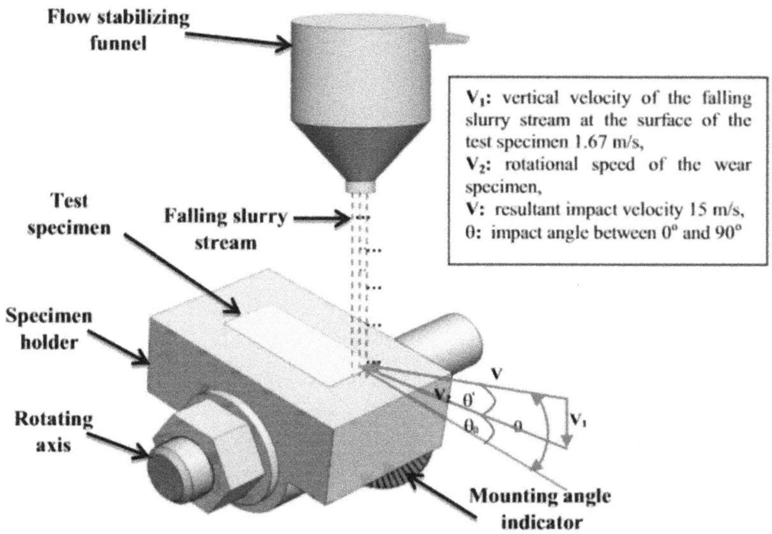

FIGURE 12.2 Schematic diagram of the impact angle and impact velocity [19].

surface hardness of the material are also important considerations. A rough surface can provide more points of contact for the abrasive particles and, therefore, higher chances for material loss. Alternatively, a smooth surface can reduce the contact area and, therefore, the erosion rate. The hardness of the surface material is also important. Harder materials are less sensitive to plastic deformation and material loss and are, therefore, better suited for the conditions of slurry erosion. However, even hard materials can be eroded if the particles are hard enough and the impact velocity is high enough [21]. The presence of corrosion can enhance the effect of slurry erosion significantly, producing a synergistic effect. Corrosion makes the surface material softer and more susceptible to erosion. It can be achieved through the creation of pits or cracks, where more attack sites are provided to abrasive particles. The corrosive condition can also accelerate the stripping of material chemically, which further enhances the erosion process. The synergistic effect is of significant interest in conditions where erosive and corrosive conditions are present simultaneously, such as in the case of the oil and gas industry [22].

Slurry erosion is a multifaceted mechanism that is dependent on velocity and angle of impact, particle shape and particle strength, surface strength and surface roughness, and corrosion. All of these factors need to be understood in an attempt to create materials and systems capable of withstanding slurry erosion conditions.

12.2.2 Material Behavior under Slurry Erosion

The behavior of materials under slurry erosion is a critical aspect of material selection and design in industries where such conditions are prevalent. Materials with high hardness, fracture toughness, and ductility generally perform better in slurry erosion environments. However, the surface conditions of the material are equally important, and post-processing techniques can be employed to tailor the material's response to erosion [23]. Figure 12.3 showing how particle size, concentration, fluid velocity, and material properties such as hardness and toughness influence slurry erosion by affecting impact energy, collision rates, and material resistance.

High-hardness materials are more resistant to material removal and plastic deformation. Hardness refers to a measure of the ability of a material to withstand scratching and indentation, which are typical modes of slurry erosion. Tungsten carbide, ceramic coatings, and hardened steels, for

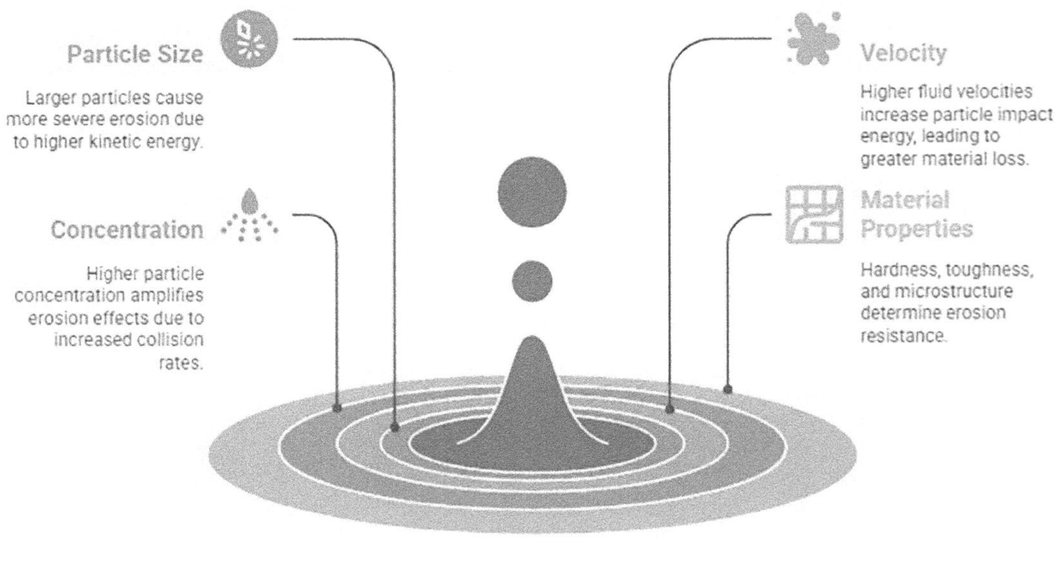

FIGURE 12.3 Key factors influencing slurry erosion mechanisms in engineering materials.

instance, are frequently applied in erosive conditions because they have high hardness. These materials will withstand repeated abrasion by hard particles without getting deformed significantly, hence minimizing material loss rate. But mere hardness is not the solution, as there are other factors such as fracture toughness and ductility that are equally important [24]. Fracture toughness refers to the property of a material to absorb energy when it's subjected to loading with no catastrophic fracture and to resist the growth of cracks. In slurry erosion, repeated particle impacts can lead to the development of micro-cracks on the surface. Fracture-tough materials can increase the resistance of the surface to crack growth and propagation, which ensures the integrity of the surface. This is especially the case in conditions where the particles are rigid and the impact velocities are high. Materials with high fracture toughness like certain stainless steels and composite materials are usually used in such applications [25].

Ductility is the capacity of a material to be deformed by stress without breaking. For slurry erosion, ductile materials are able to absorb the impact energy of particles by plastic deformation, which can act to decrease the extent of damage. Although this can result in surface roughening, it can avoid the creation of severe pits and cracks that are harder to restore. Ductile materials such as some alloys and polymers are commonly utilized in applications where the expected erosion is moderate, and there can be periodic replacement or repair of the material [26]. The surface conditions of the material are just as important in establishing the performance of the material under slurry erosion. A surface that is smooth will minimize the contact area between the particles and the material, thus minimizing the rate of erosion. Surface roughness may allow more points of attack for the abrasive particles, causing greater material removal. Surface post-processing methods like polishing, grinding, and coating are hence typically used to enhance the surface finish and lower the erosion rate [27].

12.3 SURFACE POST-PROCESSING TECHNIQUES

12.3.1 Thermal Spray Coatings

Thermal spray technologies, particularly high-velocity oxy-fuel, atmospheric plasma spray (APS), and cold spray, represent a critical class of surface post-processing techniques engineered to enhance the slurry erosion resistance of metallic substrates. These methods enable the deposition of refractory, high-hardness coatings through the propulsion of feedstock materials either in powder or wire form onto a prepared substrate at elevated velocities and controlled thermal profiles. Among the most efficacious coatings in slurry erosion environments are tungsten carbide-cobalt (WC-Co), chromium carbide-nickel chromium (Cr_3C_2-NiCr), and various oxide ceramics including alumina (Al_2O_3) and stabilized zirconia (ZrO_2-Y_2O_3). These materials are selected due to their inherent high hardness, superior fracture toughness, and thermal stability, which are indispensable for resisting erosive wear in dynamic, particulate-laden flow regimes [28].

The fundamental erosion resistance of these coatings arises from the formation of a dense, low-porosity lamellar microstructure capable of absorbing and dissipating the kinetic energy imparted by solid particle impacts. In particular, HVOF-sprayed WC-Co coatings exhibit exceptional erosion resistance due to their high coating density (>98%), fine carbide dispersion, and robust interfacial bonding between the matrix and hard phases. Cr_3C_2-NiCr coatings, meanwhile, provide not only excellent mechanical durability but also enhanced oxidation and corrosion resistance, making them particularly suited for erosive environments involving reactive media or elevated temperatures [29]. Ceramic coatings such as Al_2O_3 and ZrO_2 offer extreme hardness (up to ~2000 HV), high elastic modulus, and chemical inertness, which collectively contribute to their effectiveness in abrasive-dominated slurry erosion scenarios [30]. Figure 12.4 illustrates six sequential stages in the thermal spray process: melting material, spraying process, impact and solidification, microstructure formation, coating property development, and final performance optimization for enhanced erosion resistance in engineering applications.

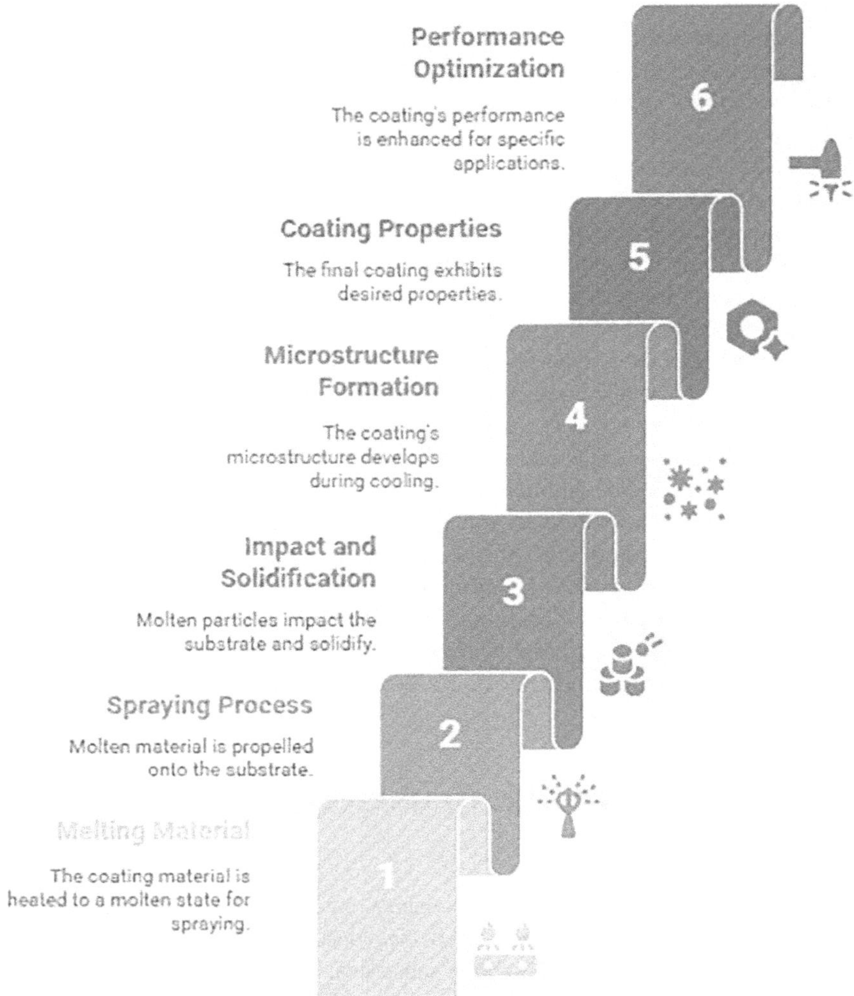

FIGURE 12.4 Stages of thermal spray coating deposition and performance enhancement.

The advantages of thermal spray coatings are manifold: their microstructure and phase composition can be finely tuned by adjusting process parameters such as feedstock velocity, combustion gas ratio, standoff distance, and substrate temperature. This allows for the tailoring of coatings with variable thicknesses (ranging from tens of microns to several millimeters), hardness, and toughness, based on the specific erosive challenge. Furthermore, these coatings can be applied to a wide variety of substrates, including stainless steels, nickel-based alloys, and titanium, provided adequate surface roughening (grit blasting) is performed to ensure mechanical interlocking and adhesion. Cold spray, which operates at lower temperatures to avoid thermal degradation of feedstock powders, also facilitates the deposition of ductile metal-based erosion-resistant coatings with minimal oxidation and negligible phase transformation [31].

12.3.2 Laser Surface Treatment

Laser surface treatment techniques, encompassing laser surface melting (LSM) and laser cladding (LC), have emerged as precise and highly controllable methods for enhancing the surface integrity of materials subjected to aggressive slurry erosion environments. These processes utilize high-energy laser beams to induce localized surface transformations or to metallurgically bond a secondary erosion-resistant material onto a substrate, thereby significantly improving wear resistance, hardness, and surface cohesion. The unique attributes of laser-material interaction—high energy density, rapid solidification rates, and minimal thermal diffusion—enable the formation of refined microstructures with superior mechanical performance. Laser surface melting involves remelting the existing surface layer without adding external material. This technique refines the grain structure via rapid solidification and directional solidification mechanisms, often resulting in equiaxed or columnar grains with reduced intergranular spacing, which contributes to improved hardness and erosion resistance. Furthermore, LSM can eliminate surface porosities, reduce pre-existing microstructural defects, and introduce compressive residual stresses due to differential cooling rates between the surface and subsurface layers factors critical in retarding crack initiation and propagation under repeated particle impacts [32, 33].

On the other hand, laser cladding incorporates a high-performance alloy, such as Inconel, Stellite, or Co-Cr-Mo, onto the substrate surface via simultaneous laser irradiation and feedstock powder injection. The localized melting and subsequent rapid solidification ensure a metallurgical bond with minimal dilution from the substrate. The resulting clad layer exhibits superior mechanical properties such as high microhardness (>700 HV), improved toughness, and enhanced corrosion resistance. The refined dendritic microstructure formed in the clad zone, coupled with hard intermetallic phases or carbides (e.g., Cr_7C_3, Co_3W_3C), significantly contributes to erosion resistance under both normal and oblique particle impact conditions [34].

One of the key benefits of laser-based treatments is the limited heat-affected zone (HAZ) due to the rapid heating and cooling cycles, which minimizes thermal distortion and preserves the mechanical properties of the bulk substrate. Moreover, enhanced metallurgical bonding at the interface, as opposed to mechanical interlocking typical of thermal spray techniques, ensures high adhesion strength, which is vital in preventing delamination during erosive loading. Additionally, laser systems offer precise control over process parameters, such as laser power, scanning speed, beam diameter, and powder feed rate, allowing tailored control over the degree of dilution, clad geometry, and final microstructural characteristics [35].

12.3.3 Shot Peening

This method involves bombarding the surface with small, high-velocity particles to induce compressive stresses in the surface layer. These compressive stresses can help to prevent the formation and propagation of cracks, thereby improving the material's resistance to erosion. Shot peening is a widely recognized mechanical surface treatment technique employed to significantly enhance the fatigue life, erosion resistance, and mechanical integrity of engineering components subjected to cyclic and impact loading, such as those exposed to slurry erosion. The process involves bombarding the material surface with high-velocity spherical media commonly made from steel, ceramic, or glass, causing localized plastic deformation. This deformation induces a layer of compressive residual stresses in the near-surface region, which serves as a formidable barrier to crack nucleation and propagation, particularly under erosive particle impact conditions [36]. Figure 12.5 shows the laser shock peening process where high-energy laser pulses induce compressive stresses to enhance surface properties.

The underlying mechanism by which shot peening enhances erosion resistance lies in the induced compressive stress field, which counteracts the tensile stresses generated by high-velocity particle impacts in slurry erosion environments. Since slurry erosion typically initiates through micro-crack formation followed by progressive propagation due to repetitive erosive impacts, the presence of compressive stresses effectively retards this damage mechanism by reducing the stress intensity factor at surface flaws. This delay in crack formation and growth translates to increased resistance to material removal and longer service life of components exposed to erosive wear [38]. Furthermore, shot peening results in grain refinement and strain hardening of the surface layer, leading to

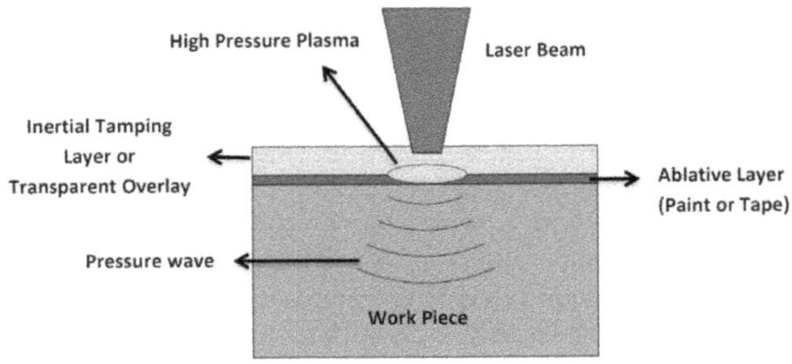

FIGURE 12.5 Schematic of laser shock peening (LSP) process [37].

an appreciable increase in micro-hardness. This hardened layer, often extending up to several hundred micrometers in depth depending on process parameters such as shot size, velocity, and duration, exhibits improved resistance to surface deformation and micro-cutting by abrasive particles. Additionally, the process modifies the surface roughness, which, while potentially increasing the real contact area between the surface and erodent, can also enhance the mechanical interlocking of protective coatings when used as a pre-treatment method [39].

Shot peening has found extensive applications in industries where slurry erosion is a persistent problem. In aerospace, the leading edges of turbine blades—especially in environments involving ingestion of particulates—are commonly peened to enhance erosion resistance and delay fatigue cracking. In the hydraulic and mining sectors, pump impellers and steel pipeline elbows, which routinely handle abrasive slurries, benefit from shot peening through reduced erosion rates and extended operational lifespans. The versatility of the technique allows for its application on complex geometries and a wide range of materials, including high-strength steels, titanium alloys, and Ni-based superalloys [40]. Importantly, the effectiveness of shot peening is contingent upon meticulous control of process variables to avoid over-peening, which can lead to surface damage or stress relaxation. Advanced techniques such as computer-controlled shot peening and laser shock peening have been developed to further improve precision and control over stress distribution profiles [41].

12.3.4 Surface Alloying and Cladding

Surface alloying and cladding are advanced post-processing methodologies aimed at tailoring the near-surface chemical and microstructural characteristics of metallic substrates to enhance their resistance against severe slurry erosion environments. These methods, including tungsten inert gas surface alloying, friction stir processing (FSP), and electron beam cladding (EBC), introduce alloying elements or reinforcement phases into the substrate surface through controlled thermal or thermomechanical interaction, facilitating the in-situ synthesis of erosion-resistant intermetallics and carbides [42]. TIG surface alloying, a fusion-based technique, utilizes a non-consumable tungsten electrode under an inert shielding atmosphere to create a localized molten pool, into which elements such as chromium (Cr), molybdenum (Mo), boron (B), and titanium (Ti) are incorporated. These elements can chemically interact with the substrate or additional feedstock materials to form hard secondary phases like Cr_7C_3, Mo_2C, TiC, and various borides. The solidification microstructure is typically refined due to high cooling rates and thermal gradients, yielding a dendritic or cellular morphology with enhanced microhardness and resistance to erosive degradation. The mechanical interlocking and metallurgical bonding produced through this process drastically improve load-bearing capacity under high-velocity slurry impact [43].

Friction stir processing represents a solid-state alternative where a rotating tool plastically deforms the surface without reaching the melting point, promoting dynamic recrystallization and grain refinement. When reinforced particles or alloying powders are introduced, FSP enables their uniform dispersion within the plastically deformed matrix, resulting in a composite surface layer with superior hardness and wear resistance. The absence of melting circumvents the formation of undesirable porosity or dilution effects commonly encountered in conventional fusion-based techniques, thereby ensuring structural integrity and performance reliability [44]. Electron beam cladding, conducted under high-vacuum conditions using a focused electron beam, offers exceptional control over thermal input and energy localization. It permits deep penetration and rapid solidification, leading to dense, metallurgically bonded clads with finely distributed carbides and intermetallics. Materials such as Ni-Cr-B-Si, Co-based alloys, and TiC-reinforced matrices have been successfully cladded using EBC to improve erosion resistance by forming high-hardness surface architectures with minimal dilution [45].

12.3.5 Heat Treatment

Heat treatment is one of the significant material science processes, which is primarily employed for enhancement of the bulk material properties, that is, strength, hardness, and ductility. It also has significant application in modifying the surface hardness gradient, which significantly enhances the erosion resistance of the material, particularly under slurry erosion. Carburizing, nitriding, and quenching are the techniques commonly employed for these surface modifications, which have their own advantages and disadvantages [46].

Carburizing is a thermal treatment process where carbon is diffused into the surface of a low-carbon steel to form a high-carbon case. The process surface hardens, and it becomes erosion and wear resistant. The depth of the carburized layer, or the case depth, is a very important parameter. A deeper case can provide higher erosion resistance but can also increase the risk of residual stresses and even cracking. The best-case depth is application and erosion type dependent. For instance, in slurry erosion, a case depth of 0.5 to 1.0 mm is generally sufficient to provide a substantial improvement in erosion resistance. In addition, microstructural stability in the carburized layer is required. Post-treatments of carburizing like quenching and tempering can stabilize the microstructure and reduce the risk of thermal fatigue in cyclic service. Thermal fatigue is a widespread issue in components under cyclic heating and cooling, and a stable microstructure can alleviate the issue [47].

Nitriding is another process of heat treatment wherein nitrogen is diffused into the surface of a steel or similar metal to form a hard, wear-resistant surface. In contrast to carburizing, which adds primarily carbon, nitriding forms a nitride surface layer with the ability to substantially increase surface hardness and wear resistance. Case depth in nitriding is typically lower than for carburizing, from 0.1 to 0.5 mm, but can be of considerable advantage in erosion resistance. Microstructure of the nitrided surface is also important. Nitriding can form a compound layer of iron nitrides, which can be very hard but brittle. Underneath this compound layer, a nitrogen-rich diffusion zone can provide a combination of hardness and ductility that renders the material resistant to wear and thermal fatigue. Stability of this microstructure under cyclic loading conditions is important, and post-nitriding treatments, such as tempering, can be employed to prevent the material from degrading over time [48].

Quenching is a heat treatment involving the rapid cooling of a material from elevated temperature to produce a hard, martensitic microstructure. Quenching can significantly increase the surface hardness and erosion resistance of a material. Quenching can produce residual stresses that will cause cracking and other modes of failure if not properly controlled. Case depth, or hardened layer depth, is important. A shallow case depth will provide excellent surface hardness with low residual stresses, but a higher case depth will provide better total erosion resistance. Microstructural stability of the material that has been quenched is important. Quenching treatments after the quench, that is, tempering, can reduce the residual stresses, stabilize the microstructure, and reduce the risk of thermal fatigue under cyclic environments [49].

In the use of heat treatments for enhancing erosion resistance, there are several considerations that need to be satisfied to realize the optimum performance. Case depth is one of these, as it determines the depth of surface modification and erosion resistance of the material. A hard case could be more protective but could also result in residual stresses that lead to cracking. Therefore, the case depth must be well controlled to balance hardness and structure [50]. Microstructural stability is also a significant consideration. Microstructure in the heat-treated layer must be stable under service conditions of the component, that is, temperature cycling and mechanical loading. For example, in carburized steels, formation of stable martensitic structure is very critical in order to maintain hardness and wear resistance. In nitrided materials, stability of the iron nitride compound layer and the diffusion zone must be ensured to prevent brittleness and cracking [51]. Resistance to thermal fatigue under cyclic loading conditions is of special interest in components that are subjected to repeated heating and cooling cycles. Thermal fatigue can lead to micro-cracking and subsequent failure of the material. Heat treatments that favor a stable microstructure and minimize residual stresses can minimize thermal fatigue. For example, tempering after quenching can minimize residual stresses and improve the ductility of the material, thus making it less susceptible to thermal fatigue [52].

12.4 PERFORMANCE ASSESSMENT AND TESTING METHODS

Assessment of surface integrity and functional performance of post-processed material with respect to slurry erosion requires the use of sophisticated microstructural, crystallographic, mechanical, and topographical characterization methods. Diagnostic methods are helpful in relating process-induced changes to erosion resistance and general surface durability. Scanning electron microscopy is a key tool for delineating the surface morphology, failure modes, and phase distribution at micro- and nanoscales. SEM allows for high-resolution imaging of erosion-induced craters, microcracks, delamination regions, and embedded slurry particulates. SEM also allows for elemental analysis by energy dispersive x-ray spectroscopy (EDS), which is commonly built into SEM platforms to determine compositional gradients and the distribution of alloying elements or hard phases over the erosion zone [53]. X-ray diffraction (XRD) is used to identify the phase constitution, crystallographic orientation, and residual stresses in the surface layers that have been modified. The non-destructive method detects the presence of favorable intermetallics (C, Cr_7C_3, Ni_3Al) or ceramic phases (Al_2O_3, ZrO_2) after processing. Peak broadening and shifts can also yield information on microstrain and lattice distortion caused by surface treatments, which act directly to impact erosion resistance through increased dislocation density and defect hardening [54]. Microhardness testing, often conducted with Vickers or Knoop indenters under controlled loads, is crucial for determining the local hardness profile of treated layers. The hardness gradient through the cross-section is a measure of the depth of hardening, grain refinement, or secondary phase dispersion. Increased microhardness is most commonly related to enhanced resistance to erosive wear by allowing for increased capacity to absorb and dissipate impact energy [55].

3D surface profilometry, either optical or stylus, gives quantitative surface roughness, waviness, and volumetric material loss information. Topographical maps produced by profilometry provide an overall picture of surface degradation after erosion and are critical when quantifying wear rates and comparing various surface treatments in a statistically significant manner [56].

12.5 ADDITIVE MANUFACTURING INTEGRATION

The integration of post-processing strategies with additive manufacturing (AM) technologies presents a transformative approach for enhancing the slurry erosion resistance of complex geometries, particularly in components used in aerospace, energy, and marine applications. AM techniques, such as selective laser melting, electron beam melting (EBM), and directed energy deposition (DED),

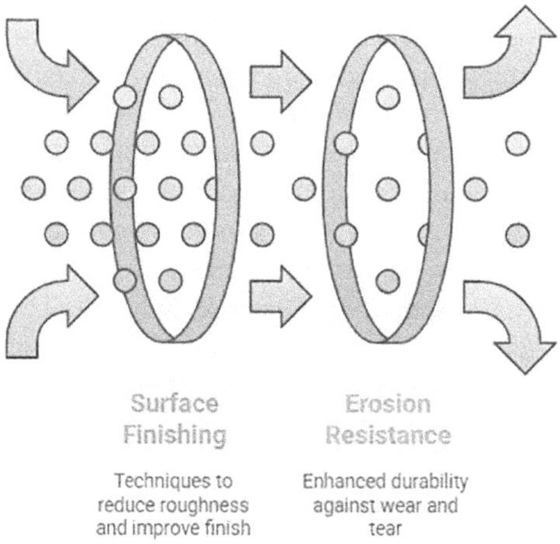

FIGURE 12.6 Role of AM surface finishing in enhancing erosion resistance of engineering materials.

enable the layer-wise fabrication of metallic and ceramic parts with tailored internal architectures and near-net-shape geometries. However, the as-built surfaces generated by AM are often characterized by high surface roughness, microstructural heterogeneity, unmelted particles, and porosity, which can significantly compromise the erosion resistance under high-velocity slurry exposure [57]. Figure 12.6 shows how surface finishing techniques reduce roughness, leading to improved erosion resistance by minimizing particle impact damage and enhancing durability. Arrows and particles represent the interaction between surface quality and external erosive forces in engineering applications.

Post-processing operations, including thermal spray coatings, laser surface modification, friction stir processing, and surface alloying, are critically essential to rectify these microstructural deficiencies and impart surface characteristics that are functionally optimized for erosive environments. These treatments not only refine surface microstructure but also promote the formation of dense, hard surface layers via metallurgical bonding or solid-state reinforcement dispersion, which serve as effective barriers against repeated particle impingement and slurry abrasion. For instance, applying high-velocity oxy-fuel coatings on SLM-produced substrates can yield WC-Co-based layers with enhanced adhesion and low porosity, dramatically improving erosion performance [58]. Moreover, AM allows localized tailoring of material compositions and surface gradients, which can be synergistically enhanced through post-processing. Techniques like laser cladding or electron beam remelting can be applied selectively to critical wear-prone zones, reinforcing them with intermetallics or hard carbides without affecting the overall geometry or integrity of the component. The coupling of additive-subtractive hybrid systems further enables simultaneous fabrication and in-situ post-processing, thus improving surface finish and microhardness directly during build cycles [59].

The convergence of AM with post-processing not only mitigates the intrinsic weaknesses of AM surfaces but also opens avenues for multifunctional surface design, where erosion resistance is achieved in tandem with corrosion resistance, fatigue life extension, or thermal stability—rendering the manufactured component suitable for next-generation, high-performance applications.

12.6 CONCLUSION

Surface post-processing is fundamental to improving slurry erosion resistance in engineering components subjected to harsh multiphase environments. The complex interaction of abrasive particles and fluid dynamics demands surface modifications that enhance hardness, toughness, and microstructural stability at the near-surface region. Techniques such as thermal spraying, laser surface treatment, shot peening, surface alloying, and cladding effectively modify surface chemistry and microstructure, thereby reducing erosion-induced material loss. The choice of an appropriate post-processing method depends critically on substrate compatibility; the nature of the slurry; and operational conditions, including temperature, pressure, and flow velocity. Economic viability and process scalability are key factors influencing industrial adoption, balancing performance gains with implementation costs. The integration of post-processing with additive manufacturing further broadens the capability to engineer complex geometries with site-specific erosion resistance, enabling functionally graded surfaces tailored to service demands. The advancement of hybrid and adaptive surface modification technologies promises enhanced multifunctionality, combining wear, corrosion, and thermal resistance in a single treatment. Moreover, coupling these techniques with real-time monitoring and data-driven process optimization will enhance control and reproducibility. Continued innovation in surface post-processing is therefore vital to extending component lifespan, reducing downtime, and supporting sustainable operation in slurry erosion-prone applications.

REFERENCES

1. Hariprasad KS, Seshadri SK, Sarma VS. Surface alloying of steels with Cr-Mo-B using TIG arc: Microstructure and slurry erosion behaviour. *Mater Des.* 2013;49:761–770.
2. Ma ZY, Mishra RS. Friction stir processing: A novel technique for fabrication of surface composite. *Mater Sci Eng A.* 2006;418(1–2):1–8.
3. Chen R, Zhang B, Li Y. Electron beam cladding of hard-facing alloys: Microstructure evolution and erosion performance. *Surf Coat Technol.* 2018;344:338–346.
4. Goldstein JI, Newbury DE, Michael JR, et al. *Scanning Electron Microscopy and X-ray Microanalysis.* 4th ed. New York: Springer; 2017.
5. Cullity BD, Stock SR. *Elements of X-ray Diffraction.* 3rd ed. Upper Saddle River, NJ: Prentice Hall; 2001.
6. Kalácska Á, Nagarajan R, Tóth LF, et al. Surface damage analysis on the application of abrasion and slurry erosion in targeted steels using an erosion test rig. *Lubricants.* 2022 Nov;10(11):316.
7. Wilson A. *Hardness Testing.* Materials Park, OH: ASM International; 2004.
8. Leach R. *Fundamental Principles of Engineering Nanometrology.* 2nd ed. London: CRC Press; 2014.
9. DebRoy T, Wei HL, Zuback JS, et al. Additive manufacturing of metallic components—process, structure and properties. *Prog Mater Sci.* 2018;92:112–224.
10. Guo N, Leu MC. Additive manufacturing: Technology, applications and research needs. *Front Mech Eng.* 2013;8(3):215–243.
11. Herzog D, Seyda V, Wycisk E, et al. Additive manufacturing of metals. *Acta Mater.* 2016;117:371–392.
12. Brown J. *Surface Engineering for Corrosion and Wear Resistance.* Cambridge: Woodhead Publishing; 2012.
13. Davis JR. *Heat Treatment of Steels.* Materials Park, OH: ASM International; 2017.
14. Johnson R. *Nitriding and Nitrocarburizing: Principles and Applications.* Cham: Springer; 2018.
15. Miller S. *Quenching and Tempering of Steel.* Boca Raton, FL: CRC Press; 2013.
16. Smith P. *Carburizing: Microstructure and Properties.* Materials Park, OH: ASM International; 2010.
17. Tejero-Martin D, Rezvani Rad M, McDonald A, et al. Beyond traditional coatings: A review on thermal sprayed functional and smart coatings. *arXiv preprint arXiv:1811.05289.* 2018.
18. Lelevic A, Walsh FC. Electrodeposition of Ni-P composite coatings: A review. *arXiv preprint arXiv:1807.04692.* 2018.
19. Aldahash SA, Abdelaal O, Abdelrhman Y. Slurry erosion-corrosion characteristics of as-built Ti-6Al-4V manufactured by selective laser melting. *Materials.* 2020 Sep 8;13(18):3967.
20. Verma PK, Minhas AS, Singh P, et al. Slurry erosion behaviour of thermal sprayed Al2O3 and Cr2O3 coatings for turbine steels. In *AIP Conference Proceedings* 2024 Feb 20 (Vol. 2986, No. 1). AIP Publishing.
21. Iquebal AS, Sagapuram D, Bukkapatnam S. Surface plastic flow in polishing of rough surfaces. *arXiv preprint arXiv:1610.09719.* 2016.
22. Krishnan M, Nalaskowsk JW, Cook LM. Chemical mechanical planarization: Slurry chemistry, materials, and mechanisms. *Chem Rev.* 2010;110:178–204.

23. Chattopadhyay R. *Advanced Thermally Assisted Surface Engineering Processes*. Kluwer Academic Publishers; 2004.
24. Chattopadhyay R. *Surface Wear: Analysis, Treatment, and Prevention*. ASM International; 2001.
25. Konda S, Mohammadi MM, Buchner RD, Lin H, Swihart MT. Flame-based synthesis and in situ functionalization of palladium alloy nanoparticles. *AIChE J*. 2018 Nov;64(11):3826–3834.
26. Singh RP, Singhal S. Rotary ultrasonic machining: A review. *Mater Manuf Processes*. 2016 Oct 25; 31(14):1795–1824.
27. Kumar J. Ultrasonic machining—A comprehensive review. *Machining Sci Technol*. 2013 Jul 3;17(3): 325–379.
28. Espallargas N, Fischer WJ, Armada S. Slurry erosion of HVOF coatings: Influence of microstructure and mechanical properties. *Wear*. 2014;319(1–2):74–85.
29. Pawlowski L. *The Science and Engineering of Thermal Spray Coatings*. 2nd ed. Chichester: Wiley; 2008.
30. Lima CRM, Trevisan RE, Marple BR. Slurry erosion behavior of alumina-based ceramic coatings. *Surface and Coatings Technology*. 2005;200(5–6):1501–1507.
31. Assadi H, Gärtner F, Stoltenhoff T, et al. Bonding mechanism in cold gas spraying. *Acta Mater*. 2003;51(15):4379–4394.
32. Steen WM, Mazumder J. *Laser Material Processing*. 4th ed. Springer; 2010.
33. Toyserkani E, Khajepour A, Corbin SF. *Laser Cladding*. CRC Press; 2004.
34. Bandyopadhyay A, Heer B. Additive manufacturing of biomaterials. *Prog Mater Sci*. 2018;93:112–224.
35. Chen Y, Li C, Yao X. Microstructure and erosion resistance of laser-cladded Co-Cr-Mo coatings. *Surf Coat Tech*. 2017;320:138–147.
36. Guzzo PL. A comparative study on ultrasonic machining of hard and brittle materials. *J Braz Soc Mech Sci Eng*. 2004;26:56–61.
37. Aldahash SA, Abdelaal O, Abdelrhman Y. Slurry erosion-corrosion characteristics of as-built Ti-6Al-4V manufactured by selective laser melting. *Materials*. 2020 Sep 8;13(18):3967.
38. Liu H, Zhang S, Liu Y, et al. Influence of shot peening on erosion resistance of 17–4PH stainless steel. *Surf Coat Tech*. 2016;289:152–158.
39. Meguid SA, Shagal G, Stranart JC. 3D FE analysis of peening of strain-rate sensitive materials using multiple impingement model. *Int J Impact Eng*. 2002;27(2):119–134.
40. Guagliano M. Relating almen intensity to residual stresses induced by shot peening: A numerical approach. *J Mater Process Technol*. 2001;110(3):277–286.
41. Ooi PPS, Shipway PH, Leen SB. Parametric effects on slurry erosion of shot-peened 316 stainless steel. *Wear*. 2013;303(1–2):325–333.
42. Hariprasad KS, Seshadri SK, Sarma VS. Surface alloying of steels with Cr-Mo-B using TIG arc: Microstructure and slurry erosion behaviour. *Mater Des*. 2013;49:761–770.
43. Ma ZY, Mishra RS. Friction stir processing: A novel technique for fabrication of surface composite. *Mater Sci Eng A*. 2006;418(1–2):1–8.
44. Chen R, Zhang B, Li Y. Electron beam cladding of hard-facing alloys: Microstructure evolution and erosion performance. *Surf Coat Tech*. 2018;344:338–346.
45. Brown J. *Surface Engineering for Corrosion and Wear Resistance*. Cambridge: Woodhead Publishing; 2012.
46. Davis JR. *Heat Treatment of Steels*. Materials Park, OH: ASM International; 2017.
47. Johnson R. *Nitriding and Nitrocarburizing: Principles and Applications*. Cham: Springer; 2018.
48. Miller S. *Quenching and Tempering of Steel*. Boca Raton, FL: CRC Press; 2013.
49. Smith P. *Carburizing: Microstructure and Properties*. Materials Park, OH: ASM International; 2010.
50. Goldstein JI, Newbury DE, Michael JR, et al. *Scanning Electron Microscopy and X-ray Microanalysis*. 4th ed. New York: Springer; 2017.
51. Cullity BD, Stock SR. *Elements of X-ray Diffraction*. 3rd ed. Upper Saddle River, NJ: Prentice Hall; 2001.
52. Wilson A. *Hardness Testing*. Metals Park, OH: ASM International; 2004.
53. Leach R. *Fundamental Principles of Engineering Nanometrology*. 2nd ed. London: CRC Press; 2014.
54. Awano Y. Carbon Nanotube (CNT) Via Interconnect Technologies: Low temperature CVD growth and chemical mechanical planarization for vertically aligned CNTs. In *Proceedings of the 2006 ICPT* (Vol. 10), 2006.
55. Voges K, Mueth A. Eco-pickled surface: An advantageous alternative to conventional acid pickling. *Iron Steel Tech*. 2008;4:81–96.

56. DebRoy T, Wei HL, Zuback JS, et al. Additive manufacturing of metallic components—process, structure and properties. *Prog Mater Sci*. 2018;92:112–224.
57. Guo N, Leu MC. Additive manufacturing: Technology, applications and research needs. *Front Mech Eng*. 2013;8(3):215–243.
58. Herzog D, Seyda V, Wycisk E, et al. Additive manufacturing of metals. *Acta Mater*. 2016;117:371–392.
59. Li H, Yin WY, Banerjee K, Mao JF. Circuit modeling and performance analysis of multi-walled carbon nanotube interconnects. *IEEE Trans Electron Devices*. 2008 May 20;55(6):1328–1337.

13 Artificial Intelligence in Slurry Erosion and Protection Techniques

Jyoti Bhattacharjee, Subhasis Roy, and Hyun Suk Jung

13.1 INTRODUCTION

Slurry erosion is characterized by the progressive material loss due to solid particles contained in a liquid medium which strike surfaces frequently and with a sufficient force. This phenomenon typically happens in a multiphase flow system, which are systems in which the kinetic energy of the high-speed particles is the eroding force against the engineering surfaces [1]. Industries that see the most predominating anecdotal evidence of this wear mechanism are hydropower (turbine blades), oil and gas (pipelines), marine structures, processing (valves, pumps), and thermal (heat exchangers), with the implications being not only premature failure but also lost uptime and increased economic expenses [2]. Historically, standard systems to manage the effects of slurry erosion have been the use of highly wear resistant materials such as high-performance alloys, ceramics, and composites, as well as coating and surface modification methods including laser cladding or thermal spraying [3]. While some strategies have shown partial success, they were inevitably limited by the complex interaction of mechanical, hydrodynamic, and material parameters.

To potentially alleviate these limitations, there has been a more recent coupling of AI and machine learning that has the potential to be a silver-bullet answer to these challenges [4]. AI leverages data to provide rapid, accurate predictions of erosion behavior, wear rates, and material response under different flow conditions. Algorithms trained on experimental and simulated data can identify hidden correlations and optimize material selection or surface treatments without conducting large amounts of testing. This predictive capability not only accelerates the design and development of erosion-resistant systems but allows the creation of adaptive, intelligent maintenance programs, as shown in Figure 13.1 [5].

Each subplot shows the pairwise relationship between the corresponding features of the horizontal and vertical axis. The three phase types are shown using different symbols: amorphous metals (AMs) are shown with black stars, single solid solution (SS) phases are indicated by red stars, and therefore, high-entropy alloy (HEA)-IM denotes the intermetallic compounds or phases which are developed as a result of high-entropy alloys or within them, thus merging the idea of high-entropy multi-element alloys with the concept of ordered intermetallic phases. The clustering behavior not only provides trend analysis outlining the distributions but also highlights potential separability of the feature set depending on phase type.

As the demand for more efficient and longer-lasting components increases in the energy, manufacturing, and infrastructure sectors, integrating the conventional field of materials engineering and new models utilizing AI creates a paradigm shift. This chapter describes the evolution of slurry erosion research, including traditional mitigation strategies as well as the new role of AI in fundamentally changing the prediction, analysis, and mitigative control of erosion effects. In this context, artificial intelligence plays a powerful role in evolving the study, prediction, and management of slurry erosion [6]. Models applying AI, and particularly models applying machine learning (ML) and deep learning (DL), have the capacity to evaluate vast data sets of material properties, fluid and particle dynamics,

DOI: 10.1201/9781003561156-15

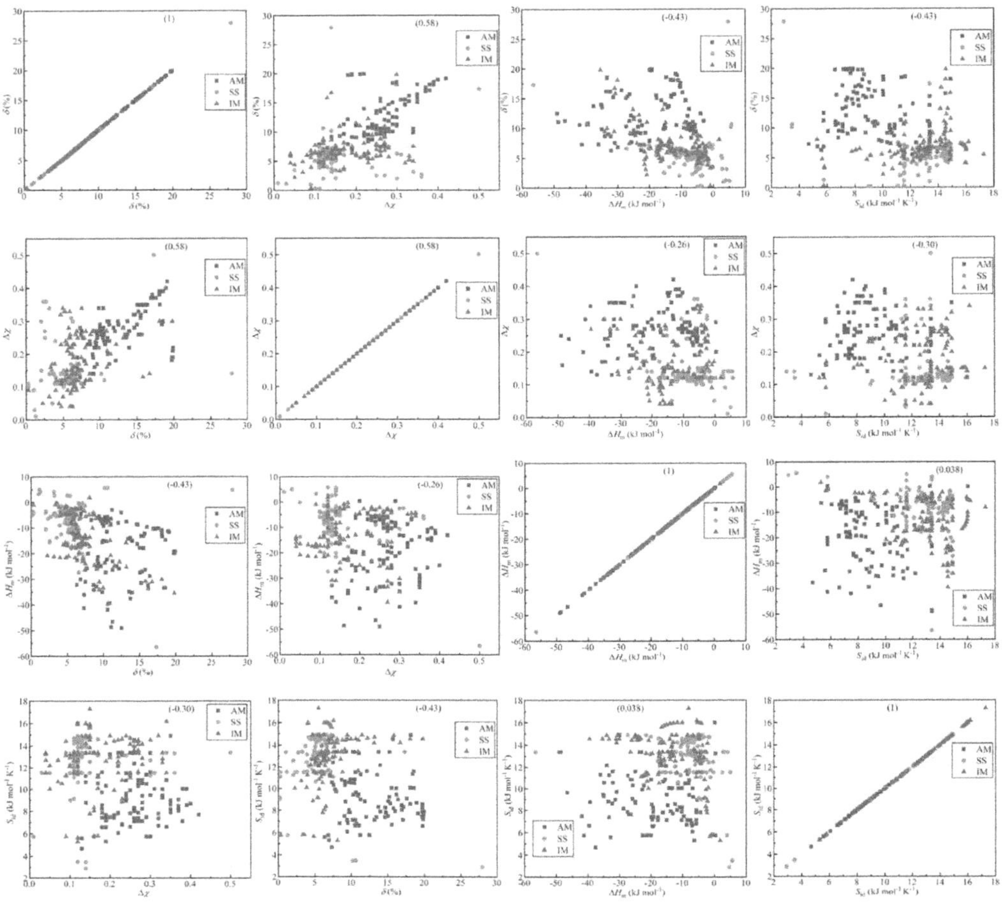

FIGURE 13.1 Scatter plot matrix showing the correlation between four input features used in phase selection analysis for 330 alloy compositions [6].

and operating conditions to identify complex relationships and successfully predict anticipated erosion behavior with high levels of certainty. Accurate prediction in this manner allows for early identification of high-risk erosion sites and can come closer to optimizing design parameters before any physical testing is performed. Moreover, AI-enabled simulations can greatly reduce dependency on expensive physical models and enable virtual screening of potential material performance across many conditions. Data models also allow for real-time monitoring and adaptive control in operational systems, creating possibilities for a new future of predictive maintenance and lifecycle management [7].

The increased need for AI stems from its potential to close the gap between nonlinear, high-order erosion mechanisms and practical engineering design. As industries transition toward intelligent manufacturing and autonomous technologies, the inclusion of AI in slurry erosion studies is not just desirable—it is essential to enable more efficiency, reliability, and sustainability across fluid handling and process technologies [8].

13.2 MECHANISM OF SLURRY EROSION

Slurry erosion is a multifaceted wear mechanism that results as solid particles suspended within a liquid carrier strike a material surface, causing gradual material loss. Slurry erosion is different from the ordinary mechanical wear process, as it consists of a multifaceted interaction between fluid flow, particle behavior, and the response of the surface material. These basic mechanisms are controlled by a set of variables that include the impact angle, particle velocity, particle shape and size, fluid properties, and surface microstructure. All of these parameters must be known in order to predict erosion behavior properly and to design materials and systems resistant to such wear [9].

13.2.1 Impact Angle

The impact angle at which particles collide with a surface is one of the controlling factors in the shape and extent of erosion. At low impact angles (15°–30°), erosion is in a cutting or plowing mode where particles flow along the surface and erode material by a shear-activated process. This is typically experienced for ductile materials like carbon steels and aluminum alloys. On the other hand, for steep impact angles (60°–90°), erosion is controlled by deformation and fracture, as particles strike the surface normally, inducing local plastic deformation or brittle fracture. Maximum erosion at or near normal incidence in very brittle materials such as ceramics and certain hardened alloys results from the inability of these materials to deform plastically [10].

13.2.2 Particle Velocity

The erosion particle impact velocity is nonlinearly related to erosion rate and typically modeled by a power law. Erosion rate increases with increasing velocity in most materials, per the equation [11]:

$$E \propto V^n \tag{13.1}$$

where the erosion rate is presented as E, particle velocity is presented as V, and exponent n is usually between 2 and 3. Greater velocities have increased particle kinetic energy and more forceful impacts, causing more material loss. Small fluid velocity increases in high-pressure slurry loops or hydro-transport pipe circuits produce disproportionately large erosion, and thus flow control and moderation need to be exercised to avoid excessive erosion, as shown in Figure 13.2 [12].

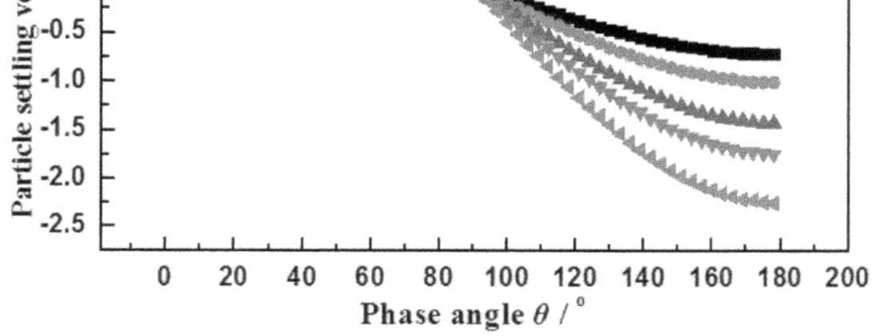

FIGURE 13.2 Association of settling velocity and phase angle for varying pipe diameters [12].

Figure 13.2 shows that with changes in pipe diameter, partially due to other complexities, the settling velocity of the particles can change, and for the phase lag of the particles in slurry transport it can also change. The settling velocity typically decreases with pipe diameter increase since the shear near the wall of the pipe is lower and turbulence is typically lower with increasing pipe diameter. Thus, the phase angle and, ultimately, the diameter also vary with increasing diameter due to inertial and drag forces, and buoyancy needs to be considered during repetitive oscillatory motion. This relationship is important for considerations of prediction of stratification of flow, pipeline designs that need to account for pressure changes due to diameter changes, and the efficiency in solid–liquid separations during unit operations in the food process and chemical industry [12].

13.2.3 Particle Size and Shape

Particle morphology has a direct impact on the nature and severity of erosion damage. Large particles have greater kinetic energy and, consequently, more severe impacts on the surface. Yet they can rebound more than smaller particles and decrease their surface residence time. Small particles, though delivering less individual energy, can produce more uniform surface wear in that they can penetrate small features and deposit in stagnation points. In form, angular or sharp-edged particles are especially destructive, with small cutting tools that cut into the surface on impact. Round particles, in contrast, produce more deformational erosion than cutting. Mixed shape and mixed size in industrial slurry add additional complexity to the exact prediction of erosion performance [13].

13.2.4 Fluid Properties

The fluid in which the particles are transported also influences erosion behavior. Characteristics like viscosity, density, and turbulence intensity influence particle momentum, transport behavior, and surface interaction time. The aforementioned fluid's high viscosity, reduced particle velocity and lower overall erosion rate are drawbacks; but on the other hand, they encourage agglomeration of particles or deposition and give rise to very localized erosion. Turbulent flows, prevalent in industrial pipes and turbines, increase the probability of particle-surface collisions because of random motion, leading to even greater erosion [14].

13.2.5 Surface Characteristics and Microstructure

The intrinsic nature of the target surface, including hardness, ductility, grain structure, and phase composition, is responsible for how material impacts energy and resists such as radiation. Hard surfaces tend to be most resistant to plowing and micro-cutting but may be susceptible to brittle fracture for severe high-stress impacts. Ductile surfaces are able to absorb impact energy through plastic deformation but wear more quickly during cyclic loading [15]. Surface roughness is also important. Highly rough surfaces retain particles and generate secondary impacts and therefore increased wear. However, highly smooth surfaces decrease adhesion and sliding friction, thus keeping erosion at a minimum. The incorporation of surface coating, oxides, or embedded reinforcement also greatly changes erosion resistance by altering energy dissipation mechanisms.

13.2.6 Cumulative and Synergistic Effects

Erosion in real systems is seldom the result of a single factor. It is likely to be a synergistic combination of fluid dynamics, particle movement [16], and material response. For example, increases in particle velocity can not only increase impact energy but influence particle path and turbulence to result in more frequent and erratic impacts. In the same way, cyclic micro-impact fatigue can relax the grain boundaries and enhance erosion over the long term. The combined effect of these variables requires multidisciplinary research and online monitoring, particularly in high-risk applications like turbine blades, downhole tools, and slurry transport pipes [17].

13.2.7 Modeling and Simulation

In order to more clearly understand the forces involved in slurry erosion, numerical approaches like computational fluid dynamics (CFD) and discrete element modeling (DEM) are used. CFD modeling allows visualization of impact patterns, such as regimes of turbulence and profiles of flows, whereas DEM is able to simulate particle-by-particle motion and contact of surfaces. Coupled with experimental data, these simulations provide rich information on hotspots of erosion, leading to more effective design of durable systems and materials. The slurry erosion mechanism is an interactive dynamic process of particle, fluid, and material properties. All variables, from impact angle and particle size to fluid turbulence and material microstructure, each contribute uniquely to the erosion process. Understanding these mechanisms collectively is required for predictive modeling and high-strength materials to resist very aggressive slurry environments. As engineering systems move toward more efficient performance and extended life [18], materials scientists, mechanical engineers, and industrial designers all must become familiar with the principles of slurry erosion.

13.3 MACHINE LEARNING METHODS IN SLURRY EROSION PREDICTION AND ANALYSIS

Implementations of machine learning in slurry erosion research are a revolutionary leap in materials technology and wear prediction. Traditional experimental designs, valuable as they are, have limitations in the presence of high complexity, nonlinear, and multivariate interactions among process parameters, material response, and environmental conditions. Machine learning offers data-driven alternatives that are capable of pattern recognition, outcome prediction, and decision aid without the need for programming or model-specific form assumptions. With experimental, simulation, or production data training, researchers and engineers are able to estimate erosion rates rapidly, propose materials, and uncover latent variables causing component failure [19].

13.3.1 Supervised Learning Models

Supervised learning involves teaching an algorithm with labeled data, wherein the input features (like particle size, velocity, impact angle, slurry concentration, and material hardness) and also the target outputs (like erosion rate, mass loss, or surface, like radiation depth) are all specified. These kinds of models have the mapping of both input and output that allows them to foresee the erosion behavior of further, not seen before data [20].

13.3.2 Regression Techniques

Arguably, the most primitive ML technique used for slurry erosion prediction is regression analysis. In its simplest form, linear regression models the erosion rate as a weighted sum of process parameters. An example of a linear regression model would model the relationship [21]:

$$\text{Erosion rate} = (a \times \text{velocity}) + (b \times \text{impact angle}) + (c \times \text{particle size}) + \text{intercept} \quad (13.2)$$

Decision tree models are another robust supervised learning algorithm. Decision trees divide the data into branches based on the features that most clearly separate erosion outcomes. A potential decision tree may divide the data first by the velocity of the particles, second by the material they are made from, and third by the angle of impact to show a hierarchy of parameters influencing erosion. To avoid over-fitting on a single decision tree, random forest algorithms develop a number of trees on various data subsets and aggregate the trees. The collective output of the group of trees provides a more reliable, accurate, generalizable, and robust output than a single decision tree. Random forests have been used with success to predict wear rates, rank the importance of parameters, and optimize conditions for erosion testing [22].

13.3.3 Support Vector Machines

The support vector machine (SVM) method is designed for datasets that have a large number of dimensions and/or nonlinear properties. In slurry erosion, SVMs can identify alternatives to complex boundary conditions that regression analyses cannot. Using a kernel function (e.g., a radial basis function), SVMs can move input variables into a newly defined feature space, whereby conditions associated with erosion-prone and erosion-resistant conditions can be effectively and precisely assigned [23]. In predictive maintenance, SVMs can be employed to identify surface wear levels from surface images, acoustic emissions, or vibration signals for real-time pipeline or rotating equipment damage assessment.

13.3.4 Unsupervised Learning Techniques

Unsupervised learning doesn't involve labeled output data. It scans the data structure to find hidden patterns, groupings, or relationships between variables. It comes in handy when the process of erosion isn't fully known or there isn't much labeled data [24].

13.3.5 Clustering of Wear Patterns

Clustering methods such as k-means or hierarchical clustering can split erosion data into clusters based on similar wear behavior, operating conditions, or material responses. K-means clustering, for example, can group surface images from profilometry or SEM into distinctive erosion modes, namely cutting, plowing, pitting, or delamination. These groups are useful in enabling engineers to relate specific process parameters to erosion signatures, enabling root cause analysis and prevention of failure. Clustering is also useful in quality control applications in identifying unusual erosion behavior during real-time monitoring [25].

13.3.6 Principal Component Analysis

Principal component analysis (PCA) is a dimensionality reduction technique that identifies the most significant variables (principal components) responsible for variance within the dataset. In slurry erosion research, datasets typically include dozens of related features, from flow conditions and particle properties to material microstructure and hardness. Through compressing the dataset into a smaller, uncorrelated set of variables, PCA simplifies modeling, noise removal, and visualization. PCA is frequently employed before clustering or supervised learning to guarantee the most important features are prioritized [26].

13.4 APPLICATIONS IN SLURRY EROSION RESEARCH AND INDUSTRY

Applications in the physical world of machine learning in slurry erosion includes a number of areas of applications—newly developed or automatized from existing methodologies into intelligent, automatic, and even predictive systems.

13.4.1 Ability to Predict Erosion Rate Based on Process Parametrics

ML models allow us to reliably predict erosion rates from input parameters including particle size, angle of impact, slurry concentration, and flow velocity. Once trained, the model completely displaces many, if not all, of the necessary subsequent physical experiments, saving immeasurable time and money. In some (e.g., hydropower, oil and gas) industries where laboratory conditions cannot be well simulated outside of laboratory, predictive models such as ML can be invaluable to risk analysis or design improvements. By integrating erosion models based on trained ML algorithms with computational fluid dynamics [27] simulation, hybridized modes are capable of highlighting erosion hotspots/areas, predicting loss of material (over time) and providing a powerful tool to system designers and maintenance planners.

13.4.2 Material Selection and Optimization

It is possible to employ machine learning algorithms to assist in selecting the optimum material for a given slurry environment. ML models can analyze past erosion test data to rank materials by their performance in specified operating conditions. Multi-objective methods can even compromise between resistance to erosion, cost, machinability, and weight. Bayesian optimization and genetic algorithms have been applied in advanced ML pipelines to optimize over material design spaces, proposing new compositions or coatings that might otherwise take a protracted amount of time to discover experimentally [28].

13.4.3 Identifying Failure Trends and Anomalies

Real-time monitoring devices fitted with sensors (vibration, acoustic, temperature, or imaging) provide unprecedented amounts of operating data. Machine learning algorithms developed from these data can search for subtle signatures of minor versions of erosion or non-standard surface motion. Anomaly detection algorithms based on unsupervised learning can help warn operators of incipient fault failure before operating failure occurs at the moment of catastrophic collapse.

The predictive maintenance system that is proposed will take care of unplanned outages, prolong the life of the equipment and improve safety in slurry systems like dredging pumps, industrial mixers, or cooling systems. While machine learning has the potential to change the way slurry erosion is researched, its use is still in its infancy. Some of the major constraints are [29]:

- Lack of Data: Good-quality, labeled data from erosion experiments are rare, and there is almost zero standardization across laboratories.
- Model Interpretability: Because most of the ML models are "black boxes", interpretation of their predictions is difficult, particularly for deep learning networks.
- Generality: Models trained on one particular material or slurry condition may not generalize to other systems without calibration [30].

Still, with the increasing availability of experimental data, simulation software, and computational resources, the outlook for ML in slurry erosion appears promising. The convergence of physics-informed machine learning, where physical principles are embedded into the model structure, offers a path to explanatory and generalizable predictions. Additionally, digital twins with real-time ML analytics have the potential to transform the way industries react to erosion threats along the entire life cycle of critical assets [31].

13.5 AI-BASED SLURRY EROSION PREDICTION MODELS

13.5.1 Introduction of AI in Slurry Erosion Modeling

Slurry erosion is a process of wear caused by the action of solid particles suspended in a fluid against a target surface, resulting in material removal. Ancient lineage empirical models typically do not have the necessary means to capture the complex and nonlinear correlations between the operating and material parameters, such as flow rate, angle of attack, and hardness. Artificial intelligence and machine learning techniques provide a new paradigm to simulate slurry erosion based on data learning from experiments and simulations. These models can predict such important results as wear depth, erosion rate, and weight loss. With the inclusion of various inputs like impact angle, flow rate, velocity, hardness of the material, and concentration of particles, AI models are able to pick up patterns and relationships that are not possible to distinguish using traditional methods. The next section outlines the development stages of AI models for slurry erosion from data acquisition and preprocessing to training, validation, and performance assessment, with a focus on applying them to predictive maintenance and process optimization [32].

13.5.2 Data Acquisition for Model Building

The reliability of any AI model is dependent on the amount and quality of the input data. In slurry erosion, information can be obtained from field sensor measurements, numerical simulations, and laboratory tests. Experimental approaches comprise slurry loop test systems and jet erosion rigs that offer controlled conditions to quantify the erosion behavior at specified flow rates and particle impacts. Numerical approaches, like computational fluid dynamics and finite element analysis (FEA), are employed to simulate a broad range of operating conditions and create artificial data. Industrial sensors implanted in reactors and pipelines also give real-time measurements of process parameters and wear. Literature-reported cases and standards, such as ASTM G76, provide wear datasets that are applied in some cases to augment experimental data that are readily available. By combining these different sources, a diverse dataset is created that builds a robust model with better learning capabilities and generalizability across a wide range of domains and materials, as shown in Figure 13.3 [33].

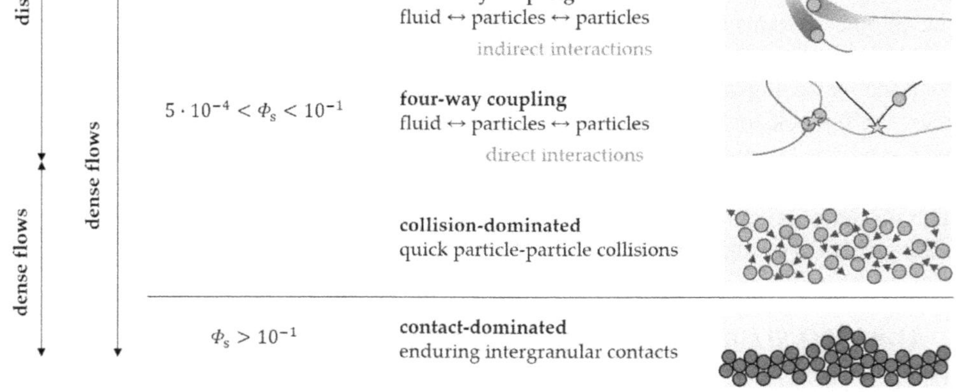

FIGURE 13.3 Classification of particle-laden flows in terms of coupling regression and dominant physical mechanisms [33].

Few researchers have characterized the classification which distinguishes between disperse and dense flows, and the two coupling types (dilute and dense) that are defined by the solid volume fraction limits, which have been determined on the flow map created by Elghobashi. The critical factors that involve particle–particle interaction, momentum transfer between particles and fluid, and feedback coupling (two-way, four-way, etc.) are the key points to this classification (R.J. emphasized a lot) [34].

13.5.3 Data Preprocessing and Feature Engineering

Unprocessed data collected from different sources usually require a great deal of preprocessing, particularly to render those data sets consistent, accurate, and usable; the first phase of transformation consists of normalization, which makes all input parameters comparable in a common range to eliminate biases produced from varying units or scales—this is typically achieved with normalization methods (MinMax or Z-score). Often it is also a necessity to remove outliers, where the extreme values that occur due to noise in measurement, noise in sensors, and so on are identified and suppressed with the aid of statistical techniques (e.g., interquartile range, Mahalanobis distance). Feature engineering can also be applied to help improve model performance, as it takes several variables and creates new variables from them, such as kinetic energy of colliding particles or erosion severity indices. In addition, the application of dimension reduction techniques (e.g., PCA) can also reduce the dimensionality of a model without sacrificing prediction capability [34].

13.5.4 Model Training, Testing, and Validation Methods

The data are divided into training, validation, and testing sets after preprocessing. Normally, 70% of the data is used for training, 15% for validation (used for hyperparameter optimization), and 15% for testing to measure generalization performance. Different AI and ML algorithms can be employed depending on the nature of the problem. Regression models such as linear regression or nonlinear polynomial regression are suitable for continuous outputs [35] such as erosion rate or wear depth. For more complicated and nonlinear data, decision trees, random forests, and support vector machines are common examples. Artificial neural networks (ANNs) with a high number of hidden layers are especially powerful in their ability to discern complex patterns among variables. Ensemble techniques, based on combining the predictions from a variety of models, remain growingly reliable. K_{Fold} cross-validation, commonly specified as $K = 5$ or 10, is performed to ensure that all available data points are used in training and validation to promote model generalizability and variance reduction [36].

13.5.5 Performance Metrics and Model Evaluation

After the model has been trained, its performance should be estimated with proper metrics to ensure accuracy and reliability. For such regression models, mean absolute error (MAE) and root mean square error (RMSE) are popular choices. RMSE penalizes greater errors more, and MAE simply gives a simple average of prediction error. The coefficient of determination (R^2) is another important measure, which indicates the proportion of the variance in the target output explained by the model—values close to 1 suggesting high predictive performance. For models based on classification, which may be applied for classifying erosion severity (such as low, medium, high, etc.), measures like the confusion matrix, precision, recall, and F1 score are utilized. More sophisticated devices like receiver operating characteristic (ROC) curves and area under the curve (AUC) [37] give better insights into classifier performance. Visualization methods such as parity plots, residual plots, and learning curves are employed to ensure consistency and reliability of prediction by the models along concerning time and changing sizes of data.

Comparison of the two modes of cross-validation: *URun*: unrestricted mode of cross-validation; *RRun:* restricted mode of cross-validation. This comparison illustrates the differences in predictive performance under these two validation constraints [37]. AI-based models for the predictions of slurry erosion are a monumental advancement over traditional empirical methods. By systematically modeling the process parameters, material characteristics, and environmental conditions, the models can enable accurate prediction of erosion rates and wear behavior for a variety of operating conditions. Deploying them effectively requires a serious development pipeline that includes data generation, data pre-processing, training on the model appropriately, and analysis of the predictive performances. The models have remarkable value in deployment under development for online monitoring, predictive maintenance, and optimal design to applications where these operate such as mining, oil and gas, and chemical processing. The ongoing progress nowadays is integrated with the concept of deep learning [38], physics-ML hybrids, and IoT-based sensing networks that would allow for more intelligent, automated, and scalable erosion predictions. This integration would contribute to extending the life of equipment, reducing downtime, and encouraging the more sustainable use of materials in engineering systems.

13.5.6 Industrial Environment Deployment

Getting applications like AI-based models of slurry erosion prediction into a genuine industrial or production environment requires the appropriate integration of expertise with existing infrastructure and operational procedures. Industrial deployments suggest several key considerations that are favorable in the commercial adoption of slurry erosion predictions; more specifically, action naturally demands a clear understanding of best practices, the appropriate use of data, and creating a safe and reliable evidence base [39]. These sensors pass real-time information to edge devices or a cloud-based AI platform, where trained models analyze the incoming data to determine erosion severity and maintenance scheduling. However, one of the challenges is that the model should remain robust in ever-changing conditions, alterations in slurry composition, geometric changes to equipment, or operating pressures. Hence, models are fine-tuned with transfer learning or updated continuously through online learning mechanisms. Further, dashboards for prediction and mobile interfaces can be linked with plant control for engineering alerts.

13.6 CHALLENGES AND LIMITATIONS IN AI-BASED EROSION AND COATING OPTIMIZATION

13.6.1 Data Limitations and Inconsistencies

The robustness and reliability of AI models in the prediction of slurry erosion and the optimization of coating parameters depend finally on the quality of datasets. In the majority of industrial and laboratory environments, data are inconsistent, sparse, or non-standardized. Problems like multiple experimental protocols, limited access to high-resolution sensors, and inconsistent recording practices result in datasets that may not be comparable. Additionally, data collection in real time remains challenging in harsh erosion-prone areas where sensor incorporation [40] is hindered by ambient temperatures, abrasiveness, and inaccessibility. This results in overreliance on static laboratory-derived data that may be invalid to simulate dynamic infield conditions. Moreover, small sample sizes might constrain the learning capability of advanced AI models, increasing the likelihood of overfitting and suppressing generalizability to novel conditions. The lack of well-annotated, freely available benchmark datasets in the erosion and coatings community also prevents reproducibility and collaboration between groups of researchers [41].

13.6.2 Generalization and Over-Fitting in Small Datasets

A frequent problem concerning training AI models is generalization. This is probably the case for specialized engineering fields. When models are trained on small or biased datasets, they are likely to overfit the data using specific input–output pairs rather than general patterns, which prevent the model from making consistent predictions when exposed to new or somewhat different inputs. When considering erosion prediction or coating optimization, for instance, operating conditions like flow speed, temperature, and particle angle of incidences may experience tremendous variation in reality. If the AI model has not been trained using a sufficiently varied dataset, its predictive capabilities with new conditions cannot be trusted. This is the worst-case scenario for industrial applications, where poor predictions could result in premature failure or inadequate protective performance. Techniques to mitigate overfitting have become more sophisticated, such as dropout (in neural networks); regularization; and the most impactful option, a high volume of rich data [42].

13.6.3 Multiphysics Modelling and Complexity

The biggest technical challenge may be in modelling the inherently complex multiphysics behavior of slurry erosion and coatings. Erosion needs an understanding of fluid behavior, solid mechanics, heat transfer, and chemical reactions that all occur simultaneously. Classical methods such as computational fluid dynamics and finite-element modeling provide an in-depth understanding of physics-based learning but require a large amount of computation time. Folding these simulation methods into an AI pipeline represents a larger challenge, which involves the type, resolution, and timescales of the data used in CFD/FEM simulation methods. AI models should be trained to account for the outputs of CFD/FEM simulations such as stress fields [43], temperature gradients, and turbulence in the flows at the cost of computational speed. This requires developing surrogate models or reduced-order representations of physical behavior without invoking full CFD/FEM simulations. However, developing surrogate models and ensuring accuracy or calibration over a wide range of operational conditions remains a significant challenge in the field today.

13.6.4 Interpretability and Trust in Models

The more sophisticated AI models become, especially with deep learning architectures, the lower the interpretability of decisions made by the models. For high-risk engineering applications such as erosion prediction and preventative coating development, users need explanations and interpretability in order to trust AI-suggested alternatives. Most AI models, especially deep neural networks, are "black boxes", offering high accuracy without much insight into how predictions are being made [44, 45]. The absence of explainability in AI systems might delay their acceptance among safety-critical sectors like aerospace, nuclear, or oil and gas, where every single decision made requires a clear and justifiable trace through regular demand and inquiry. Researchers are exploring explainable AI (XAI) techniques—such as SHAP, local interpretable model-agnostic explanations (LIME), and saliency maps—but these instruments remain under development within the scientific application. There remains a need for trading model complexity for interpretability or creating hybrid models combining physical rules and machine learning in an effort to balance both accuracy and understanding, as shown in Figure 13.4 [46].

Classification, such as regression, and clustering are the main types of ML tasks being used by papers for predicting the phase of materials, optimizing coatings, and estimating the properties of materials. Other ML types such as dimensionality reduction, anomaly detection, and reinforcement learning appear less often and are mostly limited to one or two studies. SHAP is flagged as an important interpretability tool incorporated in several papers to highlight the importance of features and improve model transparency. This distribution is simply a reflection of current trends and preferences in using ML in material design and process optimization [46].

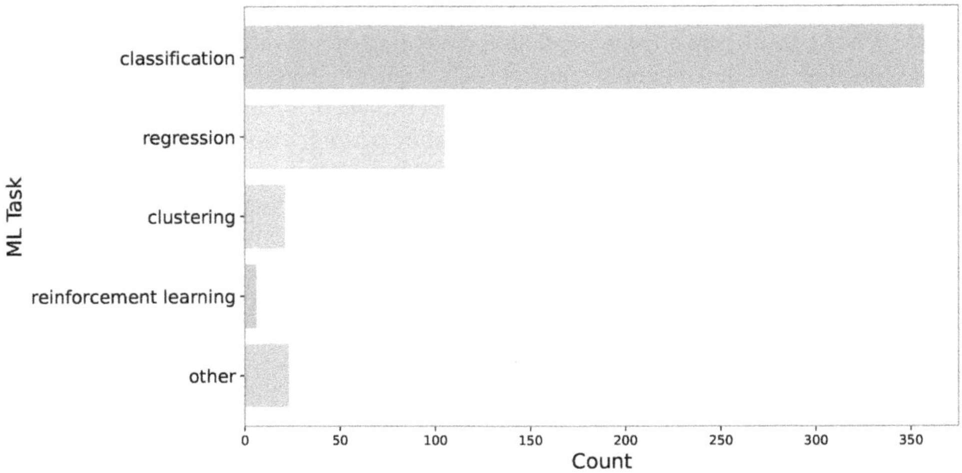

FIGURE 13.4 Summary of the major ML tasks found in the literature review [46]. **(Under CC 4.0)**

13.7 FUTURE TRENDS IN AI FOR EROSION CONTROL AND COATING TECHNOLOGIES

13.7.1 IoT and Smart Sensor Networks

Integrating with the Internet of Things (IoT) and distributed sensor networks is one of the more exciting trends in AI-based erosion and coating optimization. Today, advancements in robust, small sensors enable measurements of important process variables (pressure, temperature, wear depth, vibration, and flow profiles) in real systems in real time even in operational conditions that are challenging or extreme. These sensors can transmit real-time data to cloud or edge AI systems that use machine learning algorithms to search data for patterns, anomalies, and in some cases radiation trends. For example, predictive maintenance systems may be placed on equipment to alert operators before erosion occurs at a critical point. In coating operations, embedded in situ sensors placed in robotic spray arms can facilitate in real-time feedback information for adaptive process control. The combination of AI and IoT thus enables closed loop systems to autonomously adjust operating conditions in real time to optimize efficiency and safety and extend the life of critical equipment. A smart infrastructure initiates building strong, agile, and data-driven industrial ecosystems [47].

13.7.2 Physics-Informed AI Models

The community has moved on to physics-informed AI (PIAI) to achieve maximum performance from purely data-driven models. In this approach, physical principles, such as conservation of mass, momentum, and energy, are applied in machine learning frameworks. Such models ensure AI predictions are physically feasible even for sparse or noisy training data. PIAI models for slurry erosion can incorporate Navier–Stokes fluid motion equations, erosion rate formulations, or coating heat transfer laws for thermal coatings. Similarly, in coating processes [48], physical constraints such as diffusion barriers or thermodynamic equilibrium may be incorporated to impact the optimization of deposition conditions. PIAI models improve generalizability, reduce data requirements, and enhance credibility by linking machine learning predictions with established scientific laws. They are particularly useful for projecting predictions into unobserved and in carrying out sound sensitivity analysis. The future includes the development of open-source libraries and frameworks that allow seamless integration of physics and AI into domain-specific applications.

13.7.3 Digital Twins and Real-Time Predictive Maintenance

Digital twins—essentially real-time updates of physical systems, through live data—represent another major future avenue. For slurry erosion systems, a digital twin of a reactor, pump, or pipeline could be generated through CFD simulations updated in real time through live data from IoT sensors. AI models in the digital twin would follow the wear progression and predict failures far in advance so that maintenance can be conducted on demand. A digital twin of the spray system could predict coating thickness, surface roughness, and microstructure based on changing environmental and operating conditions in coating applications. A predictive approach reduces maintenance cost and provides a means to reduce unplanned downtimes, increase equipment longevity, and optimize operational performance. Furthermore, integrating digital twins with AI optimization software offers the potential to optimize system design and process performance continuously. These twins would provide a mechanism to train operators, plan operating scenarios, and virtually commission new coating lines or erosion-resistant systems [49].

13.7.4 Democratization and Accessibility of AI Tools

For AI to be accepted widely in erosion and coating science and for it to have impact, it is crucial to democratize AI tools so that they are accessible to nonexperts. This involves creating easy-to-use graphical user interfaces, plug-and-play AI modules, and pretrained models specific to generic materials and systems. Open-source software and cloud-based AI services allow small and medium-sized enterprises (SMEs) to leverage these technologies without creating deep in-house AI capabilities. Training material, industry–academia collaboration, and standardization of data formats drive additional adoption. As the process of use and interpretation becomes easier through AI tools, more researchers and engineers are able to get involved in developing innovative, adaptive surface protection systems [49–50].

13.7.5 Case Study: AI-Powered Erosion Predictions for Oil and Gas Pipelines

Slurry erosion is a significant pipeline integrity risk during offshore oil drilling operations due to the continuous transport of sand sediment with crude. A prominent example is an AI-powered erosion prediction system that a leading oilfield service provider used. The system used sensor data from pressure and acoustic sensors installed on pipeline elbows and bends. It used prior data on flow rate, particle concentration, and impact angle as inputs to a regression (random forest regression) model. The model successfully predicted wear depth and erosion rate with over 92% accuracy. The integration with the IoT platform enabled real-time viewing for planning maintenance. Over 18 months, the predictive model reduced unplanned shutdowns by 35%, demonstrating the potential for AI in predictive maintenance [51].

13.7.6 Case Study: PSO Optimization of Thermal Spray Coatings for Turbines

An aerospace engineering company used particle swarm optimization (PSO) to optimize high-velocity oxy-fuel spray parameters for tungsten carbide coatings on turbine blades. Feed rate, fuel–oxygen ratio, and standoff distance were optimized to produce maximum erosion resistance with minimal porosity. PSO was in a position to impart a 25% [52] improvement in the hardness of the coating and a 30% improvement in erosion resistance compared to conventional settings. Optimization was verified by jet erosion testing and SEM microstructural analysis. This project demonstrates the power of AI for complex parameter tuning in high-performance protective coatings in severe working environments, as shown in Figure 13.5.

This multi-level mapping—a direct zero order (direct relationships between process and properties), first order (particle state), and second order (microstructural evolution)—allows effective

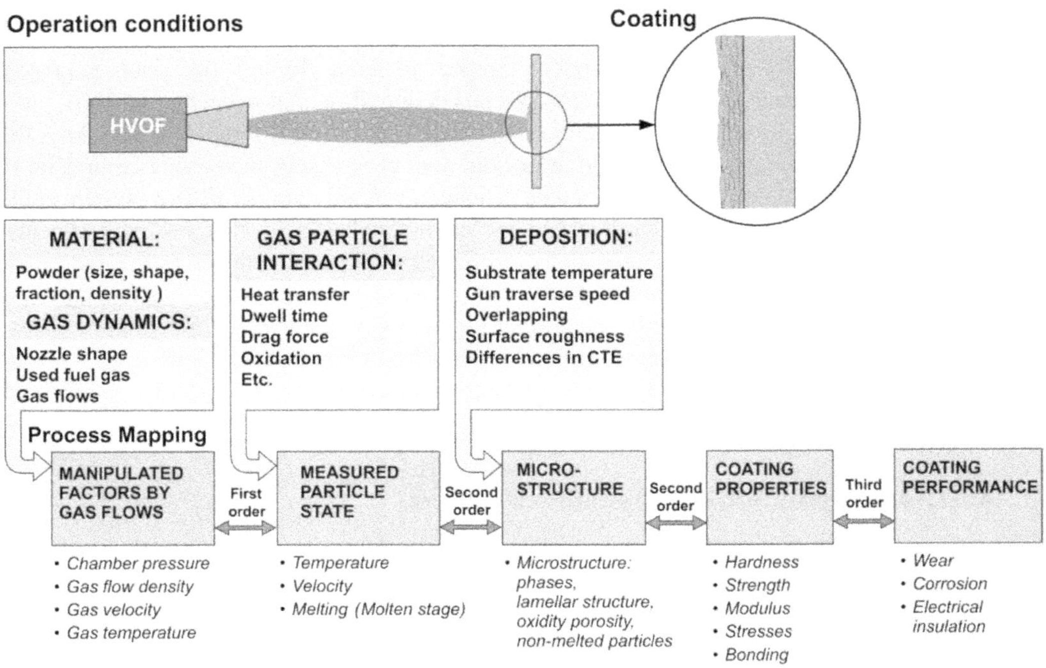

FIGURE 13.5 The conceptual framework for the HVOF process map depicts the hierarchical relationships between process parameters, in-flight particle properties, coating microstructure, and final coating characteristics [52].

mapping of how input variables (such as fuel ratio, spray distance, and particle velocity) enable structural features (such as porosity and phase composition) and subsequently affect performance properties (such as hardness, adhesion, and resistance to erosion). Process mapping aims to generate a robust process–structure–property link for the informed design and optimization of coating systems to meet application-specific needs [52].

A factory producing car parts adopted reinforcement learning (RL) in robotized spray painting operations. It targeted a maximum coating thickness on complex geometries like cylinder heads and crankcases. The RL agent utilized LIDAR and infrared feedback to learn surface properties in real time and adjusted spray path, speed, and nozzle direction as required. With repeated training cycles, the system reduced material overspray by 18% and achieved more uniform thickness with less rework. The adaptability potential of RL helped the system to learn and adapt to component shape variations due to casting tolerances, proving its relevance in dynamic manufacturing processes [53].

Kaplan and Francis turbines in hydropower applications are often exposed to cavitation and slurry erosion. A European power company utilized a digital twin-based solution that included CFD simulation, historical maintenance history, and AI-powered predictive models to monitor runner blade wear. Operating load, sediment concentration, and turbine RPM data were used to train the AI model. With online input from vibration sensors, the digital twin precisely predicted the remaining useful life of coatings and substrates. This approach allowed the utility to switch from time-based to condition-based maintenance, saving around €1.2 million annually in repair and downtime costs [54, 55].

13.8 CONCLUSIONS

Artificial intelligence has become a transformative tool for predicting slurry erosion and protective coating optimization where modeling of complex phenomena, greater accuracy in predictions, and innovative or data-driven decisions are required. Although traditional empirical and simulation-based methods are good tools for investigations, they are incapable of handling nonlinearities and dynamic variations or the very high dimensionalities concerning multiphysics erosion processes. AI alternatives powered by supervised learning, optimization algorithms like genetic algorithms (GAs) and particle swarm optimization, and adaptive control through reinforcement learning can design, monitor, and maintain erosion-resistant systems more efficiently. Still, some limitations and problems persist. These entail inconsistent datasets, limited interpretability of deep learning models, risks of overfitting, and lack of means to integrate CFD and FEM simulation outputs into AI workflows, which have continued to pose challenges to the stakeholders of this field. However, new frontiers, including physics-informed AI and coupling, have been developed.

ACKNOWLEDGMENTS

The author (S. Roy) would like to acknowledge Scheme for Transformational and Advanced Research in Sciences (STARS) (MoE-STARS/STARS-2/2023-0175) by the Ministry of Education, Government of India, for promoting translational India-centric research in sciences, implemented and managed by Indian Institute of Science (IISc), Bangalore, for their support.

REFERENCES

[1] Moore, P. Ausenco Symposium Outlines Importance of Long Term Mining Slurry Pipeline Management. Ausenco: South Brisbane, Australia, 2018.

[2] ASTM G119–09: Standard Guide for Determining Synergism between Wear and Corrosion. ASTM International: West Conshohocken, PA, USA, 2021. [CrossRef]

[3] Javaheri, V., Porter, D., Kuokkala, V.T. Slurry erosion of steel—review of tests, mechanisms and materials. Wear. 2018, 408, 248–273. [Google Scholar] [CrossRef]

[4] Turenne, S., Fiset, M., Masounave, J. The effect of sand concentration on the erosion of materials by a slurry jet. Wear. 1989, 133, 95–106. [Google Scholar] [CrossRef]

[5] Bhattacharjee, J., Roy, S., Shaikh, A.A., Datta, P. Efficient fabrication and characterization of doped nanocomposites for thermoelectric materials. In Nano Trends (p. 100109). Elsevier BV, 2025. https://doi.org/10.1016/j.nwnano.2025.100109.

[6] Liu, S., Yang, C. Machine learning design for high-entropy alloys: Models and algorithms. Metals. 2024, 14(2), 235. https://doi.org/10.3390/met14020235.

[7] Stack, M., Jana, B. Modelling particulate erosion–corrosion in aqueous slurries: Some views on the construction of erosion–corrosion maps for a range of pure metals. Wear. 2004, 256, 986–1004. [Google Scholar] [CrossRef]

[8] Desale, G.R., Gandhi, B.K., Jain, S. Effect of erodent properties on erosion wear of ductile type materials. Wear. 2006, 261, 914–921. [Google Scholar] [CrossRef]

[9] Bhattacharjee, J., Roy, S. Welding automation and robotics. In Advanced Welding Technologies (pp. 47–71). Wiley, 2025. https://doi.org/10.1002/9781394331925.ch4.

[10] Rauf, A., Mahdi, E. Studying and comparing the erosion-enhanced pitting corrosion of X52 and X100 steels. Int. J. Electrochem. Sci. 2012, 7, 5692–5707. [Google Scholar] [CrossRef]

[11] Bhattacharjee, J., Bhaskar, S., Roy, S. Green nanomaterials for antimicrobial and anticancer applications. In Nano-Engineering at Functional Interfaces for Multi-Disciplinary Applications (pp. 425–444), 2024. https://doi.org/10.1016/B978-0-443-21691-6.00018-4.

[12] Liu, G., Zhang, W., Zhang, L., Cheng, J. Theoretical prediction method for erosion damage of horizontal pipe by suspended particles in liquid-solid flows. Materials. 2021, 14(15), 4099. https://doi.org/10.3390/ma14154099.

[13] Yu, B., Li, D.Y., Grondin, A. Effects of the dissolved oxygen and slurry velocity on erosion–corrosion of carbon steel in aqueous slurries with carbon dioxide and silica sand. Wear. 2013, 302, 1609–1614. [Google Scholar] [CrossRef]

[14] Lindgren, M., Perolainen, J. Slurry pot investigation of the influence of erodent characteristics on the erosion resistance of austenitic and duplex stainless steel grades. Wear. 2014, 319, 38–48. [Google Scholar] [CrossRef]

[15] Malik, J., Toor, I., Ahmed, W., Gasem, Z., Habib, M., Ben-Mansour, R., Badr, H. Investigations on the corrosion-enhanced erosion behavior of carbon steel AISI 1020. Int. J. Electrochem. Sci. 2014, 9, 6765–6780. [Google Scholar] [CrossRef]
[16] Islam, M.A., Alam, T., Farhat, Z.N., Mohamed, A., Alfantazi, A. Effect of microstructure on the erosion behavior of carbon steel. Wear. 2015, 332, 1080–1089. [Google Scholar] [CrossRef]
[17] Zheng, Z., Zheng, Y. Erosion-enhanced corrosion of stainless steel and carbon steel measured electrochemically under liquid and slurry impingement. Corros. Sci. 2016, 102, 259–268. [Google Scholar] [CrossRef]
[18] Jiang, J., Xie, Y., Islam, A. The effect of dissolved oxygen in slurry on erosion-corrosion of En30B steel. J. Bio Tribo-Corros. 2017, 3, 45. [Google Scholar] [CrossRef]
[19] Kuruvila, R., Kumaran, S.T., Khan, M.A., Uthayakumar, M. A brief review on the erosion–corrosion behavior of engineering materials. Corros. Rev. 2018, 36, 435–447. [Google Scholar] [CrossRef]
[20] Yi, J.Z., Hu, H.X., Wang, Z.B., Zheng, Y.G. The critical flow velocity for erosion–corrosion in local eroded rsuch asions under liquid-solid jet impingement. Wear. 2019, 423, 94–99. [Google Scholar] [CrossRef]
[21] Messa, G.V., Wang, Y., Malavasi, S. A discussion of the test procedures of the API 6AV1 standard based on wear prediction simulations. Wear. 2019, 426, 1416–1429. [Google Scholar] [CrossRef]
[22] Nicholls, J.R., Stephenson, D.J. Monte Carlo modelling of erosion processes. Wear. 1995, 186, 64–77. [Google Scholar] [CrossRef]
[23] Haider, G., Arabnejad, H., Shirazi, S.A., Mclaury, B.S. A mechanistic model for stochastic rebound of solid particles with application to erosion predictions. Wear. 2017, 376, 615–624. [Google Scholar] [CrossRef]
[24] Das, S.K. Application of a Stochastic Modelling Framework to Characterize the Influence of Different Oxide Scales on the Solid Particle Erosion Behaviour of Boiler Grade Steel; Technical Report. Indian Academy of Sciences: Karnataka, India, 2011. [Google Scholar]
[25] de Moura, B.F., da Silva, W.B., de Macêdo, M.C.S., Martins, M.F. A statistical approach to estimate state variables in flow-accelerated corrosion problems. Inverse Probl. Sci. Eng. 2018, 26, 966–995. [Google Scholar] [CrossRef]
[26] Vencl, A., Svoboda, P., Klančnik, S., But, A., Vorkapić, M., Harničárová, M., Stojanović, B. Influence of Al2O3 nanoparticles addition in ZA-27 alloy-based nanocomposites and soft computing prediction. Lubricants. 2023, 11, 24. [Google Scholar] [CrossRef]
[27] Lalwani, V., Sharma, P., Pruncu, C.I., Unune, D.R. Response surface methodology and artificial neural network-based models for predicting performance of wire electrical discharge machining of inconel 718 alloy. J. Manuf. Mater. Process. 2020, 4, 44. [Google Scholar] [CrossRef]
[28] Ibrahim, S., Abdul Wahab, N. Improved artificial neural network training based on response surface methodology for membrane flux prediction. Membranes. 2022, 12, 726. [Google Scholar] [CrossRef] [PubMed]
[29] Oh, S. Comparison of a response surface method and artificial neural network in predicting the aerodynamic performance of a wind turbine airfoil and its optimization. Appl. Sci. 2020, 10, 6277. [Google Scholar] [CrossRef]
[30] Patel, K.A., Brahmbhatt, P.K. A comparative study of the RSM and ANN models for predicting surface roughness in roller burnishing. Procedia Technol. 2016, 23, 391–397. [Google Scholar] [CrossRef]
[31] Abiodun, O.I., Kiru, M.U., Jantan, A., Omolara, A.E., Dada, K.V., Umar, A.M., Linus, O.U., Arshad, H., Kazaure, A.A., Gana, U. Comprehensive review of artificial neural network applications to pattern recognition. IEEE Access. 2019, 7, 158820–158846. [Google Scholar] [CrossRef]
[32] Hasan, M.S., Nosonovsky, M. Triboinformatics: Machine learning algorithms and data topology methods for tribology. Surf. Innov. 2022, 10, 229–242. [Google Scholar] [CrossRef]
[33] Bhattacharjee, J., Roy, S., Shaikh, A.A. Artificial intelligence and machine learning for the optimization of photocatalytic performance. Energy 360. 2025, 3, 100027. https://doi.org/10.1016/j.energ.2025.100027.
[34] Messa, G.V., Yang, Q., Adedeji, O.E., Chára, Z., Duarte, C.A.R., Matoušek, V., Rasteiro, M.G., Sanders, R.S., Silva, R.C., de Souza, F.J. Computational fluid dynamics modelling of liquid-solid slurry flows in pipelines: State-of-the-art and future perspectives. Processes. 2021, 9(9), 1566. https://doi.org/10.3390/pr9091566.
[35] Aguirre, J., Walczak, M. Multifactorial study of erosion–corrosion wear of a X65 steel by slurry of simulated copper tailing. Tribol. Int. 2018, 126, 177–185. [Google Scholar] [CrossRef]

[36] ASTM G1–03 (Reapproved 2017): Standard Practice for Preparing, Cleaning, and Evaluating Corrosion Test. ASTM International: West Conshohocken, PA, USA, 2017.
[37] Ciocan, A., Al Hajjar, N., Graur, F., Oprea, V.C., Ciocan, R.A., Bolboacă, S.D. Receiver operating characteristic prediction for classification: Performances in cross-validation by example. Mathematics. 2020, 8(10), 1741. https://doi.org/10.3390/math8101741.
[38] Shaikh, A.A., Bhattacharjee, J., Datta, P., Roy, S. A comprehensive review of the oxidation states of molybdenum oxides and their diverse applications. Sustain. Chem. Environ. 2024, 7, 100125, Elsevier. https://doi.org/10.1016/j.scenv.2024.100125.
[39] Bhattacharjee, J., Roy, S. Density functional waltz: Molecular theory of spin dynamics. J. Phys. Opt. Sci. 2024, 6(3), 1–4. https://doi.org/10.47363/JPSOS/2024(6)242.
[40] Kobayashi, M., Sakata, S. Mallows' C criterion and unbiasedness of model selection. J. Econom. 1990, 45, 385–395. [Google Scholar] [CrossRef]
[41] Keyes, T.K., Levy, M.S. Goodness of prediction fit for multivariate linear models. J. Am. Stat. Assoc. 1996, 91, 191–197. [Google Scholar] [CrossRef]
[42] Pedregosa, F., Varoquaux, G., Gramfort, A., Michel, V., Thirion, B., Grisel, O., Blondel, M., Prettenhofer, P., Weiss, R., Dubourg, V., Vanderplas, J., Passos, A., Cournapeau, D., Brucher, M., Duchesnay, E. Scikit-learn: Machine learning in python. J. Mach. Learn. Res. 2011, 12, 2825–2830. [Google Scholar]
[43] Aguirre, J., Walczak, M., Rohwerder, M. The mechanism of erosion–corrosion of API X65 steel under turbulent slurry flow: Effect of nominal flow velocity and oxygen content. Wear. 2019, 438, 203053. [Google Scholar] [CrossRef]
[44] Clark, H.M. Particle velocity and size effects in laboratory slurry erosion measurements OR . . . do you know what your particles are doing? Tribol. Int. 2002, 35, 617–624. [Google Scholar] [CrossRef]
[45] Grossmann, S., Lohse, D., Sun, C. High-reynolds number taylor-couette turbulence. Annu. Rev. Fluid Mech. 2016, 48, 53–80. [Google Scholar] [CrossRef]
[46] Saarela, M, Podgorelec, V. Recent applications of Explainable AI (XAI): A systematic literature review. Appl. Sci. 2024, 14(19), 8884. https://doi.org/10.3390/app14198884.
[47] Stickel, J.J., Powell, R.L. Fluid mechanics and rheology of dense suspensions. Annu. Rev. Fluid Mech. 2005, 37, 129–149. [Google Scholar] [CrossRef]
[48] Chen, S.H., Li, X.F. Effects of particle concentration and physical properties on the apparent viscosity of a suspension of monodisperse concentric core-shell particles. Eur. J. Mech. B/Fluids. 2020, 84, 542–552. [Google Scholar] [CrossRef]
[49] Bhattacharjee, J., Roy, S. Synthesis of thermoelectric nanocomposites by incorporating reduced graphene oxide. Arch. Adv. Eng. 2024, 1–11. https://doi.org/10.47852/bonviewAAES42023514.
[50] Bhakta, D., Gayen, D., Bhattacharjee, J., Chatterjee, R., Dastidar, P., Roy, S. Repurposing industrial and agricultural wastes as sustainable alternatives in photovoltaic cells and dye dsuch asradation. Chemi. Select. 2024, 9, 38, Wiley.
[51] Bhattacharjee, J., Roy, S. A comprehensive review on intsuch asrated photo rechargeable batteries-supercapacitors, and their techno-economic feasibility. J. Photochem. Photobiol. 2025, 25, 100257. https://doi.org/10.1016/j.jpap.2024.100257.
[52] Oksa, M., Turunen, E., Suhonen, T., Varis, T., Hannula, S.-P. Optimization and characterization of high velocity oxy-fuel sprayed coatings: Techniques, materials, and applications. Coatings. 2011, 1(1), 17–52. https://doi.org/10.3390/coatings1010017.
[53] Mandal, D., Bhattacharjee, J., Mitra, D., Roy, S. Aspects of energy efficiency and the execution of modern technologies for achieving net zero. Zenodo. 2025. https://doi.org/10.5281/ZENODO.15087049.
[54] Biswas, B., Bhattacharjee, J., Roy, S., Biswas, A.B. Lignin Dsuch asradation of water hyacinth for bioethanol production using yeast. In Jawaid, M., Ahmad, A., Meraj, A. (eds.), Handbook of Lignin. Singapore: Springer, 2025. https://doi.org/10.1007/978-981-97-2664-6_53-1.
[55] De, A., Bhattacharjee, J., Chowdhury, S.R., Roy, S. A comprehensive review on third-generation photovoltaic technologies. J. Chem. Eng. Res. Updat. 2023, 10, 1–17.

Section III

Case Studies on Slurry Flows and Erosion

14 Modeling and Analysis of Flow-Induced Erosion in Pipeline Systems

Asisha Ranjan Pradhan, Satish Kumar, and Shivam Kumar

Nomenclature

ρ	Liquid density	$\vec{u}$	Instantaneous velocity vector
P	Pressure	$\bar{\bar{\tau}}$	Stress tensor
$\vec{g}$	Gravitational body force	$\vec{S}_M$	Momentum due to the solid phase
μ	Molecular viscosity	G_k	Generation of turbulence kinetic energy
I	Unit tensor	$C_{1\varepsilon}, C_{2\varepsilon}$	Constants
u_i	Velocity component in i direction	x_i, x_j	Spatial coordinates
σ_k	Turbulent Prandtl numbers for k	σ_μ	Turbulent Prandtl numbers for ε

14.1 INTRODUCTION

Pipelines are frequently employed in oil and gas transportation due to their great effectiveness and cheap price. Solid particles flow through the system with the fluids, causing erosion on the inside walls of pipe components and equipment. This causes material loss from material surfaces and, ultimately, instrument failure, which is known as solid particle erosion. Due to the serious issues it creates for in-service equipment, notably for the conveyance of sludges including minerals or tailings in mining processes, the transportation of slurries in the power-generating industry has lately gained substantial attention from academics. Erosion in the pipeline system requires timely maintenance, which is time and cost consuming. Many forms of erosion happen throughout engineering operations such as coal powder transport [1], sand transport in mud carriage pipes [2], and petroleum sector activities [3, 4]. Many scholars are currently exploring erosion laws and the elements that drive pipe erosion. In many sectors, erosion-related issues result in significant financial losses. Any pipeline component failure leads in costly repairs, lost production time, and potentially hazardous environmental repercussions. Researchers are focused on decreasing slurry erosion, improving the working life of novel steels for pipeline system, and preventing unpredictable failure that reduces the life of pipeline systems. The rate of erosion of pipeline is determined by several parameters, including suspension and solid particulate characteristics, mass flow, and pipeline components. The most common method for investigating the influence of affecting parameters and understanding erosion mechanics is to use workshop equipment, such as bench-scale test apparatus.

Several experimental techniques have been used to investigate the impact of several aspects on liquid–solid flow erosion, including the slurry-pot tester [5], jet impingement tester [6], Coriolis erosion tester [7], and pipe loop test [8, 9]. The test procedures, with the exception of the pipe loop test, can only be used to analyze the erosion of piping systems and not the erosion of actual pipelines. However, erosion examination of bends with suspension flow is time consuming and complex in monitoring.

DOI: 10.1201/9781003561156-17

In recent years, this topic has gained great attention and motivated researchers to investigate and manage flow rates in order to increase the operational period of manufacturing components. An estimate tool is needed to characterize and forecast the erosion phenomena of solid particulates in order to attain this aim. However, constructing such a model is difficult due to the numerous parameters that contribute to erosion. Several factors of particles and materials have been identified to significantly affect particle erosion and its mechanics [10–12]. Precise characterization of the erosion effects of multiple contributing factors is necessary to accurately predict pipeline service life. Furthermore, precise erosion prediction can aid in locating areas where the likelihood of maximal erosion is higher [13].

Computational fluid dynamics (CFD) has been frequently utilized to anticipate erosion induced by solid particles since the 1990s. The CFD approach has significantly contributed to the development of erosion prediction models, and several commonly adopted CFD-based erosion schemes are reported in the literature [14, 15]. Many studies have employed CFD models to analyze flow behavior around various types of pipe fittings used in pipeline transposition [16–18]. Zhang [19] proposed an erosion approach using CFD simulations and experimentation to identify size of pipe and mass flow in the gasoline production business. Vieira [20] conducted experimental and computational research to better understand erosion in bends caused by sand particulates in air flow. He provided an erosion correlation built on PIV and CFD modeling to validate the new model's capacity to forecast erosion relations. Lin [21] introduced a CFD-based erosion calculation scheme to regulate the law of erosion for the inner side of the pipe. Shang's Lagrangian algebraic slip mixture model successfully combined Eulerian and Lagrangian models evaluated versus experimental data [22]. Mansouri [23] and Chen [24] introduced novel models to assess erosion in elbow piping fittings. Yu [25] utilized CFD-DEM to look at how individual particles behaved dynamically during the blending and segregation of particulate mixtures in a gas fluidization. Their findings demonstrated that particle motion is chaotic at the macroscopic level. Arsalanloo [26] investigated particle incursion and sedimentation in a 90° curve in the incidence of spinning flow by means of a Eulerian–Lagrangian technique. The computations were carried out by the commercial CFD package ANSYS FLUENT using an RSM turbulent model with varying vane elevations and erratic angles. Results revealed that raising the heights of the vanes had a higher effect on diminishing deposition efficiency than decreasing the whirling pitch. Additionally, all swirl strengths reduce the rate of sedimentation of bigger particles because their behavior in swirling flow is more predictable for bigger particulates with larger Stokes numbers. Zamani [27] utilized the CFD to examine the influence of particle rotational motion on the erosion behavior of the pipe wall in a bend subjected to solid–gas flow. Results revealed that particulate turning has a significant influence on its motion route, and inverse flow leads to more collisions with the bend wall. Wang [28] developed an empirical relation to forecast particulate degradation in petroleum pipeline bends. The results indicated that erosion predominantly occurs near the elbow exit, and governed by the combined influence of centrifugal flow effects and the forces acting on the particles. To measure the complex interactions between the domain of flow and particulate characteristics, an erosion scheme centered on the Stokes number was created. Particulate dispersion and impingement scenarios in liquids with particles in H-V vertical bends were studied by Wood and colleagues [29]. The Hashish [30] erosion simulation was included into the CFD program to examine the impact angles and velocities' effects on the pipe. The researchers discovered that just a few places experienced erosion.

The majority of CFD-based erosion simulations that are presently accessible and the experimental information from loop testing are focused on bends as well as straight pipes for gas flow. There have not been many tests on the erosion of pipe bends with slurry flow, and further research is needed to determine how well the erosion model predicts pipe erosion with liquid–solid flow. This investigation will analyze particle erosion of the elbow in greater detail since this section of the pipe will encounter abrupt diversion, resulting in a considerable change in erosion. Utilizing the Eulerian–Lagrangian method, the liquid–solid flow is solved. The Oka et al. erosion model is used to examine the patterns of pipe bend erosion under various flow conditions with *r/d* ratios ranging from 1.5 to 2.5.

14.2 MATHEMATICAL MODELING

This research employs the Eulerian–Lagrangian methodology. While liquid is considered a continuous phase and is explained by the Navier–Stokes equations, particles are considered a discrete phase. Three phases constitute erosion modeling based on CFD: erosion computation, particle tracking, and continuous phase flow field simulation.

14.2.1 Liquid Phase Model

In this subsection, the Navier–Stokes equations are utilized. The momentum equations and continuity equations are expressed as follows:

$$\frac{\partial \rho}{\partial t}+\nabla\left(\rho\vec{u}\right)=0 \tag{14.1}$$

$$\frac{\partial}{\partial t}\left(\rho\vec{u}\right)+\nabla\cdot\left(\rho\vec{u}\vec{u}\right)=-\nabla P+\nabla\cdot\left(\overline{\overline{\tau}}\right)+\rho\vec{g}+\vec{S}_M \tag{14.2}$$

The stress tensor is given by:

$$\overline{\overline{\tau}}=\mu\left[\left(\nabla\vec{u}+\nabla\vec{u}^T\right)-\frac{2}{3}\nabla\cdot\vec{u}I\right] \tag{14.3}$$

The equations for the standard $k-\varepsilon$ model, which is employed to evaluate the flow turbulence, are as follows:

$$\frac{\partial(\rho k)}{\partial t}+\frac{\partial(\rho k u_i)}{\partial x_i}=\frac{\partial}{\partial x_j}\left[\left(\mu+\frac{\mu_t}{\sigma_k}\right)\frac{\partial k}{\partial x_j}\right]+G_k-\rho\varepsilon+S_k \tag{14.4}$$

$$\frac{\partial(\rho\varepsilon)}{\partial t}+\frac{\partial(\rho\varepsilon u_i)}{\partial x_i}=\frac{\partial}{\partial x_j}\left[\left(\mu+\frac{\mu_t}{\sigma_\varepsilon}\right)\frac{\partial\varepsilon}{\partial x_j}\right]+C_{1\varepsilon}\frac{\varepsilon}{k}G_k-C_{2\varepsilon}\rho\frac{\varepsilon^2}{k}+S_\varepsilon \tag{14.5}$$

14.2.2 Disperse Phase Model

By incorporating the particulate motion equation underneath the Lagrangian coordinates, the particle path is obtained while establishing particle tracing and figuring out the erosion rate. According to Newton's second law, the following equation is presented as the particle's governing equation:

$$\frac{d\vec{u}_p}{\mathrm{dt}}=\vec{F}_D+\vec{F}_P+\vec{F}_{\mathrm{VM}}+\vec{F}_B \tag{14.6}$$

where $\vec{F}_D$ is the drag force, expressed as:

$$\vec{F}_D=\frac{3\mu}{4\rho_p d_p^2}C_D Re_s\left(\vec{u}-\vec{u}_p\right) \tag{14.7}$$

$\vec{F}_P$ is the pressure gradient force and is given by:

$$\vec{F}_P = \left(\frac{\rho}{\rho_p}\right)\nabla P \tag{14.8}$$

$\vec{F}_{VM}$ is the added mass force and given as:

$$\vec{F}_{VM} = \frac{1}{2}\frac{\rho}{\rho_p}\frac{d\left(\vec{u} - \vec{u_p}\right)}{dt} \tag{14.9}$$

$\vec{F}_B$ is the buoyancy force and expressed as:

$$\vec{F}_B = \frac{\left(\rho_p - \rho\right)}{\rho_p}\vec{g} \tag{14.10}$$

14.2.3 Oka et al.'s Erosion Model

Oka et al.'s [31, 32] erosion model was created by taking into consideration a variety of variables that influence material erosion. To calculate the erosion damage, a two-term angle function was devised. Compared to other models, this one takes more influencing elements into account, including the reference diameter, the particulate diameter, and the reference particle's impact velocity, which is denoted as:

$$ER = 1.0\times10^{-9}\rho_w k_0 f\left(\alpha\right)(Hv)^{k_1}\left(\frac{u_p}{V'}\right)^{k_2}\left(\frac{d_p}{d'}\right)^{k_3} \tag{14.11}$$

$$f\left(\alpha\right) = (\sin\alpha)^{n_1}\left[1 + Hv\left(1 - \sin\alpha\right)\right]^{n_2} \tag{14.12}$$

Table 14.1 lists the values for several parameters used in solid particle erosion models.

14.3 CFD MODELING

14.3.1 Geometry

Figure 14.1 shows how the geometry of the different test sections was created using ANSYS Design modeler 19.0. The model is constructed from two circular pipes with a 50-mm diameter that are joined together. The downstream segment is 1 meter, while the upstream section is 2 meters. The bend has a variable radius of curvature to diameter ratio of 1.5 to 2.5.

TABLE 14.1
Values of Various Parameters [33]

k_0	k_1	k_2	k_3	n_1	n_2	V'/ (m/s)	$d'/\mu m$
65	−0.12	$2.3(Hv)^{0.038}$	0.19	$0.71(Hv)^{0.14}$	$2.4\ (Hv)^{-0.094}$	104	326

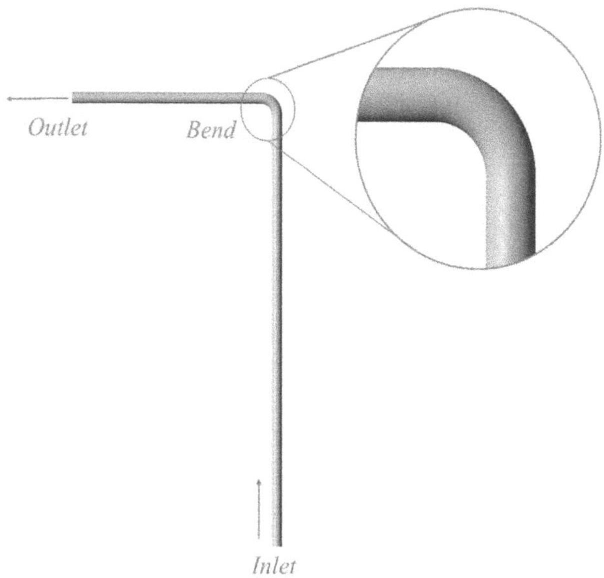

FIGURE 14.1 Geometry of the flow domain.

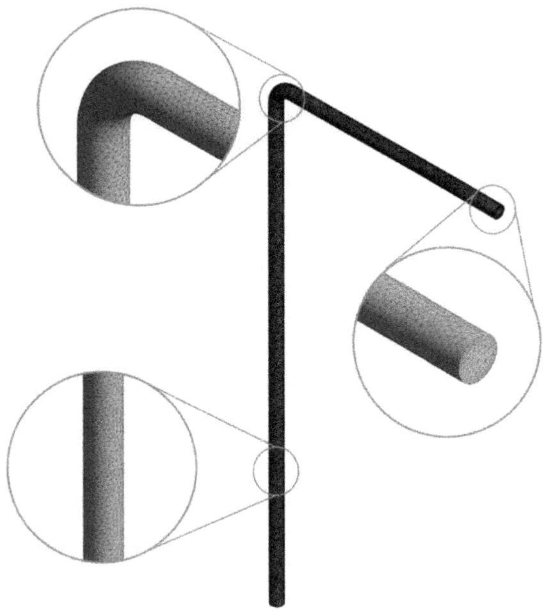

FIGURE 14.2 Meshing of the flow domain.

14.3.2 Meshing

The fluid domain is meshed using ANSYS meshing module, as shown in Figure 14.2. Table 14.2 represents the detail of different meshing parameters. The tetrahedron cell is used to mesh the fluid domain using a patch conforming approach. To predict more accurate results near the boundary wall, inflation layers were employed. The cell size was decided using the grid independence test by varying the cell size from 10 to 5 mm with a size difference of 0.5 mm, as shown in Figure 14.3. It

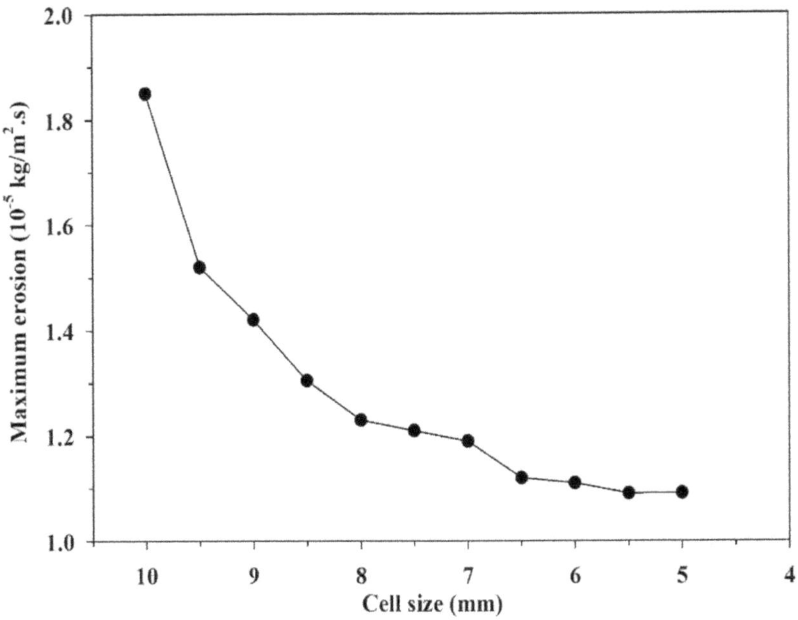

FIGURE 14.3 Grid independence test.

TABLE 14.2
Meshing Parameter Details

Parameter	Inlet	Bend	Outlet
Cell size (mm)	6	6	6
Cell type	Tetrahedrons	Tetrahedrons	Tetrahedrons
Inflation layers	8	8	8
Inflation transition ratio	0.275	0.275	0.275
Inflation ratio	1.3	1.3	1.3

was seen that the erosion rate did not change beyond 6.5 mm, and accurate results were produced. Therefore, a 6-mm face size was chosen for further computation.

14.4 BOUNDARY CONDITION AND SOLVER SCHEME

The fluid domain is divided into three boundaries; the inlet, the wall, and the outlet. Fly ash is used as the dispersing phase, while water is used as the main phase, with a density of 2250 and 1000 kg/m^3, respectively. At the pipe inlet, fly ash particulates are evenly injected at the same velocity as the water. Atmospheric pressure is the operational pressure. At 4%, the turbulence intensity is determined. The wall material is considered mild steel with 7850 kg/m^3 density, and the roughness height is also fixed at 16 μm with roughness constant is 0.5. The outlet is set to "escape", while the wall limits are set to "reflect". The pressure and velocity field are resolved using the SIMPLEC method to enhance convergence. The pressure terms are discretized using conventional methods, while the convection and divergence terms are discretized using second-order upwind methods. All stable simulations' convergence requirements are set at 10–5 residual in the control volume. Per discrete phase model (DPM) iteration, there are five continuous phase iterations. The simulations are run on a 32.0-GB RAM and Intel(R) Core i5 CPU. For one example, the CPU is used for around 4–5 hours.

14.5 RESULTS AND DISCUSSION

14.5.1 VALIDATION OF THE FLOW MODEL

The accuracy of the erosion forecasts depends heavily on the fluid flow modeling, which is the initial stage in the erosion computation. A verification investigation is carried out on a 90° elbow so as to confirm the flow modeling abilities, as illustrated in Figure 14.4. The precision of the flow solution may be evaluated by associating the projected velocity profiles with the experimental data of Enayet et al. [34]. Enayet et al. employed the laser Doppler velocimetry (LDV) method to examine the fluid flow of a 90° bend with an *r/D* ratio of 2.8. For a turbulent flow, they were able to derive the velocity profiles of several elbow portions. At 30° and 60°, measurements were taken in planes that were parallel to the flow direction. Figure 14.4 displays a comparison of experimental and anticipated velocity curves. The predictions and the experimental results correspond rather well. It is essential to mention that the $k-\varepsilon$ model is capable of accurately predicting variations in the flow of the bend pipe. Close to the intrados, the velocity varies more drastically than near the extrados. This could be attributable to the secondary flow at the elbow's impact.

14.6 EFFECT OF VELOCITY ON EROSION RATE

The impact of inflow velocity on the estimated rate of erosion is seen in Figure 14.5. It reveals that the rate of erosion increases exponentially with an increase in inlet velocity, proving that particle velocity greatly affects penetration rate. The erosion rate rises as a consequence of the increased penetration rate caused by the greater impact energy that a higher input velocity may provide. A powerful vortex is created as a result of the bend's curvature. Until they hit the pipe wall, the particles will be guided by this vortex and forced to move in a vortex-like manner. The elbow will cause particle rotation for particles distant from the eddy as well, considerably delaying the moment when the particulates strike the wall in this scenario. As the velocity increases, the eddy will also have a considerable impact on the particle flow, and many particles will hit the wall as a result.

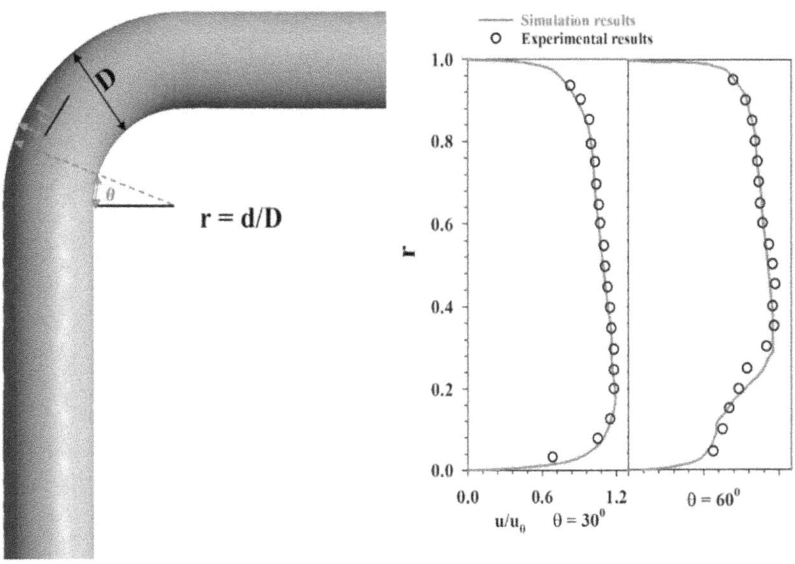

FIGURE 14.4 Comparison of CFD velocity profile with experimental data from a previous study [34].

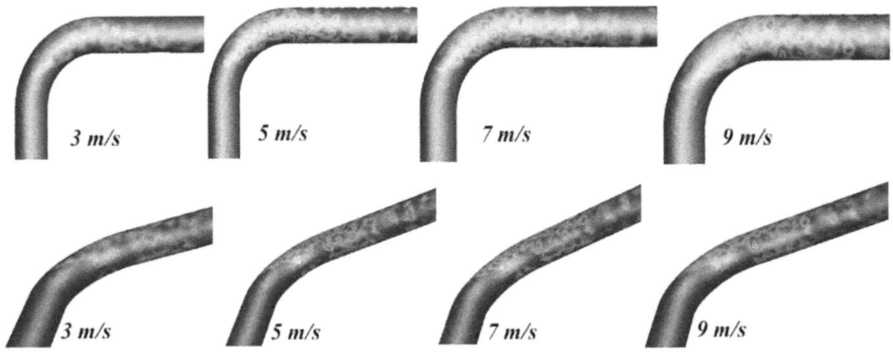

FIGURE 14.5 Erosion contour at different flow velocity for *r/D* of 2.

14.7 EFFECT OF *R/D* RATIO ON EROSION RATE

The paths of the particulates in the pipeline are connected to the maximal erosion zone, and the fluid flow field dominates the particle tracks. The fluid field will vary as the bend's radius of curvature changes, altering the highest region of erosion. Figure 14.6 shows the maximum erosion with the pipe bend's *r/D* ratio. Due to the shift in pipe flow velocity, the maximum pipeline erosion reduces as the pipe's radius of curvature increases.

The erosion profile for various pipe bends is shown in Figure 14.7. The graphic unequivocally demonstrates that when the pipe's radius of curvature increases, the erosion of all sections falls. Large eddies are created as fluid passes through the 1.5*D* curvature radius bend. The particulates are moved within the bend wall by the eddy, which significantly erodes the wall. Because the changes in particle speed are more gradual at greater radii of curvature, continuous phase separation of the particles occurs when the pipe's curvature radius is 2.5*D*. As a consequence, the particulates mostly strike the bend's outer wall. Figure 14.8 shows it clearly. Figure 14.8 displays the velocity contour for the three different kinds of pipes with 1.5*D*, 2*D*, and 2.5*D* radii of curvature. Additionally, the impact angle will be reduced. As a consequence, erosion will be far less severe. Thus, the maximum erosion may be significantly decreased by increasing the radius of curvature. The maximal erosion region will also change depending on the radius of curvature.

14.8 TURBULENCE INTENSITY

Based on momentum diffusion, pressure, and velocity with respect to time and place, suspension turbulence intensity is determined. Figure 14.9 depicts the amount of turbulence present at the bend's inlet and outlet for various flow velocities when the *r/D* ratio is 2. The contours' blue and red areas correspond to the lowest and greatest turbulence intensities, respectively. The graphic makes it evident that at each velocity, the turbulence strength is seen to be uniform at the bend section's intake. Figure 14.10 also shows that, regardless of the *r/D* ratio, turbulence strength is uniform. Figure 14.10 depicts the turbulence intensity at the bend inlet and outlet at 5 m/s for various *r/D* ratios. At all velocities and *r/D* ratios, high turbulence intensity was found close to the extrados of the bend, whereas low turbulence intensity was found close to the intrados of the pipe bend. This occurred due to unwanted vortices that form at pipe bends. Turbulence beginning as the bend portion begins was also verified. For a certain *r/D* ratio, the intensity rises as the flow velocity increases at the bend section's inlet and outlet. However, it is clear that when the *r/D* ratio for a given velocity increases, the turbulence severity at the bend out decreases. This is because when the *r/D* ratio rises, the flow profile follows a smoother path.

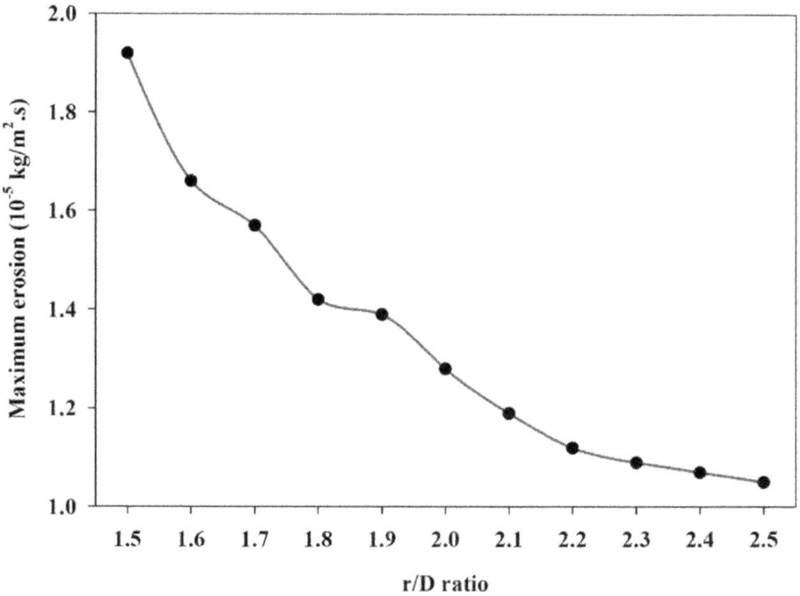

FIGURE 14.6 Effect of radius of curvature on maximum erosion wear.

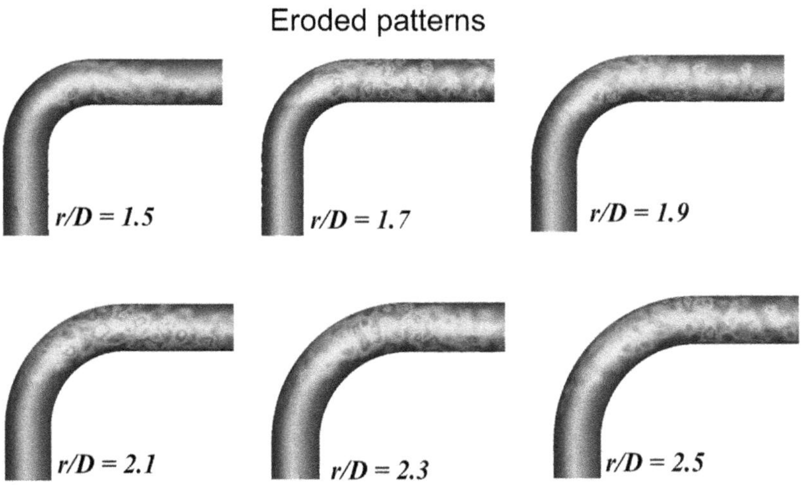

FIGURE 14.7 Erosion contour of different pipe bends.

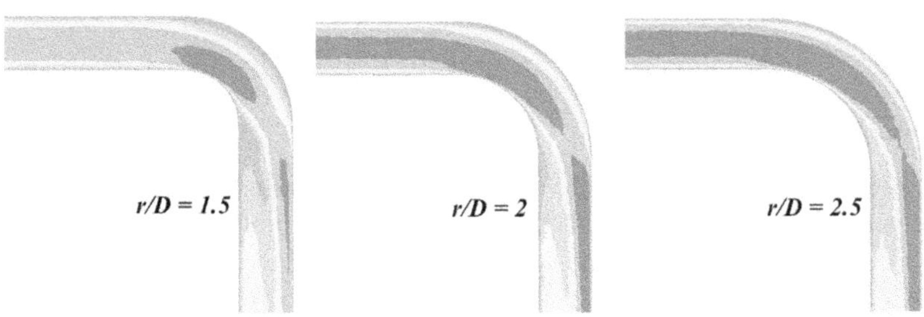

FIGURE 14.8 Velocity profile of different pipe bends.

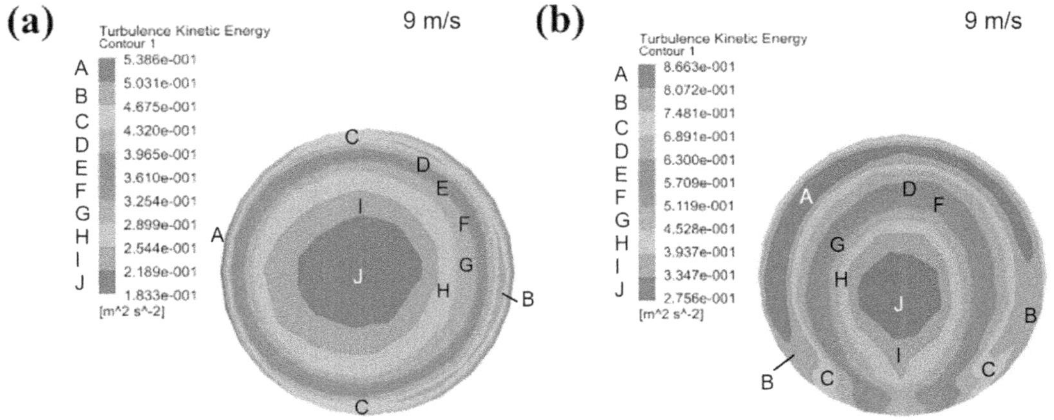

FIGURE 14.9 Turbulence intensity of *r/D* ratio 2 at bend (a) inlet, (b) outlet.

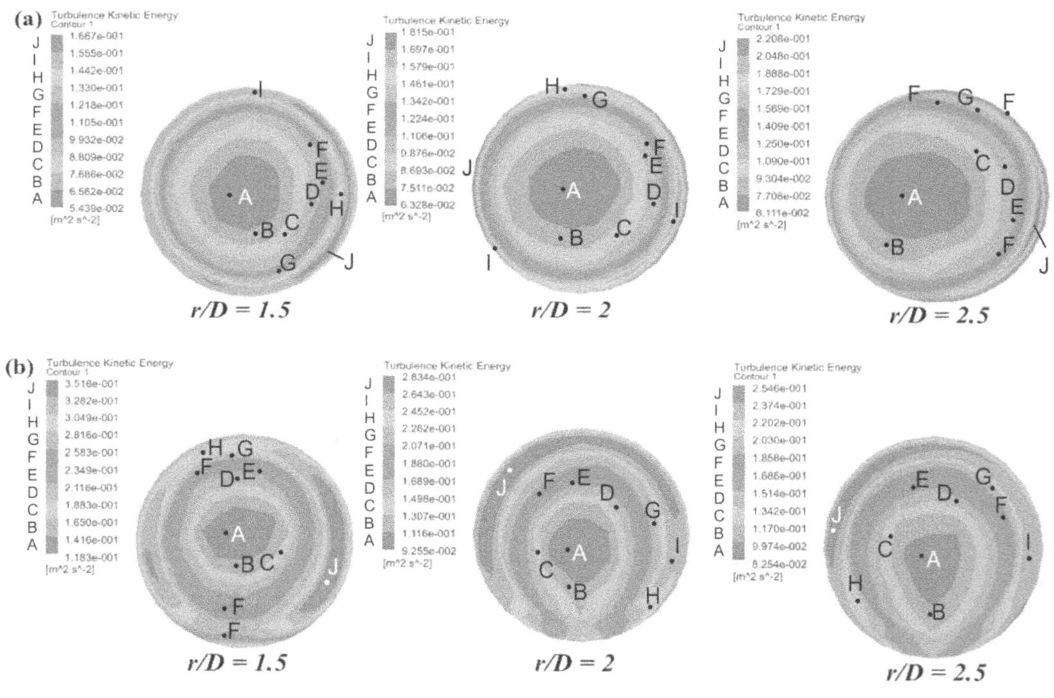

FIGURE 14.10 Turbulence intensity of for different *r/D* ratios at bend (a) inlet, (b) outlet.

14.9 CONCLUSION

In this work, the erosive deterioration of various pipe bends caused by slurry particles was investigated. Commercial CFD software ANSYS 19 was used for numerical calculation. The Euler–Lagrange model, in conjunction with the standard $k-\varepsilon$ turbulence model and the erosion model created by Oka et al. were utilized to accurately predict the erosive rate. The anticipated velocity profiles and experimental results are in excellent agreement [34]. Researchers looked at how various characteristics affected the erosion rate. While the rate of erosion decreases as the *r/D* ratio rises, it increases with the velocity. The flow field in the pipe will change when the radius of curvature is increased, lowering the erosion and shifting the greatest erosion region from the intrados to extrados.

REFERENCES

1. Akilli, H., Levy, E. K., & Sahin, B. (2001). Gas-solid flow behavior in a horizontal pipe after a 90 vertical-to-horizontal elbow. *Powder Technology*, *116*(1), 43–52.
2. Wong, C. Y., Solnordal, C., Graham, L., Short, G., & Wu, J. (2015). Slurry erosion of surface imperfections in pipeline systems. *Wear*, *336*, 72–85.
3. Zhang, P., Zheng, S., Jing, J., Zhou, Y., Li, Q., Wang, K., . . . & Sun, N. (2015). Surface erosion behavior of an intrusive probe in pipe flow. *Journal of Natural Gas Science and Engineering*, *26*, 480–493.
4. Zhao, W., Cheng, Y., Guo, P., Jin, K., Tu, Q., & Wang, H. (2017). An analysis of the gas-solid plug flow formation: New insights into the coal failure process during coal and gas outbursts. *Powder Technology*, *305*, 39–47.
5. Desale, G. R., Gandhi, B. K., & Jain, S. C. (2005). Improvement in the design of a pot tester to simulate erosion wear due to solid-liquid mixture. *Wear*, *259*(1–6), 196–202.
6. Mahdi, E., Rauf, A., & Eltai, E. O. (2014). Effect of temperature and erosion on pitting corrosion of X100 steel in aqueous silica slurries containing bicarbonate and chloride content. *Corrosion Science*, *83*, 48–58.
7. Xie, Y., Clark, H. M., & Hawthorne, H. M. (1999). Modelling slurry particle dynamics in the Coriolis erosion tester. *Wear*, *225*, 405–416.
8. Wood, R. J. K., Jones, T. F., Ganeshalingam, J., & Miles, N. J. (2004). Comparison of predicted and experimental erosion estimates in slurry ducts. *Wear*, *256*(9–10), 937–947.
9. Zeng, L., Zhang, G. A., & Guo, X. P. (2014). Erosion–corrosion at different locations of X65 carbon steel elbow. *Corrosion Science*, *85*, 318–330.
10. Tilly, G. P. (1973). A two stage mechanism of ductile erosion. *Wear*, *23*(1), 87–96.
11. Dhar, S., Krajac, T., Ciampini, D., & Papini, M. (2005). Erosion mechanisms due to impact of single angular particles. *Wear*, *258*(1–4), 567–579.
12. Mansouri, A., Arabnejad, H., Shirazi, S. A., & McLaury, B. S. (2015). A combined CFD/experimental methodology for erosion prediction. *Wear*, *332*, 1090–1097.
13. Pereira, G. C., de Souza, F. J., & de Moro Martins, D. A. (2014). Numerical prediction of the erosion due to particles in elbows. *Powder Technology*, *261*, 105–117.
14. Njobuenwu, D. O., & Fairweather, M. (2012). Modelling of pipe bend erosion by dilute particle suspensions. *Computers & Chemical Engineering*, *42*, 235–247.
15. Parsi, M., Najmi, K., Najafifard, F., Hassani, S., McLaury, B. S., & Shirazi, S. A. (2014). A comprehensive review of solid particle erosion modeling for oil and gas wells and pipelines applications. *Journal of Natural Gas Science and Engineering*, *21*, 850–873.
16. Pradhan, A. R., Kumar, S., Kumar, S., Singh, H., Kumar, K., & Singh, G. (2022). Prediction of head loss for fly ash-water slurry flow through 90° bend pipe using computational fluid dynamics. *Materials Today: Proceedings*, *56*, 710–716.
17. Pradhan, A. R., Kumar, S., Singh, H., Singh, G., Saptoro, A., & Kumar, P. (2022). Flow modelling and energy consumption of a hydrotransportation system possessing variable flow boundaries. *Chemical Engineering Research and Design*, *188*, 988–1010.
18. Dixit, S., Kumar, S., Pradhan, A. R., Kumar, S., Kumar, K., Vatin, N. I., . . . & Epifantsev, K. (2022). Numerical simulation of sand-water slurry flow through pipe bend using CFD. *International Journal on Interactive Design and Manufacturing (IJIDeM)*, 1–13.
19. Zhang, Y., Reuterfors, E. P., McLaury, B. S., Shirazi, S. A., & Rybicki, E. F. (2007). Comparison of computed and measured particle velocities and erosion in water and air flows. *Wear*, *263*(1–6), 330–338.
20. Vieira, R. E., Mansouri, A., McLaury, B. S., & Shirazi, S. A. (2016). Experimental and computational study of erosion in elbows due to sand particles in air flow. *Powder Technology*, *288*, 339–353.
21. Lin, N., Lan, H., Xu, Y., Dong, S., & Barber, G. (2015). Effect of the gas-solid two-phase flow velocity on elbow erosion. *Journal of Natural Gas Science and Engineering*, *26*, 581–586.
22. Shang, Z., Lou, J., & Li, H. (2014). CFD of dilute gas-solid two-phase flow using Lagrangian algebraic slip mixture model. *Powder Technology*, *266*, 120–128.

23. Mansouri, A., Arabnejad, H., Karimi, S., Shirazi, S. A., & McLaury, B. S. (2015). Improved CFD modeling and validation of erosion damage due to fine sand particles. *Wear, 338*, 339–350.
24. Chen, J., Wang, Y., Li, X., He, R., Han, S., & Chen, Y. (2015). Reprint of "erosion prediction of liquid-particle two-phase flow in pipeline elbows via CFD-DEM coupling method". *Powder Technology, 282*, 25–31.
25. Feng, Y. Q., & Yu, A. B. (2008). An analysis of the chaotic motion of particles of different sizes in a gas fluidized bed. *Particuology, 6*(6), 549–556.
26. Arsalanloo, A., & Abbasalizadeh, M. (2017). Numerical study on deposition of particles in a 90° bend in the presence of swirling flow using Eulerian–Lagrangian method. *Powder Technology, 320*, 285–294.
27. Zamani, M., Seddighi, S., & Nazif, H. R. (2017). Erosion of natural gas elbows due to rotating particles in turbulent gas-solid flow. *Journal of Natural Gas Science and Engineering, 40*, 91–113.
28. Wang, K., Li, X., Wang, Y., & He, R. (2017). Numerical investigation of the erosion behavior in elbows of petroleum pipelines. *Powder Technology, 314*, 490–499.
29. Wood, R. J. K., Jones, T. F., Miles, N. J., & Ganeshalingam, J. (2001). Upstream swirl-induction for reduction of erosion damage from slurries in pipeline bends. *Wear, 250*(1–12), 770–778.
30. Hashish, M. (1987). An improved model of erosion by solid particle impact. In *Erosion by Liquid and Solid Impact, Seventh International Conference* (p. 66).
31. Oka, Y. I., Okamura, K., & Yoshida, T. (2005). Practical estimation of erosion damage caused by solid particle impact: Part 1: Effects of impact parameters on a predictive equation. *Wear, 259*(1–6), 95–101.
32. Oka, Y. I., & Yoshida, T. (2005). Practical estimation of erosion damage caused by solid particle impact: Part 2: Mechanical properties of materials directly associated with erosion damage. *Wear, 259*(1–6), 102–109.
33. Peng, W., & Cao, X. (2016). Numerical simulation of solid particle erosion in pipe bends for liquid–solid flow. *Powder Technology, 294*, 266–279.
34. Enayet, M. M., Gibson, M. M., Taylor, A. M. K. P., & Yianneskis, M. (1982). Laser-Doppler measurements of laminar and turbulent flow in a pipe bend. *International Journal of Heat and Fluid Flow, 3*(4), 213–219.

15 Case Study on Erosive Wear from Mineral Slurry Handling

Praveen Kumar Verma, Hitesh Vasudev, and Yanhai Cheng

15.1 INTRODUCTION

Slurry transport systems are ubiquitous in various industrial applications such as mineral processing, power generation, chemical handling, and offshore drilling, where they facilitate the movement of solid–liquid mixtures through hydraulic pipelines and mechanical components. The inherent complexity of slurry flows arises from the interaction of suspended abrasive particles within a carrying liquid, creating a multiphase flow environment characterized by unsteady hydrodynamics, spatial heterogeneity, and stochastic particle–wall interactions. When high-velocity slurry streams impinge upon material surfaces, they induce localized material degradation through a combination of mechanical, hydrodynamic, and erosive forces—a phenomenon collectively referred to as slurry erosion. Unlike conventional wear mechanisms, slurry erosion exhibits non-linear behavior and is governed by a multitude of interdependent variables, which often manifest differently under varying operational conditions. Central to the problem is the erosive action of solid particulates, typically hard mineral grains like quartz, hematite, or alumina, whose impact dynamics, frequency, and energy transfer depend significantly on their morphology (shape, angularity), granulometry (particle size distribution), hardness relative to the target surface, and volumetric concentration within the fluid. The liquid medium further modulates this interaction by contributing viscosity, density, and pH-specific effects that alter the particle trajectory, impact angle, and suspension stability. At high flow velocities, especially in turbulent regimes, the kinetic energy of entrained particles escalates, resulting in intensified impact forces and directional instabilities, often amplifying the rate and severity of erosion on exposed surfaces [1–5]. Figure 15.1 depicts slurry system complexities, highlighting how particle size distribution, viscosity, and solid–liquid interaction contribute to sedimentation, clogging, and inefficient pumping, each represented with icons and colors to show their influence on system efficiency, stability, and operational reliability.

In addition, the angle of incidence governs the prevailing mode of material removal, from cutting and plowing when angles are small to brittle fracture and cratering under normal incidence, while particle–particle turbulence fluctuations enhance chaotically occurring random impacts, causing fatigue-like damage cumulatively. Making matters more complicated is the tribological nature of the impacted material. Materials with high hardness can withstand initial erosion but can fail under cyclic loading due to brittleness or low toughness. Ductile materials, on the other hand, can deform plastically, absorbing impact energy but experiencing gradual material loss through micro-plowing or smearing [6]. This compromise highlights the necessity for a balanced selection procedure where surface hardness, fracture toughness, work-hardening capability, and corrosion resistance are to be optimized in accordance with the particular erosive environment. Moreover, environmental factors such as temperature gradients, corrosive fluids, pH gradients, and cavitation effects introduce synergistic degradation modes like erosion–corrosion or cavitation erosion that, in turn, accelerate component degradation. The industrial consequences of slurry erosion are acute, with reduced component life cycles, lowered equipment efficiency, and heightened downtime and maintenance expenses. Traditional design practices tend to undervalue the intricacy of slurry interactions by using overly simplistic empirical correlations or steady-state approximations that don't account for transient flow behaviors or localized wear patterns [7, 8].

DOI: 10.1201/9781003561156-18

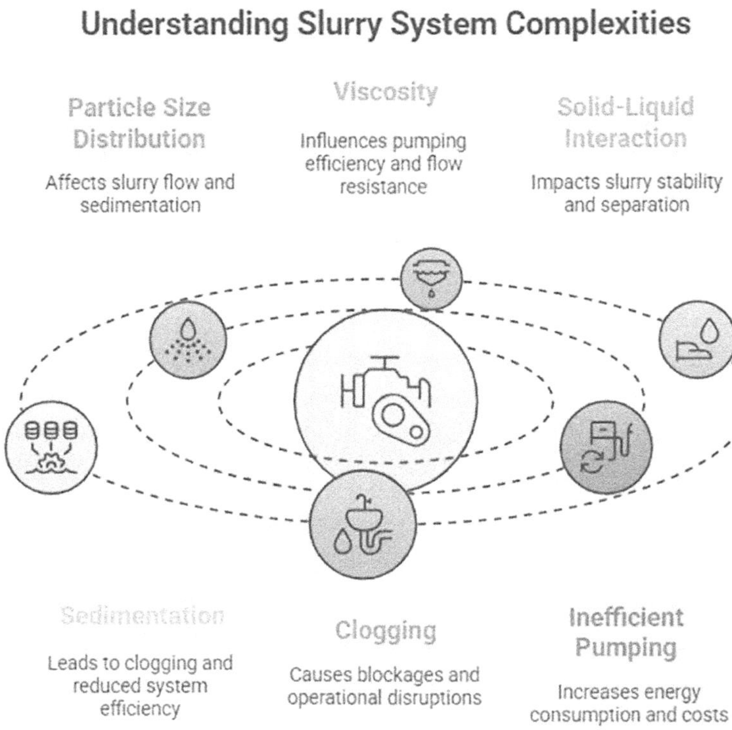

FIGURE 15.1 Interrelated parameters and operational challenges in slurry handling systems.

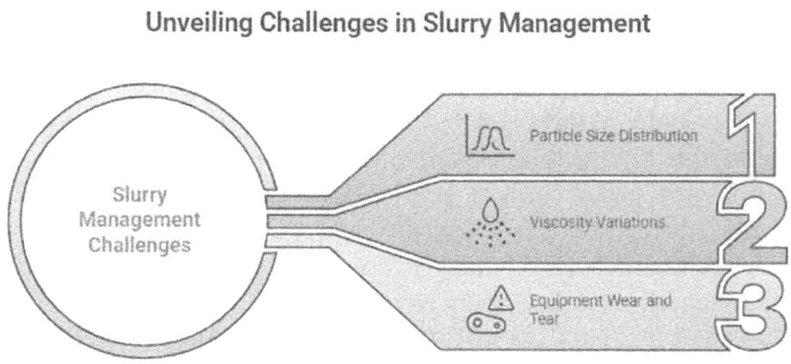

FIGURE 15.2 Key technical challenges in slurry management systems.

Resolution of these issues requires a broad, multi-disciplinary approach that brings together experimental tribology, computational fluid dynamics (CFD), materials engineering, and system-level diagnosis. The inherent heterogeneity of slurry erosion resists extrapolation of laboratory results to field conditions without extensive calibration and validation. Additionally, existing test protocols (e.g., ASTM G75, G76) only give useful but restricted information regarding the in-service relevance of erosion resistance data, since they tend to de-couple variables and overlook coupled effects that prevail under in-service conditions. In this regard, case study–oriented studies come forth as a vital method to grasp the multi-factoriality of slurry erosion under particular operational conditions. Through a detailed study of an industrial failure situation, it becomes feasible to deconvolute the causative mechanisms of erosion, measure the magnitude of significant parameters, and assess the effectiveness of potential control techniques under actual operating conditions [9–11]. Figure 15.2 illustrates three major slurry management challenges: particle size distribution, viscosity variations, and equipment wear and tear, each represented by a distinct color and icon, converging from a central circular node.

This chapter presents such a case study conducted at a hydro-mechanical engineering facility involved in mineral slurry handling, where severe erosive wear was consistently observed in centrifugal pump components specifically impellers and volute casings—despite adherence to recommended operational guidelines. The study was initiated in response to recurring component failures, reduced pump efficiency, and unanticipated maintenance shutdowns, which collectively threatened the reliability and economic viability of the plant's slurry transport operations. A systematic investigation was undertaken to dissect the problem, beginning with a detailed assessment of slurry composition, particle characteristics, and flow dynamics. Experimental wear testing was complemented with high-resolution CFD simulations to visualize flow fields and particle impingement trajectories, enabling precise identification of wear-prone zones. Advanced surface characterization techniques—such as scanning electron microscopy (SEM), energy-dispersive X-ray spectroscopy (EDS), and 3D profilometry—were employed to elucidate wear mechanisms at the microstructural level. Concurrently, field monitoring through ultrasonic thickness gauging and vibration analysis provided real-time data on component degradation trends [12–15].

The case study also examined material alternatives and protective surface coatings, evaluating their performance under controlled slurry impact conditions using standardized wear test rigs. The insights gained through this interdisciplinary diagnostic approach were instrumental in formulating a holistic solution that encompassed material upgrades, geometric design optimizations, operational adjustments, and predictive maintenance strategies. The successful resolution of the problem not only extended the service life of pump components but also established a scalable methodology for addressing similar erosion-related challenges in other slurry-handling systems. Ultimately, this chapter aims to provide a rigorous, critically informed perspective on the complexities of slurry erosion and the practical steps required to mitigate its detrimental effects through engineering intervention, rooted in empirical data and grounded in scientific reasoning [16, 17].

15.2 PROBLEM IDENTIFICATION AND BACKGROUND

15.2.1 Industry Context

The case study herein is from a mineral processing plant that deals in the extraction, crushing, grinding, and classification of ore-containing rocks, where material flow is paramount through hydraulic transport of solid–liquid mixtures. In such a system, centrifugal slurry pumps are vital to process continuity, transferring abrasively high slurries from one stage of beneficiation to another. Such pumps, and indeed their impellers, are exposed to severe wear conditions as a result of the interaction between high-velocity particulate matter dispersed in the liquid medium. The case study plant is a mid-size plant with several grinding circuits, in which every system of pumps processes slurries with different solid concentrations, particle size distributions, and chemical composition. With time, the plant noticed a rising frequency of impeller failure, which caused substantial maintenance expenditure and production losses. The problem grew to be crucial as downtime incidents interrupted the whole mineral processing sequence, which compelled an intensive diagnostic and mitigation program [18, 19].

The operating conditions of the plant were especially tough from a tribological aspect. Slurry concentrations in many cases were above 40 wt%, particle sizes between 50 μm (fine fraction of clay and silt) and 2–3 mm (coarse sand and quartz fragments), and the solid phase consisted of angular and hard particles with Mohs hardness numbers greater than 6. The carrier fluid, often process water adjusted with reagents, was in a pH range of 7–9 but sometimes exhibited local variations according to upstream chemical dosages. The slurry temperature varied with process energy input, reaching as high as 45–50°C when operating under maximum loading conditions. Such a thermomechanical condition further increased the difficulty in forecasting and preventing wear, as the interaction among temperature, slurry rheology, and erosive impact increased surface deterioration [20]. Plant centrifugal slurry pumps employed high-chromium white cast iron (Cr > 25%) impellers, chosen due to their purported hardness and erosion resistance. Nonetheless, this selection was not driven by legacy design principles and did not take into consideration changes in ore characteristics because of deeper mining blocks that added more abrasive and angular particles into the system. The slurry circuits were also not optimized for hydrodynamic stability. Pumps were usually oversized or undersized for specific duties, running well away from their best efficiency point (BEP), and this caused varying flow regimes—fluctuating between turbulent and transitional inducing instabilities such as vortexing, recirculation, and even transient cavitation. These aberrations encouraged localized areas of high-energy particle impingement, which further enhanced wear [21]. A key to the foregoing problems was the lack of an integrated erosion management approach. The current practice was largely reactionary—centered on periodic visual inspections and time-based component replacement—with minimal focus on predictive diagnostic and root cause analysis. Even with SCADA (Supervisory Control and Data Acquisition) systems in place, the facility did not possess real-time wear monitoring technologies like vibration analytics, acoustic emission sensing, or flow-induced stress measurement. In addition, there was very little feedback between design changes and maintenance events with resulting repeated utilization of inefficient impeller geometries under tough and dynamic slurry conditions [22].

In effect, the industry context described here shows a multi-faceted challenge underpinned by both operation complexity and engineering inertia. An integration of extremely abrasive operating conditions, poorly aligned material specifications, compromised hydraulic optimization, and insufficient monitoring infrastructure together translated to an unsustainable tendency towards pump impeller failures. This called for a thorough investigation aimed at diagnosing failure mechanisms; correlating operational parameters with damage modes observed; and suggesting targeted interventions underpinned by materials science, fluid mechanics, and process optimization [23].

15.2.2 Observed Symptoms

The experienced symptoms of slurry pump impeller failure in the mineral processing facility were multifaceted and conveyed a complex and aggressive wear condition. The most dominant symptom, one reported uniformly across all operating units, was vanishing thinning of impeller vanes. Impellers with design life of 3–6 months were failing in less than 4 weeks of operation at full load. Thickness measurements performed with ultrasonic test equipment indicated appreciable material loss, specifically on the front edges and mid-spans of the vanes—regions that are generally subject to direct particle impacts and turbulent eddy flows. The material removal rate was determined to be greater than 0.5 mm/week in some instances, much higher than acceptable wear levels for the alloy in question [24].

Microscopic and macroscopic examinations of the failed components unveiled distinct signs of erosion-governing wear mechanisms. SEM examination of the worn surfaces displayed micro-cutting, plowing, and cratering characteristic of high-angle solid particle impingement. Surface fatigue characteristics and indications of crack initiation in subsurface inclusions further substantiated a synergistic erosion-fatigue mechanism, probable promotion by process flow pulsations. Moreover, irregular wear patterns were regularly encountered, indicating non-uniform slurry flow throughout the impeller radius. Individual impeller quadrants experienced up to 40–50% more wear than others, indicating asymmetrical flow distribution or mishandled pump installation and alignment [25]. These physical manifestations translated into palpable performance degradation at the system level. One of the first indicators of pump health decline was decreased volumetric efficiency, as seen by declining flow rates and head delivery at constant motor power inputs. Operators complained that pumps were becoming increasingly incapable of achieving throughput targets, and process engineers had to compensate by modifying upstream and downstream equipment, influencing circuit stability. Internal flow measurements and simulations (performed subsequently in the case study) validated that profile distortion of impellers caused by erosion increased turbulence, recirculation zones, and more hydraulic losses—all resulting in the decrease in efficiency [26].

A secondary but more serious symptom was the increased level of vibration and noise of the involved pumps. These were measured with portable vibration analysis equipment and accelerometers during regular maintenance inspections. The spectra of vibration data indicated dominant frequencies due to impeller imbalance, bearing condition, and mechanical looseness. Cyclical amplitude variations indicating rotating component damage were evidenced by time-domain analysis. Increased levels of vibration not only indicated continuous deterioration but also induced collateral damage to the neighboring pump parts including seals, bearings, and couplings, thus aggravating maintenance costs [27].

The noise levels in the malfunctioning units were also significantly higher, with pronounced pitch changes during operation, most commonly understood as signs of cavitation or flow instability. Such audible abnormalities were usually preceded by abrupt efficiency losses and emergency shutdowns. Acoustic emission analysis later verified that cavitation was occurring intermittently, particularly upon transient operation or passage through slurry density spikes. These cavitation zones created localized pressure reductions and collapsed violently as vapor bubbles, which caused surface pitting and fatigue damage to the impeller surfaces. Figure 15.3 shows six stages of erosion damage in slurry systems: impeller vane thinning, erosion wear, performance drop, increased vibration, cavitation events, and missed early-warning indicators, each linked to specific mechanical symptoms and represented with distinct icons and colors [28].

The post-failure analysis also revealed several early warning indicators that had been overlooked due to the lack of a proactive monitoring protocol. For example, thermal imaging showed localized hot spots near pump casings and impeller hubs—early signs of frictional heating and energy losses due to impeller-rubbing or flow-induced misalignment. Additionally, visual inspections during pump disassembly showed signs of rubber liner delamination in certain models and weld spatter detachment in hardfaced impellers, indicating material and bonding issues under repetitive stress cycles [29]. Collectively, these observed symptoms provided a clear diagnostic baseline that pointed to an aggressive erosive environment, exacerbated by poor flow distribution, material mismatch, and inadequate system-level monitoring. The clear identification and correlation of these symptoms served as a foundation for a detailed root cause analysis and guided the development of robust design, material, and operational countermeasures in subsequent sections of the study [30].

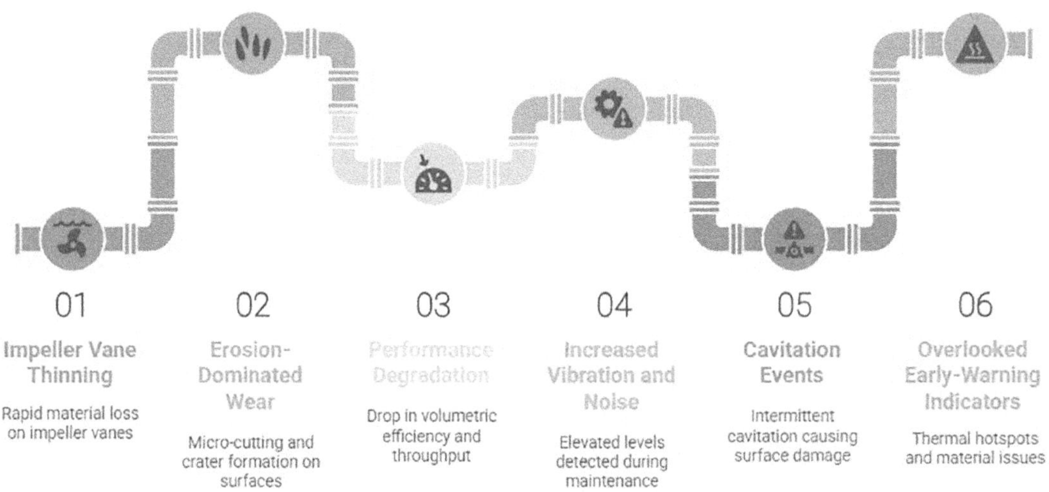

FIGURE 15.3 Progressive stages of erosion-related degradation in slurry-handling systems.

15.2.3 Initial Hypothesis

The observed erosion damage was initially hypothesized to result from a complex interplay of material, operational, and fluid dynamic factors. First, the primary contributor was identified as the highly abrasive nature of the slurry, which contained a high concentration of angular quartz particles. Quartz, as one of the hardest naturally occurring minerals, with a Mohs hardness of 7, can induce significant micro-cutting, plowing, and surface fatigue when impinging repeatedly on metallic or polymeric surfaces [31]. The angular morphology of the particles exacerbates this effect by creating localized stress concentrations, leading to accelerated material loss [2]. Second, the operational design was suspected to include suboptimal flow alignment within the hydraulic system, resulting in turbulent impingement zones. In well-designed slurry transport systems, laminar or controlled flow conditions are crucial for minimizing particle impact velocity and angle variability. However, in this case, the misalignment likely created regions of flow separation, vortices, and eddy currents, causing high-velocity, randomly oriented impacts that intensified erosive wear [32]. Such turbulent conditions not only increase the frequency and severity of particle-wall interactions but also produce localized pressure fluctuations that can contribute to surface fatigue and subsurface crack initiation [33]. Finally, the material selection for the impeller was considered inadequate for the prevailing erosive environment. The impeller material, potentially austenitic stainless steel or a similar alloy with moderate hardness and toughness, may not have been suitable for resisting high-energy particle impacts. Materials for slurry applications must balance hardness (to resist surface cutting) and toughness (to absorb impact energy and resist cracking) [34]. Figure 15.4 compares the reactive and proactive erosion management strategies in mineral processing, showing visual inspections on the left and real-time monitoring on the right. Highlights the transition from manual checks to sensor-based techniques using vibration, acoustic, and stress analysis tools.

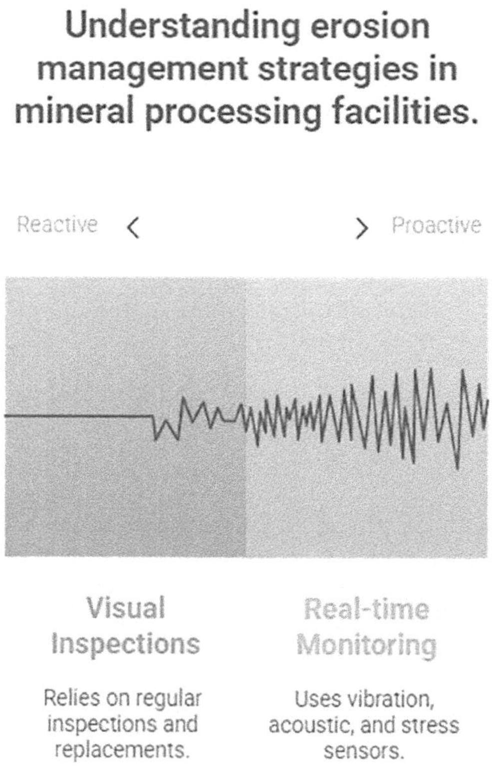

FIGURE 15.4 Comparative approaches to erosion management in mineral processing systems.

In this case, the impeller material likely lacked sufficient surface hardness and may have exhibited poor strain hardening capability, making it susceptible to repeated micro-deformation and fatigue. Furthermore, if the material had a heterogeneous microstructure with second-phase particles or inclusions, it could have become an initiation site for crack propagation under cyclic loading [35]. The combination of a harsh slurry environment, turbulent impact conditions, and a mechanically suboptimal impeller material likely created a synergistic degradation mechanism, leading to the observed erosion. This hypothesis underscores the importance of holistic design encompassing material compatibility, hydrodynamic optimization, and abrasive characterization for enhancing erosion resistance in slurry-handling systems [36].

15.3 MULTIMODAL APPROACHES FOR CHARACTERIZING AND MITIGATING SLURRY EROSION

15.3.1 Advanced Material Characterization Techniques for Erosion-Resistant Alloys

Accurate material characterization forms the foundation for understanding and improving erosion resistance in slurry-handling systems. Alloys such as ASTM A532 high-chromium white cast iron have emerged as leading candidates owing to their high hardness, carbide content, and corrosion resistance in hostile environments. Advanced characterization techniques are essential to quantify microstructural attributes that govern material response under erosive conditions [37]. Figure 15.5 categorizes the advanced techniques for characterizing erosion-resistant materials, including mechanical testing (e.g., hardness, AFM), spectroscopy (EDS, XRD), microscopy (OM, SEM), and advanced microscopy (TEM), showing the multi-disciplinary approach required to evaluate material behavior under erosive conditions.

Optical microscopy (OM) and scanning electron microscopy (SEM) are widely employed to examine the distribution, morphology, and orientation of primary and secondary carbide phases. SEM, especially in backscattered electron (BSE) mode, offers high-resolution imaging of carbides (e.g., M_7C_3, M_3C) embedded within martensitic or austenitic matrices. Energy dispersive X-ray spectroscopy (EDS), integrated with SEM, facilitates compositional analysis of both matrix and carbide constituents, revealing elemental segregation or enrichment that may impact local hardness and toughness. X-ray diffraction (XRD) is utilized to identify crystallographic phases, measure lattice strains, and evaluate residual stress states. In high-Cr white irons, phase quantification using Rietveld refinement is instrumental in determining the proportion of retained austenite and

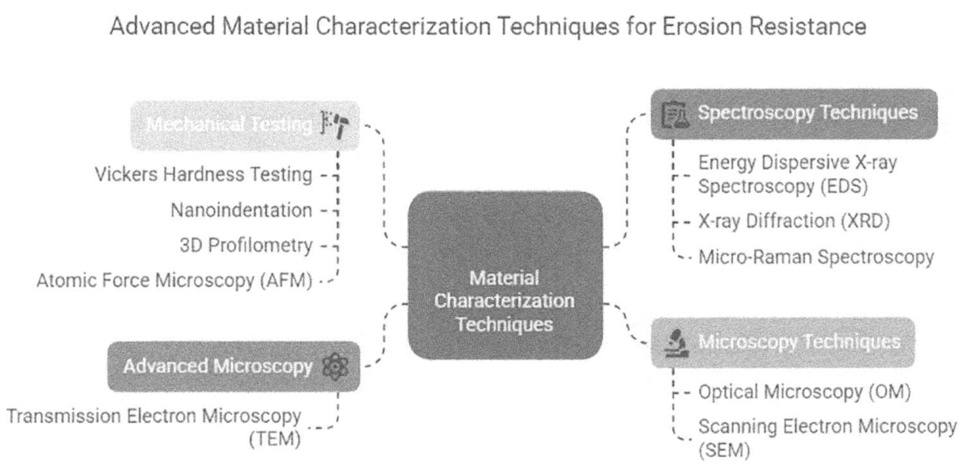

FIGURE 15.5 Overview of advanced material characterization techniques for erosion-resistant alloys.

martensite, which directly correlates with impact resistance and crack propagation behavior. Additionally, micro-Raman spectroscopy can be deployed for localized phase mapping, especially in assessing transformation-induced hardening mechanisms [38, 39].

Hardness testing using the Vickers method is standard for assessing surface resistance, with typical values for ASTM A532 ranging between 600–750 HV. Nanoindentation provides localized mechanical property gradients, particularly useful in layered coatings or multi-phase matrices. Furthermore, 3D profilometry and atomic force microscopy (AFM) enable precise surface topography mapping, offering insights into post-erosion wear features and micro-fracture propagation patterns [40]. Transmission electron microscopy (TEM), although less commonly used due to sample preparation complexity, offers atomic-scale insights into dislocation structures, carbide/matrix interfaces, and phase transformation zones. These detailed assessments are valuable when correlating nanoscale phenomena with macro-scale erosion performance [41].

Advanced material characterization thus enables predictive correlation between microstructure, hardness, and erosion resistance, allowing optimization of heat treatments, alloy design, and surface modifications. A robust understanding of these properties is essential in selecting materials tailored to specific slurry dynamics, particle characteristics, and operational regimes. Such multi-scale analysis ensures the development of erosion-resistant alloys capable of enduring aggressive slurry environments with minimized material loss and extended service life [42].

15.3.2 Laboratory-Based Erosion Testing Protocols

Laboratory-scale erosion testing protocols play a critical role in evaluating the performance of materials under controlled yet representative conditions. Standardized procedures such as ASTM G75 and ASTM G76 are widely adopted for assessing the erosive wear behavior of materials subjected to slurry and air jet impacts, respectively [43]. ASTM G75, also known as the Miller Slurry Abrasion Test, simulates low-velocity erosive environments characterized by rolling and sliding of abrasive particles. In this test, a specimen is pressed against a rotating rubber wheel in the presence of an abrasive slurry—typically comprising quartz particles suspended in water. Material loss is measured periodically by weight, allowing the calculation of wear rates over extended test durations. This test provides critical insight into ductile wear mechanisms such as plowing and micro-cutting, particularly in materials with martensitic matrices and hard carbides [44].

ASTM G76, in contrast, simulates high-velocity, angle-sensitive erosion scenarios using a dry, compressed air stream to propel abrasive particles at the specimen surface. The erosion rate is measured as a function of impact angle, revealing the angular dependency of material loss. Ductile materials tend to show maximum erosion at shallow angles (30–45°), whereas brittle materials peak at normal incidence (90°). This test is invaluable for understanding the behavior of materials under dynamic impingement, replicating field conditions such as those in pump impellers and slurry pipelines [45]. Complementary tests like slurry pot erosion testing offer a dynamic fluid–particle interaction environment closer to real-world applications. Coupons are mounted in rotating frames within a slurry bath, enabling evaluation of erosion under complex flow patterns and varying impact orientations. Wear depth is quantified using profilometry and correlated with flow velocity and particle energy distribution [46].

Post-test characterization using SEM, profilometry, and microhardness mapping further clarifies the wear mechanisms, indicating whether erosion occurred via micro-fracture, delamination, or abrasive grooving. These laboratory tests provide a quantitative and mechanistic understanding of material response, which is essential for screening erosion-resistant candidates, validating coatings, and calibrating computational models. When performed systematically, such protocols serve as a predictive toolset for designing materials and geometries optimized for aggressive slurry conditions [47].

15.3.3 Computational Fluid Dynamics as a Predictive Tool for Erosion Mapping

Computational fluid dynamics has emerged as a powerful predictive tool in the study of erosion phenomena within slurry-handling systems. It enables the visualization and quantification of fluid–particle interactions, turbulent energy fields, and wall impact forces, all of which contribute to material degradation under service conditions [48]. Figure 15.6 shows CFD applications in erosion analysis, highlighting erosion modeling, particle trajectory dynamics, erosion contour mapping, validation and sensitivity analysis, and design optimization for predictive maintenance in slurry-handling systems, supporting accurate wear prediction and performance improvement in industrial components.

In erosion modeling, a three-dimensional geometry of the pump or pipeline component is developed using CAD software and imported into solvers such as ANSYS Fluent. The fluid domain is meshed using high-resolution hexahedral or polyhedral grids to capture complex flow behavior near boundaries and obstructions. A multiphase Eulerian–Lagrangian approach is typically adopted, wherein the liquid phase is treated as a continuum, and solid particles are tracked as discrete entities using the discrete phase model (DPM). Particle trajectories are governed by Newtonian dynamics incorporating drag, gravitational, and inertial forces. Turbulence is resolved using models like realizable k–ε or SST k-ω, with turbulence–particle interaction further refined using stochastic tracking or particle time-scale models. Wall–particle interactions are modeled through rebound conditions, accounting for restitution coefficients and energy dissipation during collisions [49].

CFD simulations yield erosion contours that visualize predicted wear locations based on particle impact frequency, velocity, and angle. These maps often highlight vulnerable zones such as impeller leading edges, volute bends, and flow reattachment points. In addition to qualitative mapping, quantitative erosion rates can be derived using empirical erosion models (e.g., Oka, Finnie)

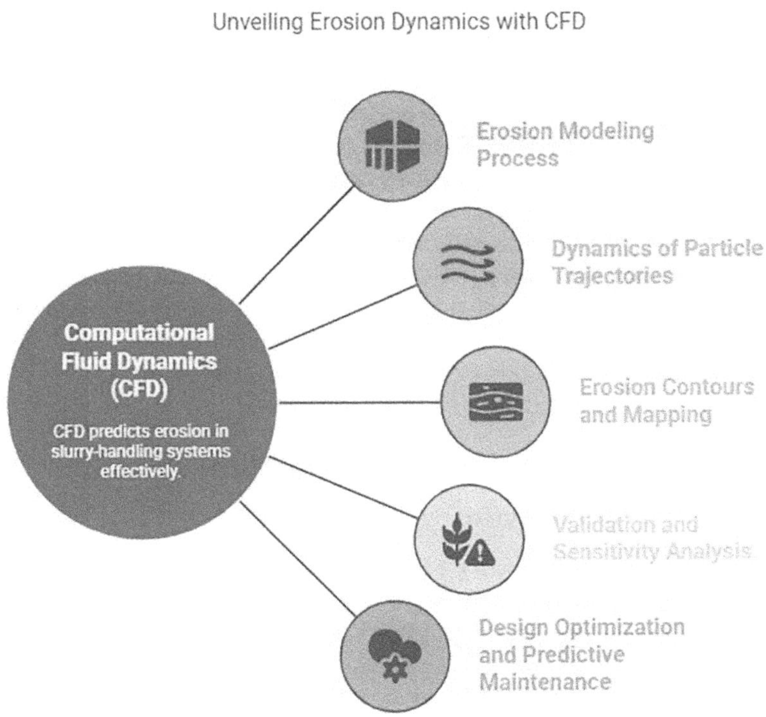

FIGURE 15.6 Key aspects of computational fluid dynamics in erosion analysis of slurry systems.

embedded within CFD codes. Model accuracy is validated through comparison with laboratory and field data, ensuring consistency in erosion depth trends and spatial distribution. Sensitivity analysis can be performed by varying flow rates, particle sizes, and impact angles to evaluate design robustness. Notably, simulations reveal the influence of geometric tolerances and installation errors; for instance, a minor misalignment in inlet geometry can double particle impact velocities in localized regions. CFD thus enables iterative design optimization by informing engineers where to apply hardfacing, redesign flow paths, or introduce flow-modifying vanes. Its integration with erosion data from ASTM and field tests makes it indispensable in predictive maintenance, material selection, and failure analysis of slurry systems [50, 51].

15.3.4 Field-Scale Validation and Real-Time Monitoring Strategies

Field-scale validation bridges the gap between laboratory predictions and real-world erosion behavior, providing empirical evidence for the effectiveness of materials, coatings, and design improvements in slurry-handling components. Real-time monitoring strategies are increasingly integrated into operational frameworks to enable condition-based maintenance and mitigate unplanned downtimes. In-situ thickness monitoring is a primary method for tracking material loss over time. Ultrasonic thickness gauges offer high-resolution (±0.01 mm) non-destructive measurements at critical zones of impellers, casings, and piping. Baseline readings are recorded pre-installation, with subsequent measurements collected at regular intervals to map erosion progression. Variations in material loss across different regions correlate closely with CFD-predicted high-erosion zones, validating model fidelity [52].

Vibration monitoring complements thickness measurements by providing dynamic insights into mechanical integrity. Accelerometers mounted on bearing housings and pump casings capture time and frequency domain data. Increases in amplitude, particularly in the 200–500 Hz range, often signal impeller imbalance or cavitation linked to asymmetric erosion. Trending this data allows early detection of performance degradation before catastrophic failure occurs. Acoustic emission (AE) sensors offer additional sensitivity to sub-surface wear and crack initiation. These sensors detect high-frequency acoustic waves generated by micro-fracturing, particle impacts, or delamination events. In slurry pumps, AE patterns are particularly useful for identifying transient changes in wear zones, often undetectable through visual inspection alone [53].

Visual inspections during scheduled shutdowns remain critical for corroborating sensor data. Surface morphology, pitting, and lip formation are documented to evaluate wear mechanisms, validate laboratory test findings, and assess repair feasibility. Digital image correlation and 3D scanning techniques are increasingly used to quantify surface erosion profiles with high spatial resolution. The integration of real-time monitoring tools supports the deployment of predictive maintenance models. These models, trained on operational and erosion data, can forecast remaining useful life and suggest optimal replacement intervals. Additionally, combining field data with CFD and lab findings ensures that design changes are evidence-based and targeted [54, 55].

15.4 CONTEXTUALIZING KEY FINDINGS IN LITERATURE

15.4.1 Wear Zones in Slurry Systems

Slurry erosion in rotating and stationary equipment is often localized and heavily influenced by component geometry and flow dynamics. In most documented cases, particularly in pumps and turbines, the leading edges of impeller blades and the curved regions of the volute casing are identified as the most vulnerable zones to erosion damage. These wear patterns are attributed to the concentration of particle-laden flow and abrupt changes in direction that amplify impingement forces. Literature consistently reports that wear localization in these areas is due to high-pressure gradients and flow separation, which causes particles to impact surfaces at high velocities and non-normal angles [56, 57]. This pattern has been confirmed through both laboratory testing and field data, including in systems operating with quartz-rich slurries, where angular particles cause pronounced surface degradation. In recent case studies using ultrasonic thickness mapping and SEM analysis, material loss was visibly concentrated at the forward-facing surfaces of the impeller blades and inner curvature of volute casings. These observations match CFD predictions and prior experimental data, highlighting the repeatability of erosion locations across different operating environments. Furthermore, regions downstream of flow obstructions or near flow reattachment points also show elevated erosion due to localized turbulence and vortex shedding [58].

From a surface engineering perspective, recognizing these wear zones is critical for deploying targeted treatments such as hard coatings, sacrificial liners, or functionally graded materials. Protective strategies are most effective when aligned with erosion "hotspots", thus minimizing unnecessary weight or cost in less vulnerable areas. The findings support the need for component-level erosion maps during design stages, which can be generated via simulation and validated by empirical data. By correlating structural features with erosion susceptibility, designers and maintenance engineers can anticipate service life and optimize inspection intervals. Overall, understanding wear localization is essential not just for damage mitigation but for predictive maintenance and life-cycle cost reduction in slurry-handling systems [59].

15.4.2 Particle Behavior and Turbulence Effects

The behavior of solid particles in slurry flows is governed by a complex set of interactions involving hydrodynamic forces, turbulence intensity, particle inertia, and wall-surface conditions. In high-turbulence regions, particle trajectories become chaotic, leading to increased rebound frequency and angular impingement, both of which significantly accelerate material loss. These effects are particularly critical in curved flow passages, blade trailing edges, and regions with recirculation zones. Numerous CFD and experimental studies have confirmed that turbulence enhances erosion not only by raising particle velocity but also by increasing the randomness of impact angles, thereby transitioning wear from pure cutting or plowing modes to mixed-mode and fatigue-based mechanisms [60, 61].

In recent evaluations of impeller components subjected to quartz slurry, high-speed imaging and discrete particle tracking simulations revealed that particles undergoing initial impact often rebounded and struck the same surface or adjacent surfaces at oblique angles. These secondary impacts introduced shear components to the erosive force, creating micro-cracks, craters, and delamination zones. Furthermore, in systems with improperly aligned flow channels or sharp bends, turbulent vortices formed, suspending particles in a rotating zone that caused repeated wall impacts. This behavior has also been documented in literature as a primary cause of accelerated wear in pump volutes and hydrocyclone walls [62].

The particle's physical properties size, shape, and hardness further modulate its erosive impact under turbulent conditions. Angular particles, especially in the 200–400 μm range, exhibit stronger rebound and transfer higher stress concentrations upon collision compared to spherical particles. Quartz, with a Mohs hardness of 7, frequently causes sharp indentations and lateral micro-fractures on metal surfaces. Studies on particle-laden turbulent jet flows have shown that particles above 200 μm with high aspect ratios tend to align with shear flows and impact at critical tangential zones on the metal surface, contributing to groove formation and fatigue wear [63]. Turbulence-induced particle rebound also complicates wear modeling. Conventional erosion models that consider only single-pass impingement often underpredict real-world erosion. As such, erosion modeling in turbulent slurry systems must account for multiphase flow, rebound dynamics, and angular re-impacts, especially in CFD simulations intended for design optimization. In practice, suppressing turbulence via flow conditioning or geometry modification can reduce rebound and improve erosion resistance in critical zones [64].

15.4.3 Material Limitations under Cyclic Impact Conditions

Despite their widespread use, high-chromium white cast irons such as ASTM A532 exhibit distinct limitations when subjected to cyclic, high-energy particle impacts commonly encountered in slurry environments. These materials derive their exceptional hardness and wear resistance from their microstructure, which typically includes a martensitic or austenitic matrix reinforced by eutectic and secondary carbides predominantly M_7C_3. While this carbide-rich structure offers impressive resistance to abrasion, its performance under repetitive impact conditions is compromised by its inherent brittleness and susceptibility to micro-crack formation at carbide-matrix interfaces [65].

Under cyclic impingement, especially at intermediate angles where tangential stresses dominate, repeated plastic strain accumulates in the metallic matrix surrounding hard carbides. As a result, micro-cracks initiate at phase boundaries, particularly where there is a mismatch in modulus or thermal expansion. Over time, these cracks propagate, coalesce, and lead to pitting or fragment detachment—mechanisms observed in both SEM post-erosion analysis and field inspections of failed components. Notably, erosion tests following ASTM G76 with quartz particles have shown that even with surface hardness exceeding 700 HV, A532 alloys begin showing surface degradation within the first few hours of exposure, especially at 30°–45° impact angles [66].

Another contributing factor is the low fracture toughness of the material, which limits its ability to absorb and redistribute impact energy. Unlike tougher metals such as duplex stainless steel or Ni-based superalloys, white cast iron lacks ductility and strain-hardening capacity, making it prone to brittle cracking under dynamic loading. Thermal cycling or pressure pulsations—common in slurry systems—further exacerbate the issue by creating fluctuating stress states that promote subcritical crack growth. Studies have shown that the addition of molybdenum or rare-earth elements can enhance matrix toughness and reduce carbide embrittlement, though at increased cost [67]. Overall, while high-chromium white cast irons remain a cost-effective and well-characterized material for moderate erosion scenarios, their limitations under cyclic, high-angle impacts necessitate strategic use of surface coatings, composite overlays, or hybrid microstructures. Future research may focus on optimizing carbide morphology, introducing nano-scale reinforcements, or developing new cast iron chemistries with improved resistance to fatigue-driven erosion in slurry systems [68].

15.4.4 CFD as a Predictive Tool for Erosion Mapping

Computational fluid dynamics has emerged as a vital tool in the prediction and visualization of erosion behavior within slurry-transporting equipment. The complexity of multiphase flows—particularly those involving high solid content, turbulence, and irregular geometries—makes experimental erosion mapping both costly and limited in scope. CFD offers a powerful alternative by enabling simulation of fluid–particle interactions under various operating conditions, thus allowing engineers to identify critical wear zones before actual deployment. Modern CFD platforms, such as ANSYS Fluent or STAR-CCM+, utilize Eulerian–Lagrangian or discrete phase modeling (DPM) to track individual particle trajectories while simultaneously resolving fluid dynamics using turbulence models like k–ε or Reynolds stress models (RSM) [69, 70].

Recent case studies have shown a high degree of correlation between CFD-predicted erosion profiles and experimental wear patterns on impellers, valves, and elbows. In slurry pump applications, CFD simulations identified high particle concentration and impact velocity at impeller leading edges, volute tongue, and outlet regions—all of which matched observed material loss in long-term field trials. These simulations revealed that secondary flows, including eddies and recirculation zones, significantly influence the residence time and impact frequency of solid particles, intensifying localized erosion. Furthermore, CFD enables visualization of impact angles, shear stress, and turbulent kinetic energy distribution, providing insights into the mechanisms driving material degradation. Beyond visualization, CFD supports design optimization by allowing parametric studies involving flow rate, particle concentration, blade curvature, and casing geometry. By iteratively modifying these parameters, designers can minimize peak erosion rates and redistribute wear across broader, less critical surface areas. For example, altering blade angles by as little as 5° in simulations has shown to reduce erosion intensity by 20–30% in high-impact zones. Some researchers have further integrated erosion prediction models, such as Finnie's or Bitter's models, within CFD solvers to estimate wear rate quantitatively based on particle properties and impact mechanics. Validation of CFD results with laboratory data remains crucial. Techniques like particle image velocimetry (PIV), erosion coupons, and 3D profilometry are often used to confirm model accuracy. When properly calibrated, CFD enables predictive maintenance scheduling by identifying surfaces at high risk of failure and informing targeted reinforcement strategies such as coatings or material substitution. With the integration of AI-based model tuning and digital twins, CFD is poised to become not only a simulation tool but also a real-time erosion management platform in smart industrial systems [71–73].

15.5 SOLUTION STRATEGY

In response to the erosion-related challenges identified through experimental, computational, and field-based evaluations, a comprehensive multi-pronged solution strategy was developed. This approach aimed not only to mitigate material loss but also to extend component lifespan, improve system efficiency, and reduce unplanned downtimes. The four pillars of this solution, material upgrade, design modifications, operational changes, and advanced monitoring, were integrated to provide synergistic erosion control across both mechanical and process domains [74].

One of the most significant interventions involved replacing the existing ASTM A532 white cast iron components with tungsten carbide-coated steel surfaces in high-erosion zones such as impeller leading edges and volute casing inlets. Tungsten carbide (WC), with its extreme hardness (~1600–2000 HV) and excellent toughness, is well known for its superior resistance to particle impact and micro-crack propagation. The coatings were applied using the high-velocity oxy-fuel (HVOF) technique, which provides a dense, low-porosity, high-adhesion coating that maintains structural integrity under cyclic impact loading. HVOF's relatively low thermal input minimizes substrate distortion while ensuring strong metallurgical bonding between the coating and the steel substrate, making it ideal for dynamic slurry environments [75]. To address turbulent flow conditions and recirculation-induced wear, impeller blade curvature was redesigned to promote streamlined flow and reduce impact angles. Computational fluid dynamics simulations guided the development of a revised impeller profile with smoother entry and exit flow transitions, thereby minimizing high-velocity collisions. Additionally, the volute casing was modified by altering curvature and increasing throat area to reduce pressure gradients and recirculation intensity. These modifications collectively contributed to more uniform particle trajectories, reducing localized erosion [76].

Operational parameters were also adjusted to minimize erosive forces. Specifically, the slurry concentration was reduced from 35% to 25% for operations where material throughput could be moderated. This significantly decreased the frequency and energy of particle collisions. Moreover, flow rates were optimized to operate closer to the laminar-to-transition flow regime, where turbulence-induced wear is less severe. These changes helped in preserving component surfaces without compromising hydraulic performance significantly [77].

15.6 RETURN ON INVESTMENT AND SUSTAINABILITY

Implementing erosion control strategies in slurry-handling systems not only enhances technical performance but also delivers measurable economic and environmental returns. A comprehensive cost-benefit analysis revealed a payback period of approximately 6 months for the material and design upgrades, making the investment both technically sound and financially justifiable. The bulk of the initial costs was attributed to the application of high-performance tungsten carbide coatings via the HVOF process and precision redesign of the impeller and volute casing. However, the rapid

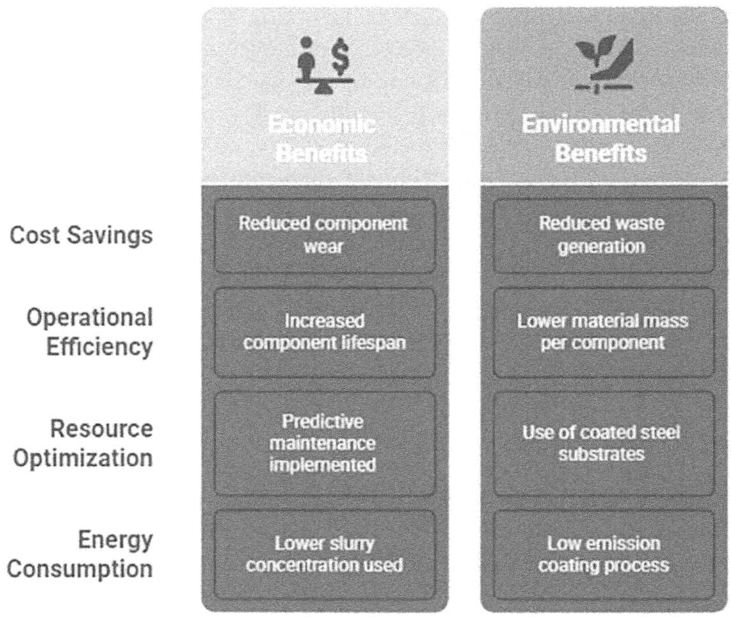

FIGURE 15.7 Economic and environmental benefits of erosion control.

reduction in component wear, decrease in unscheduled downtime, and lower frequency of replacement resulted in significant cost savings within a short operational cycle [78]. Figure 15.7 presents a visual representation of the economic and environmental benefits of erosion control, emphasizing reduced maintenance costs, extended equipment life, improved energy efficiency, minimized material waste, decreased unplanned downtimes, and lower environmental impact through sustainable operation in slurry-handling and mineral processing systems.

From an operational perspective, the lifecycle of critical components such as impellers and casings increased by more than 3 times, drastically reducing the cost of spare parts and maintenance labor. Predictive maintenance, facilitated by wireless ultrasonic erosion sensors, further optimized resource allocation by ensuring that parts were serviced or replaced only when necessary, eliminating inefficiencies associated with reactive repair models. Additionally, reduction in the slurry concentration to 25% for specific operations translated into lower energy consumption for slurry agitation and pumping, enhancing overall energy efficiency without compromising throughput in non-critical processes [79].

Beyond the economic metrics, the sustainability benefits were equally compelling. One of the most impactful environmental outcomes was the significant reduction in waste generation. Previously, frequent erosion-induced failures resulted in the premature disposal of heavy metallic components, contributing to both solid waste and embodied energy losses. By extending component lifespan and enabling refurbishment over replacement, the strategy reduced material disposal volume by approximately 60% over a 12-month cycle. This directly contributes to circular economy principles by conserving raw materials and minimizing environmental burdens associated with casting and machining new components [80]. Moreover, switching to coated steel substrates instead of monolithic white cast iron led to lower total material mass per component, further reducing the carbon footprint associated with material extraction, processing, and logistics. The HVOF coating process, being relatively low in emissions compared to some other thermal spraying techniques, also contributed positively to the system's environmental profile [81]. The integration of advanced materials, improved hydraulic design, operational optimization, and intelligent monitoring not only delivered rapid financial returns but also aligned with modern sustainability goals. The reduced environmental impact through minimized material waste, energy-efficient operations, and extended equipment service life represents a model approach for erosion control that balances performance, cost, and ecological responsibility. This sets a precedent for broader adoption of sustainable engineering practices in slurry-intensive industries such as mining, power generation, and wastewater management [82].

15.7 CONCLUSION

This case study exemplifies a rigorous, multidisciplinary methodology to mitigate slurry erosion through the synergistic integration of empirical experimentation, computational fluid dynamics simulations, and real-world operational analytics. The confluence of advanced materials engineering employing high-velocity oxy-fuel tungsten carbide coatings with precision hydraulic redesign substantiated a quantifiable enhancement in component durability and system resilience. Empirical wear characterization coupled with predictive modeling facilitated optimization of impeller and volute geometries, thereby attenuating erosive forces and flow-induced stress concentrations. The deployment of condition-based monitoring, leveraging ultrasonic sensor data, enabled prognostic maintenance schedules that minimized unplanned downtime and operational expenditure. This holistic framework underscores the imperative of convergent engineering disciplines to systematically decipher and remediate complex slurry erosion phenomena. Ultimately, the study substantiates a paradigm wherein material innovation, fluid dynamic optimization, and data-driven maintenance coalesce to elevate lifecycle performance, operational efficiency, and economic viability in slurry-intensive industrial applications.

REFERENCES

1. Pagalthivarthi KV, Gupta PK, Tyagi V, Ravi MR. CFD predictions of dense slurry flow in centrifugal pump casings. Int J Aerospace Mech Eng. 2011 Mar 1;5(4):16–28.
2. Rao PN. Manufacturing Technology: Foundry, Forming and Welding. McGraw-Hill; 2018.
3. Bitter JGA. A study of erosion phenomena: Part 2. Wear. 1963;6(2):169–190.
4. Levy AV. Solid Particle Erosion and Erosion-Corrosion of Materials. Materials Park, OH: ASM International; 1995.
5. Zhang D, Kashani HR. Advances in slurry erosion–corrosion behavior of materials: A review. Wear. 2021;480–481:203941.
6. Finnie I. Erosion of surfaces by solid particles. Wear. 1960;3(2):87–103.
7. Neilson JH, Gilchrist A. Erosion by a stream of solid particles. Wear. 1968;11(2):111–122.
8. ASTM G75–15. Standard Test Method for Determination of Slurry Abrasivity (Miller Number) and Slurry Abrasion Response of Materials (SAR Number). West Conshohocken, PA: ASTM International; 2015.
9. Tan X, Neville A. Erosion–corrosion of engineering steels: A review. Wear. 2015;334–335:73–88.
10. Karunamurthy P, Pandey PM. CFD modeling and analysis of solid-liquid slurry erosion in hydro components. Eng Fail Anal. 2018;84:344–358.
11. Stack MM, Abd El-Badia M. Mapping erosion–corrosion of carbon steels in oil sands slurries. Wear. 2006;260(7–8):816–822.
12. Clark DE, Hawthorne HM. Slurry erosion: Mechanisms and test methods. Tribol Int. 1992;25(5):297–307.
13. Tiwari S, Singh S, Singh R. Performance of different coating materials against slurry erosion. Wear. 2020;460–461:203456.
14. Zhang Y, Li X, Wang Y, et al. Research progress on coping strategies for the fluid-solid erosion wear of pipelines. Powder Technol. 2023;418:118–129.
15. Gupta N, Singh R, Singh S. An overview of slurry erosion control by the application of high-performance coatings. Proc Inst Mech Eng Part J. 2011;225(6):577–588.
16. Gaurav V, Sankaranarayanan S, Kumareshbabu SP, et al. Slurry erosion behavior of cast 25 Cr austenitic-ferritic steel with niobium addition for subsea pipelines. J Mater Eng Perform. 2021;30:6224–6234.

17. Hall N, Elenany M, Zhu DZ, et al. Experimental study of sand and slurry jets in water. J Hydraul Eng. 2010;136(10):727–738.
18. Arabnejad H, Mansouri A, Shirazi SA, et al. Abrasion erosion modeling in particulate flow. Wear. 2017;376–377:1194–1199.
19. Desale GR, Gandhi BK, Jain SC. Development of correlations for predicting the slurry erosion of ductile materials. J Tribol. 2011;133(1):011601.
20. Kumar R, Goyal K, Bhandari D. Slurry erosion behavior of thermally sprayed ceramic nanocomposite coatings on turbine steel. Int J Appl Ceram Technol. 2022;19(6):3049–3061.
21. Ndumia JN, Kang M, Gbenontin BV, et al. A review on the wear, corrosion and high-temperature resistant properties of wire arc-sprayed Fe-based coatings. Nanomaterials. 2021;11(10):2527.
22. Kumar R, Goyal K, Bhandari D. Slurry erosion behavior of thermally sprayed ceramic nanocomposite coatings on turbine steel. Int J Appl Ceram Technol. 2022;19(6):3049–3061.
23. Singh G, Ash B. A review on erosion wear of different types of slurry pump impeller materials. Mater Today Proc. 2021;37:2298–2301.
24. Arabnejad H, Mansouri A, Shirazi SA, et al. Abrasion erosion modeling in particulate flow. Wear. 2017;376–377:1194–1199.
25. ASTM G75–15. Standard Test Method for Determination of Slurry Abrasion Response of Materials (Miller Number). ASTM International; 2015.
26. Hariprasad KS, Seshadri SK, Sarma VS. Surface alloying of steels with Cr-Mo-B using TIG arc: Microstructure and slurry erosion behaviour. Mater Des. 2013;49:761–770.
27. Dehghan-Manshadi A, Hashemi R, Sohi MH. Erosive wear behaviour of white cast iron alloyed with vanadium and titanium. Wear. 2013;302(1–2):1361–1368.
28. Gandhi BK, Sarkar R, Desale GR. Parametric study of slurry erosion of ductile materials. Wear. 2007;263(1–6):240–248.
29. Shirazi SA, Shadley JR, McLaury BS. Erosion modeling in multiphase flow: A review. J Energy Resour Technol. 2004;126(1):1–12.
30. McLaury BS, Shirazi SA, Shadley JR. CFD prediction of erosion damage in elbows for gas/solid flow. ASME J Energy Resour Technol. 1997;119:105–113.
31. Lim SC, Brunton JH, Ashby MF. Wear mechanism maps. Acta Metall. 1987;35(6):1343–1348.
32. Venkatakrishnan P, Meenakshisundaram N. Application of CFD and experimental methods in erosion prediction of hydro-transport systems: A review. Powder Technol. 2021;388:525–538.
33. Rajkumar YR, Shirazi SA, McLaury BS. A combined CFD and experimental methodology for erosion prediction. Wear. 2015;332–333:1090–1097.
34. Keshavamurthy R, Naveena BE, Ramesh CS, et al. Evaluation of slurry erosive wear performance of plasma-sprayed flyash-TiO_2 composite coatings. J Bio Tribo Corros. 2021;7:92.
35. Kalácska Á, Nagarajan R, Tóth LF, et al. Surface damage analysis on the application of abrasion and slurry erosion in targeted steels using an erosion test rig. Lubricants. 2022;10(11):316.
36. Guo N, Leu MC. Additive manufacturing: Technology, applications and research needs. Front Mech Eng. 2013;8(3):215–243.
37. Herzberg M, Salama M, Shirazi SA, et al. Development of real-time erosion monitoring techniques using acoustic emission and vibration analysis. Wear. 2020;450–451:203281.
38. Debnath S, Reddy VR, Basak AK, et al. Vibration-based condition monitoring and fault diagnosis techniques for slurry pumps: A review. Measurement. 2023;207:112396.
39. Gandhi BK, Sarkar R, Desale GR. Parametric study of slurry erosion of ductile materials. Wear. 2007;263(1–6):240–248.
40. Goyal K, Das S, Kumar S. A comprehensive review of slurry erosion and its correlation with impingement angle. Mater Today Proc. 2022;62:1847–1852.
41. Arabnejad H, Mansouri A, Shirazi SA, et al. Abrasion erosion modeling in particulate flow. Wear. 2017;376–377:1194–1199.
42. Rajkumar YR, Shirazi SA, McLaury BS. A combined CFD and experimental methodology for erosion prediction. Wear. 2015;332–333:1090–1097.
43. Shirazi SA, McLaury BS. Erosion modelling in multiphase flow: Current status and future work. Wear. 1998;233–235:486–493.
44. Dehghan-Manshadi A, Hashemi R, Sohi MH. Erosive wear behaviour of white cast iron alloyed with vanadium and titanium. Wear. 2013;302(1–2):1361–1368.

45. Singh G, Ash B. A review on erosion wear of different types of slurry pump impeller materials. Mater Today Proc. 2021;37:2298–2301.
46. Shirazi SA, Rybicki EF, McLaury BS. Modeling erosion in elbows for gas/solid flow. Wear. 1995;186–187:395–401.
47. Venkatakrishnan P, Meenakshisundaram N. Application of CFD and experimental methods in erosion prediction of hydro-transport systems: A review. Powder Technol. 2021;388:525–538.
48. Kalácska Á, Nagarajan R, Tóth LF, et al. Surface damage analysis on the application of abrasion and slurry erosion in targeted steels using an erosion test rig. Lubricants. 2022;10(11):316.
49. Lim SC, Brunton JH, Ashby MF. Wear mechanism maps. Acta Metall. 1987;35(6):1343–1368.
50. Kumar R, Goyal K, Bhandari D. Slurry erosion behavior of thermally sprayed ceramic nanocomposite coatings on turbine steel. Int J Appl Ceram Technol. 2022;19(6):3049–3061.
51. Wilson A. Hardness Testing. Materials Park, OH: ASM International; 2004.
52. Keshavamurthy R, Naveena BE, Ramesh CS, et al. Evaluation of slurry erosive wear performance of plasma-sprayed flyash-TiO_2 composite coatings. J Bio TriboCorros. 2021;7:92.
53. Guo N, Leu MC. Additive manufacturing: Technology, applications and research needs. Front Mech Eng. 2013;8(3):215–243.
54. Herzberg M, Salama M, Shirazi SA, et al. Development of real-time erosion monitoring techniques using acoustic emission and vibration analysis. Wear. 2020;450–451:203281.
55. Debnath S, Reddy VR, Basak AK, et al. Vibration-based condition monitoring and fault diagnosis techniques for slurry pumps: A review. Measurement. 2023;207:112396.
56. Assadi H, Gärtner F, Stoltenhoff T, et al. High-velocity oxy-fuel (HVOF) spraying: Fundamentals and applications. J Therm Spray Technol. 2003;12(1):93–102.
57. Liang M, Lu H, Liu J, et al. Effect of HVOF-sprayed tungsten carbide coatings on slurry erosion–corrosion behavior of pump components. Wear. 2017;380–381:94–103.
58. Suresh S, Rajeswari S, Ravichandran K. Cost benefit analysis and life cycle assessment of thermal sprayed coatings for industrial components. Surf Coat Technol. 2020;388:125536.
59. Jardine AKS, Lin D, Banjevic D. A review on machinery diagnostics and prognostics implementing condition-based maintenance. Mech Syst Signal Process. 2006;20(7):1483–1510.
60. Liu Z, Chen W, Chen J. Energy efficiency improvement by slurry concentration control in mineral processing plants: A case study. J Cleaner Prod. 2019;208:1398–1406.
61. Zhang J, Cui Y, Zhang Y. Life cycle assessment of heavy-duty metallic components in slurry pumping systems with focus on material selection and coating. J Cleaner Prod. 2021;311:127593.
62. Nasr EM, Kim S, Kim KH. Sustainability assessment of thermal spray coatings: Environmental impacts and circular economy approach. J Mater Res Technol. 2020;9(6):13880–13892.
63. Almeida M, Saramago B, Ferreira F, et al. Environmental and economic impacts of maintenance optimization: A case study on predictive maintenance using wireless sensors. J Ind Eng Manag. 2020;13(2):247–262.
64. Singh R, Singh H, Dhiman SS. Circular economy in the mining industry: Strategies for sustainability. J Cleaner Prod. 2022;347:131182.
65. Ghods P, Amirkhanlou S, Fathi MH. Advances in slurry erosion resistance and coatings: A review with future perspectives. Wear. 2023;518–519:204246.
66. Cruz N, Forster J, Bobicki ER. Slurry rheology in mineral processing unit operations: A critical review. Can J Chem Eng. 2019 Jul;97(7):2102–2120.
67. Heywood NI. Stop your slurries from stirring up trouble. Chem Eng Prog. 1999 Sep 1;95:21–42.
68. Gesing F. The material politics of slurry: Mobilisations and transformations along the waste-fertiliser continuum. Political Geogr. 2023 Mar 1;101:102832.
69. El Moçayd N, Seaid M. Data-driven polynomial chaos expansions for characterization of complex fluid rheology: Case study of phosphate slurry. Reliab Eng Syst Saf. 2021 Dec 1;216:107923.
70. Cumby TR. Slurry mixing with impellers: Part 1, theory and previous research. J Agric Eng Res. 1990 Jan 1;45:157–173.
71. Hjorth M, Christensen KV, Christensen ML, et al. Solid-liquid separation of animal slurry in theory and practice. Sustain Agric. 2011;2:953–986.
72. Torvik R, Svendsen HF. Modelling of slurry reactors: A fundamental approach. Chem Eng Sci. 1990 Jan 1;45(8):2325–2332.

73. Paterson AJ. Pipeline transport of high density slurries: A historical review of past mistakes, lessons learned and current technologies. Min Technol. 2012 Jan;121(1):37–45.
74. Elkarii M, Boukharfane R, Benjelloun S, et al. A CFD-based surrogate model for predicting slurry pipe flow pressure drops. Particulate Sci Technol. 2023 Apr 3;41(3):432–442.
75. Shariq M, Nemec D. Modelling approaches of the dispersion process for conductive slurries in chemical process industries. Adv Eng Mater. 2024 Nov;26(21):2401089.
76. Clark HM. Slurry erosion: Macro-and micro-aspects. In Fundamentals of Tribology and Bridging the Gap between the Macro-and Micro/Nanoscales (pp. 439–444). Dordrecht: Springer Netherlands; 2001 Mar 31.
77. Dauzère-Pérès S, Nordli A, Olstad A, et al. Omya hustadmarmor optimizes its supply chain for delivering calcium carbonate slurry to European paper manufacturers. Interfaces. 2007 Feb;37(1):39–51.
78. He M, Wang Y, Forssberg E. Slurry rheology in wet ultrafine grinding of industrial minerals: A review. Powder Technol. 2004 Oct 11;147(1–3):94–112.
79. Lahiri SK, Ghanta KC. Development of an artificial neural network correlation for prediction of hold-up of slurry transport in pipelines. Chem Eng Sci. 2008 Mar 1;63(6):1497–1509.
80. Verma PK, Minhas AS, Singh P, et al. Slurry erosion behaviour of thermal sprayed Al2O3 and Cr2O3 coatings for turbine steels. In AIP Conference Proceedings (Vol. 2986, No. 1). AIP Publishing; 2024 Feb 20.
81. Choudhury PR. Some aspects of particle agglomeration in simulated slurry fuels. In ASME International Mechanical Engineering Congress and Exposition (Vol. 18411, pp. 179–186). American Society of Mechanical Engineers; 1997 Nov 16.
82. Cui H, Grace JR. Flow of pulp fibre suspension and slurries: A review. Int J Multiphase Flow. 2007 Sep 1;33(9):921–934.

16 Case Studies on Optimizing Industrial Slurry Systems

Jyoti Bhattacharjee, Subhasis Roy, and Hyun Suk Jung

16.1 INTRODUCTION

Slurry is a complicated mixture of solid particles suspended in a liquid medium, and its movement presents unique challenges in engineering across industries like mining, wastewater treatment, and chemical processing. Since slurries tend to be abrasive and vary with particle size, concentration, and flow velocity characteristics, adverse wear results on pipelines, pumps [1], and other related infrastructure. The issues worsen given operating efficiency, environmental considerations, and cost constraints. One needs to develop innovative solutions to balance the durability of materials and dominant hydrodynamic aspects [2]. The demand for advanced [3] predictive tools and new-generation materials is rising because industries are geared toward higher throughput and longer asset performance. This chapter presents real-world case studies through which it will show that integrated approaches involving materials science, design modification, and computational modeling can solve problems, as shown in Figure 16.1.

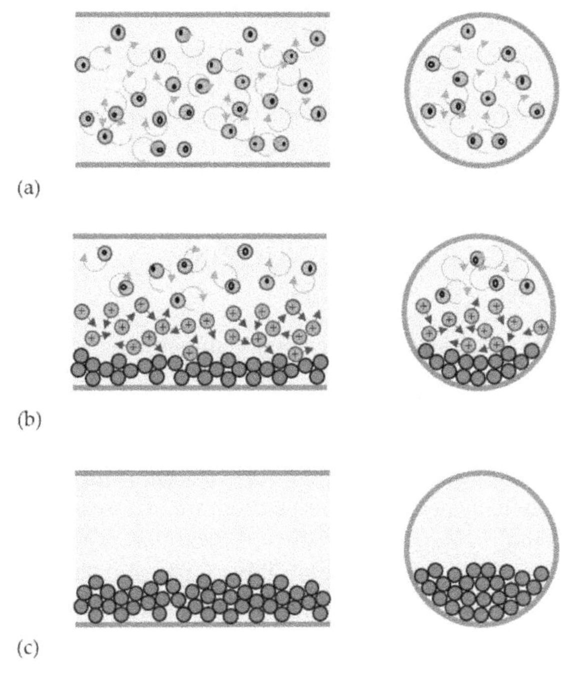

FIGURE 16.1 Flow configurations for flow of solid–liquid slurry streams in horizontal pipes: (a) pseudo-homogeneous flow, (b) heterogeneous flow, and (c) fully stratified flow. In each diagram, the color of the particles indicates their primary mechanism of transport: gray indicates particles transported primarily by interacting with the turbulent-fluid motion; green particles are subjected to rapid collision processes between particles; red particles, meanwhile, bear long-term contact between each other [4].

DOI: 10.1201/9781003561156-19

16.2 CASE STUDY 1: REDUCING PIPELINE EROSION IN TRANSPORT (MINING SECTOR)

A complete engineering solution was implemented to improve pipeline life and operating reliability on a copper mine plant plagued by repeated pipeline failures due to the erosive effects of high-velocity abrasive slurries. The original carbon steel pipes of the plant, subjected to 3.5–4.2 m/s slurry velocities and 150–400 μm-sized abrasive particles, developed localized erosion rates above 12 mm/year [5], particularly at the bottom and bends, causing recurrent unplanned stoppages and rising maintenance costs. To address the aforementioned issues, the project team adopted a multi-faceted approach, which began with replacing the regular carbon steel with superior ceramic-lined composite pipes, which possessed higher abrasion resistance due to their high alumina content and hardness, and resulted in a substantial wear reduction at critical locations. Concurrently, pipeline design was improved by including smoother bends and elbows, minimizing turbulence and particle-laden flows' erosive impact. Computational fluid dynamics (CFD) modeling played a pivotal role in the solution, enabling engineers to simulate the patterns of slurry flow, identify erosion hotspots, and simulate the effectiveness of proposed design changes before actual implementation. This forward-looking approach made interventions more accurate and reduced the chances of sudden operational interruptions [6]. Throughout the project, challenges such as thermally expanding differences between ceramic and steel components, possessing robust joining processes, and accurately regulating flow regime transitions were addressed through cautious engineering and phasing of rollout. The returns on these improvements were spectacular: maximum erosion rates decreased by nearly 90%, bend life in service increased by a factor of almost 5, and annual maintenance shutdown time was reduced by over three-quarters. In addition, the optimized system exhibited further advantages of reduced energy usage, less pipeline vibration, and increased pump life. In an economic sense, the increased initial cost of ceramic-lined pipes and CFD modeling was balanced by significant operational savings and cost avoidance in production losses, with a return on investment of less than 3 years. This case study illustrates how an interdisciplinary integrated approach (by combining innovative materials, hydrodynamic design, and predictive modeling) has the potential to transform slurry transport systems, setting a new reliability and efficiency standard for mining and offering valuable lessons to other industries facing similar challenges, as shown in Figure 16.2 [7].

16.3 CASE STUDY 2: WEAR AND MONITORING IN OIL SANDS SLURRY PIPELINES

Slurry pipelines are a vital component of conveying bitumen-charged sand to processing, but the slurry's coarse and abrasive character results in tremendous wear and operational concerns. Conventional pipelines experienced rapid deterioration and high maintenance costs due to the erosive action by coarse, hard particles suspended in the slurry, leading to recurrent unscheduled outages and progressively escalating repair costs [8, 9]. Having come full circle on these issues, plant operators such as Syncrude launched an extensive program of assessment and innovation to optimize pipeline life while improving the reliability of operation. The first mass-scale application was installing non-metallic piping equipment in rubber hoses and elastomer-lined pipes that were more abrasion resistant than traditional steel pipes. Field experience over ten years supported that such materials aided in securing increased wear life and safer and more reliable pipeline operation; currently, more than a hundred rubber hoses and some 15 kilometers of elastomer-lined pipelines are in operation at Syncrude. However, new problems developed during this period, most typically limited wear monitoring possibilities and localized liner wear in pipe unions that tended to create premature pipeline retirement and increased maintenance expenses. To address these issues, the plant developed and implemented advanced monitoring technologies, including a new wear monitoring system based on radio frequency identification (RFID) technology [10, 11], which has been implemented successfully in rubber hoses and continues to be developed for elastomer-lined

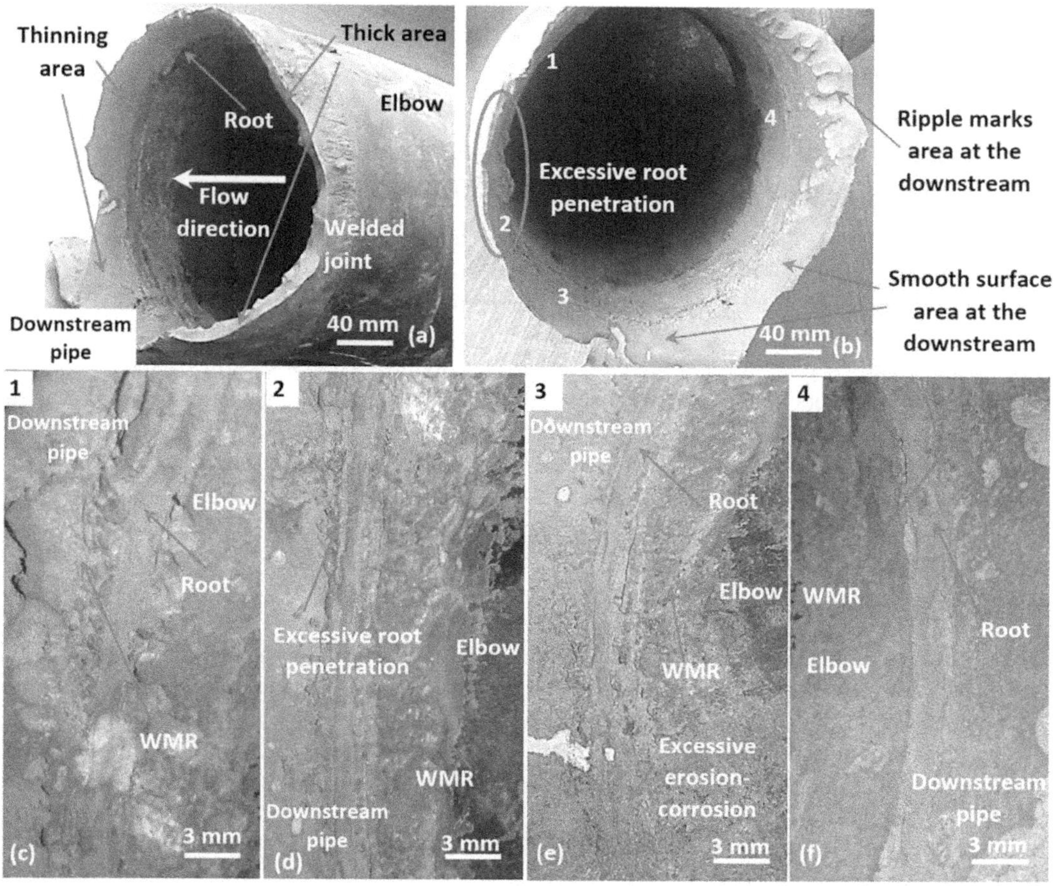

FIGURE 16.2 Fractography of the weld in the elbow and downstream pipe: (a) Failure at the downstream pipe, immediately adjacent to the weld. (b) The front view of the welded joint shows a region of severe root penetration. (c–f) Magnified views of the root region at points 1 to 4 marked in (b), showing microstructural features and crack initiation sites [7].

pipes. The technology enables real-time monitoring of wear patterns and facilitates proactive maintenance before critical failure. In addition, in-line inspection (ILI) tools such as mechanical caliper tools have been thought to provide smooth pipeline condition data and contribute to higher accuracy inspection. The second recurring issue was the internal diameter mismatch between pipe joints, leading to liner wear in particular areas and requiring repeated replacements. To prevent this, a "replaceable ring" concept was developed wherein sacrificial rings at key joints could be turned or replaced to extend the total wear life of non-metallic piping parts. On loss of liner adhesion, a novel repair technology with anchors attached to the carbon steel base was developed to improve adhesion and further extend pipeline life significantly. These combined interventions—material design, advanced monitoring, and strategic repair practices—have dramatically reduced pipeline failure, lowered maintenance costs, and improved operating uptime. The case underscores the necessity for continuous process improvement and the use of new technologies in confronting the complexities of slurry transport, particularly in settings where abrasive wear remains a constant threat. The success of these approaches in the oil sands industry offers valuable lessons to other industries faced with analogous problems, proving that an integrated approach of material selection, online monitoring, and advanced repair techniques can transform slurry pipeline performance and reliability [12, 13].

16.4 CASE STUDY 3: MULTI-SIZED SLURRY PIPELINE DIAMETER OPTIMIZATION AND COST OPTIMIZATION

The choice of pipeline diameter and operating velocity in multi-sized slurry transport pipeline systems significantly affects both capital and operating costs in industrial slurry pipeline systems. One common problem when conveying multi-sized slurries is that a diameter that is too small results in high velocities, excessive wear, and high pumping costs. In contrast, a diameter that is too large involves high initial capital expenditure without necessarily increasing operating efficiency. A specific example was a pipeline conveying a slurry with a volume solids concentration of 15% with solid density 2650 kg/m³ and carrier fluid (water) density 999 kg/m³ [14, 15]. The solid throughput required was 60 kg/hr, and the particle size range was given as indicated in Table 16.1. The problem was finding the most economical pipe diameter and operating velocity to reduce overall cost, trading off manufacturing and power costs, as shown in Figure 16.3.

16.4.1 Solution Approach

The project team implemented a conceptual optimization model accounting for manufacturing cost (pipe installation and fabrication) and power cost (power demand for pumping) over various pipe diameters and velocities. The model considered particle size gradation, which affects slurry rheology and therefore pressure drop and pumping demand. The predicted pressure drops and velocities were verified with field data and computational models to ensure accuracy in the optimization results [17].

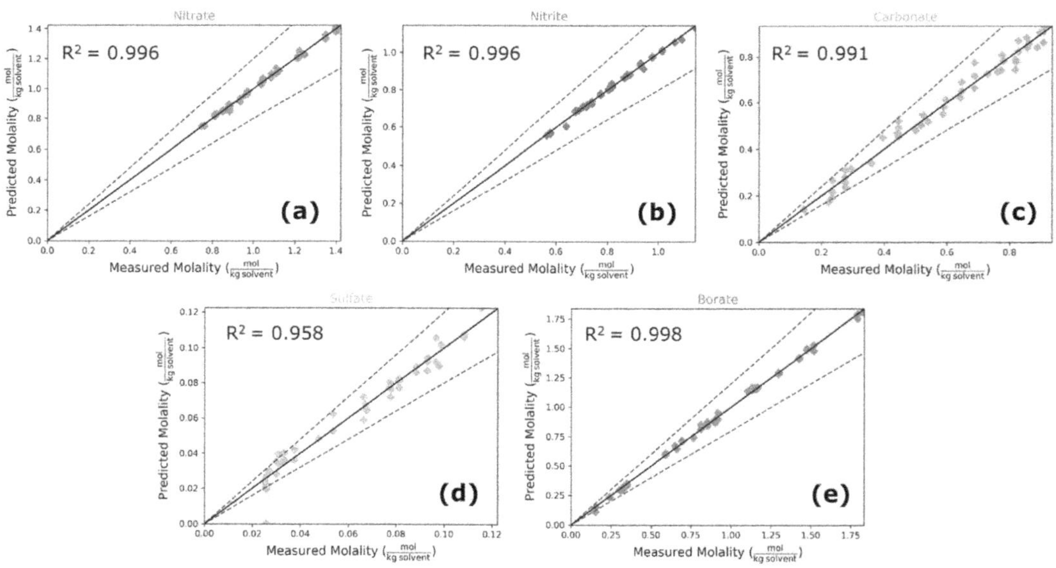

FIGURE 16.3 Schematic showing the parity plots (with accuracies within ±20%) for quantification of soluble anions in the simulant slurry, with ATR-FTIR spectral data used as input into the PLSR model with 15 latent variables: (a) nitrate (NO_3^-), (b) nitrite (NO_2^-), (c) carbonate (CO_3^{2-}), (d) sulfate (SO_4^{2-}), and (e) borate (BO_3^{3-}) [16].

TABLE 16.1
Various Pipe Diameters and Their Costs

Parameter	Baseline (Pre-Optimization)	Post-Optimization/Implementation	References
Pipeline Length	~50 km	~50 km	[19]
Slurry Type	Iron ore tailings, thickened	Iron ore tailings, thickened	[20]
Maximum Surge Pressure	Near design limit	Well below design limit	[21, 22]
Leak Detection System	Model-based	Statistical volume balance	[23]
False Alarm Rate	High	Low	[24]
Downtime Due to Surges/Blockages	Occasional	Rare	[25]
Compliance with ASME B31.3	Partial	Full	[26]

16.4.2 Implementation and Outcomes

The optimization process identified prominent trends: as pipe diameter rose, the operating velocity needed for a specific throughput fell, as did total head loss and power use. But bigger pipe diameters also meant higher manufacturing costs. The optimal diameter was 0.12 meters for this scenario, balancing capital and operating costs, as shown in Table 16.1 [18].

16.4.3 Discussion

The findings prove that the gradation of particle size and the pipe diameter are essential parameters for evaluating the most cost-effective slurry transportation system. The best diameter (0.12 m in this example) has the minimum total cost by saving on the greater manufacturing cost of large pipes against the lower power cost incurred with lower velocity and head loss. These results are consistent with general industry experience, where it is stressed that the consideration of both initial capital and continuing operating costs is paramount in slurry pipeline design [27]. This case study illustrates the effectiveness of optimization models in designing slurry pipelines, especially for multi-sized slurries. Systematically analyzing the trade-offs among pipe diameter, velocity, and cost, engineers can save substantially and enhance the integrity of slurry transport systems. The methodology applies to mining, mineral processing, and industrial waste disposal, where slurry transportation forms a vital part of operations [28].

16.5 CASE STUDY 4: RELIABILITY AND SURGE MITIGATION IN LONG-DISTANCE IRON ORE SLURRY PIPELINES

16.5.1 Background and Problem Statement

Long-distance slurry pipelines are increasingly being used in the mining sector to pipeline mineral concentrates such as iron ore, coal, and copper across hundreds of kilometers [2, 6, 29]. These pipelines offer significant cost, safety, and environmental advantages over traditional bulk transport. However, operation in hostile environments—such as cold climates or remote, challenging terrain—raises unique technical challenges, including the risk of clogging of the pipeline due to particle settling, the danger of hydraulic surges in the event of pump failure or valve closure, and the requirement for efficient leak detection systems [30].

A good example is Tata Steel's iron ore slurry pipeline network in India, which transports thickened tailings across challenging terrain and elevation gradients. The network is exposed to fluctuating operating conditions, including frequent shutdowns and startups of the pumps, and must provide reliability while minimizing downtime and environmental risk [31].

16.5.2 Solution Approach

Tata Steel adopted an integrated approach to pipeline reliability and surge mitigation to address these challenges.

- *Data Collection and System Assessment:* The project began with a rigorous review of working parameters, pipeline configuration, and slurry behavior. Key factors to be considered were slurry viscosity, density, pipeline elevation gradients, pump discharge points, and valve positions.
- *Hydraulic Simulation and Surge Analysis:* Advanced hydraulic modeling software was utilized to simulate steady-state and transient behavior, including emergency conditions such as pump trips, instantaneous valve closures, and loss of power. Analysis was verified for pipeline stress, potential cavitation, and compliance with industry standards (ASME B31.3) [32–35].
- *Leak Detection and Monitoring:* The pipeline deployed a statistical volume balance leak detection system with a higher precision than the traditional model-based systems for slurry pipelines. The technology minimized false alarms and the chance of spurious shutdowns, which can lead to costly blockages, as listed in Table 16.2.
- *Operational Best Practices:* The gradual activation of valves, regulated pump operation, and accurate pipeline sizing were employed to manage surge events and maintain operational stability [36].

The hydraulic surge analysis confirmed transient pressures within design limits even in the worst-case scenarios. Surge reduction techniques other than those needed were not required because of the pipeline design optimization and operating controls. The statistical leak detection system reduced shutdowns and typically improved overall reliability [42–44].

This case study focuses on integrating systems design and advanced modeling in long-distance slurry pipeline reliability, especially under severe environmental conditions. Statistical leak detection and robust surge analysis software have set a new standard for operating safety and effectiveness in mining [45].

TABLE 16.2
Tata Steel's Iron Ore Slurry Pipeline Key Parameters and Outcomes

Parameter	Rubber-Lined Steel	Polyurethane-Lined Steel	References
Wear Rate	High	Very low	[37]
Service Life (Key Sections)	12–18 months	5+ years	[38]
Maintenance Cycles	Frequent	Rare	[38]
Monitoring Capability	Manual inspection	Real-time digital monitoring	[39]
Downtime	High	Low	[40]
Corrosion Resistance	Moderate	Excellent	[40]
Customization	Limited	High	[41]

16.6 CASE STUDY 5: ENHANCING FLOW IN THE PROCESSING OF FOOD SLURRY

In response to chronic issues with erratic viscosity and settling in a tomato puree slurry processing line—issues that detracted from product quality [40–42], line capacity, and customer satisfaction—a food processor plant employed a systemic, technology-intensive solution that integrated process simulation, advanced temperature control, and real-time monitoring. Using Aspen HYSYS process simulation software, the facility simulated the entire slurry flow system, identified the location of critical bottlenecks and sedimentation hotspots, and maximized pipeline design and operating parameters for more efficient flow [43, 44]. The plant used zoned heat along the pipeline to supply uniform slurry characteristics and preclude temperature-induced viscosity shifts to ensure the tomato puree remained within its ideal thermal range throughout processing. Furthermore, inline viscosity sensors incorporated into the process provided instant feedback regarding slurry consistency, and real-time pump speed adjustment, heating, and flow rate could be implemented [45–48]. The likelihood of off-spec batches was significantly reduced. As a result, consistency across all product batches was considerably enhanced, line throughput increased by 25%, and product quality customer complaints fell by 60%. This example shows the synergistic combination of process simulation, precise temperature control, and real-time monitoring, empowering a revolution in food processing of slurry, providing quantifiable improvements in product quality, plant effectiveness, and customer satisfaction, as shown in Figure 16.4.

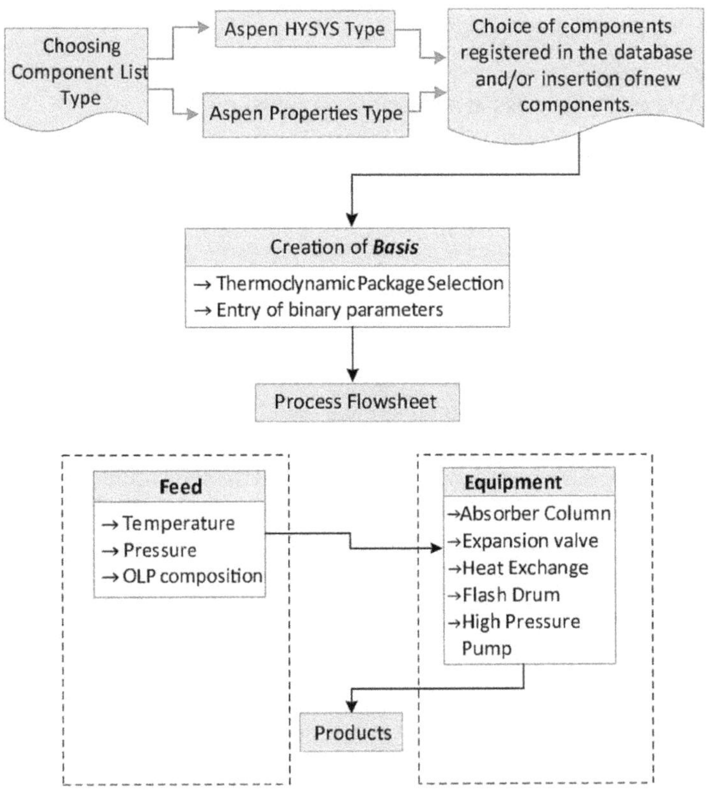

FIGURE 16.4 Detailed design process to simulate the deacidification/deoxygenation of OLP (oilfield liquids processing) utilizing Aspen HYSYS 8.4 based on the everyday simulation tasks that occur for this process. Step-by-step simulation workflow for deacidification and deoxygenation of Oilfield Liquids Processing (OLP) in Aspen HYSYS 8.4, illustrating key stages such as component selection, property method setup, process flow diagram creation, stream and unit operation specification, reaction modeling, and convergence monitoring.

(Developed by authors).

The simulation revealed the following key findings:

- Flow Bottlenecks: Multiple pipe elbows and low-flow sections were prone to particle settling.
- Temperature Sensitivity: Viscosity was highly sensitive to temperature variation, especially at seasonal changes.
- Residence Time Effects: Increased residence in holding tanks increased the risk of sedimentation [45–51].

16.6.1 Temperature Control: Zoned Heating

Based on simulation findings, the plant integrated zoned heating along the most critical sections of the slurry pipeline. This configuration allowed precise control of temperature so that the slurry could maintain ideal thermal properties for the duration of the process. Some of the critical elements utilized were:

- Localized Heaters: Installed at critical points (e.g., pre-pumps, post-holding tanks) to prevent temperature loss.
- Thermal Insulation: Added insulation to the pipes and tanks to minimize heat loss.
- Automated Control: Interconnected with the plant control system to adjust heating based on current temperature readings [52].

16.6.2 Inline Viscosity Sensors and Feedback Loops

Besides the previous, the plant put inline viscosity sensors along the slurry flow path at several points. They provided real-time data on slurry viscosity, which enabled [53, 54]:

- Continuous Monitoring: Operators could track viscosity trends and detect deviations in real time.
- Automated Corrections: The control system made real-time adjustments in pump speed, heat, and flow rates using viscosity information [54].
- Process Optimization: Patterns were recognized, and operating parameters were optimized for different batches based on historical data from sensors [55].

Alane (aluminum hydride, AlH_3), ammonia borane (NH_3BH_3), and similarly composed compounds like lithium aluminum hydride ($LiAlH_4$) and magnesium aluminum hydride ($Mg(AlH_4)_2$) present a distinct paradigm for hydrogen storage in solid form. In these substances, hydrogen is trapped in a metastable solid state, providing extremely high hydrogen densities at ambient pressure and temperature, far higher than liquid hydrogen's density. One of the advantages is their comparatively low hydrogen release temperature, generally below 150°C, making them appealing to applications involving practical utilization, such as on-board hydrogen storage. Additionally, such hydrides can produce very high hydrogen pressures—over 1 kilobar—under moderate temperatures without external compressors. Nevertheless, the use of such metastable materials for hydrogen storage poses two formidable challenges: in the first place [56], governing the rate of hydrogen release from a system that is intrinsically very far from equilibrium; second, and possibly more importantly, regenerating the hydrogen-poor material—that is to say, recharging the hydride with hydrogen despite its low formation enthalpy.

16.7 SYNTHESIS AND STRUCTURES OF ALUMINUM HYDRIDE

Aluminum hydride (ALANE) is a promising metastable hydride due to its high gravimetric hydrogen content (10.1 wt%) and volumetric hydrogen content (148 g H_2 per liter). It releases hydrogen quickly at low temperatures (below 150°C) in a single, clean step, yielding solely aluminum and hydrogen gas with no nasty intermediates or unwanted side reactions. The preparation of AlH_3 usually consists of a reaction between lithium aluminum hydride ($LiAlH_4$) and aluminum chloride ($AlCl_3$) in diethyl ether to produce an alane-etherate adduct. Other tetrahydroaluminates, binary hydrides, and aluminum salts can also be employed. Following the reaction, a salt byproduct such as lithium chloride (LiCl) [56] is filtered away. The solution of etherate is subsequently subjected to dynamic vacuum between 65 and 75°C to desolvate and recuperate crystalline AlH_3. With an excess of $LiAlH_4$ alone or sometimes in association with $LiBH_4$, the time and temperature of desolvation can be minimized. Following desolvation, the excess hydrides are removed by washing with diethyl ether. Vacuum desolvation of the ethereal solution directly results in sub-micron (200–500 nm) irregular AlH_3 particles. For larger production, a continuous process known as Brower and co-workers can be used, where desolvation is done in toluene at around 77°C to form large cubic crystallites (50–100 μm). The large crystals can be washed with a dilute aqueous acid solution (e.g., dilute HCl), which sweeps out impurities and passivates the hydride surface to enhance shelf life, and can be utilized to set the dehydrogenation temperature finely. Aluminum hydride also occurs in several crystalline phases, or polymorphs, α, α′, β, γ, δ, ε, and ζ. The most stable and well known of these is the α-AlH_3 polymorph. Its structure includes AlH_6 octahedra sharing their corners with hydrogen atoms, having a trigonal space group ($R\bar{3}c$). The other polymorphs contain the same structural units but are less stable and usually considered impurities during α-AlH_3 synthesis. These more unstable forms will decompose at room temperature, so they are unsuitable for most practical uses [57].

16.7.1 Thermodynamics and Kinetics of Alane

Dehydrogenation of α-AlH_3 to aluminum and hydrogen is an entropically favored, endothermic reaction that occurs in one step (α-AlH_3 → Al + 3/2 H_2). The reaction enthalpy is about 7 kJ per mole of H_2, with a reaction entropy of about 129 J per K per mole of H_2. The free energy change is about −31 kJ per mole of H_2 at 300 K. The other polymorphs are less stable than α-AlH_3 by 1–2 kJ per H_2 but possess comparable entropies. While the hydrogen pressure in equilibrium with α-AlH_3 is achieved at 1 bar at an extremely low temperature (−218°C), the release of hydrogen is kinetically resisted until temperatures reach 100°C. Conversely, hydrogenation of aluminum metal to AlH_3 is an exothermic process. Still, it is only facilitated with exceptionally high hydrogen pressure (about 7 kilobar at ambient temperature) to overcome the barrier of entropic nature [58].

Hydrogen release kinetics from α-AlH_3 are characteristic of solid-state hydrides. The hydrogen evolution rate obeys an Arrhenius relationship with temperature but also varies with the composition of the material (aluminum/AlH_3 ratio). The isothermal desorption curve is sigmoidal, with an initial induction period (nucleation of aluminum), an acceleratory period (three-dimensional growth of aluminum), and a decay period. The lower stability polymorphs of AlH_3 tend to make an exothermic phase transition to the α-form at high temperatures (above 100°C). Although this phase transition releases a limited amount of heat, no hydrogen is evolved in the process, and thus the decomposition kinetics for all polymorphs are comparable above 100°C. At lower temperatures, decomposition of less stable polymorphs is more complicated: part of the material decomposes to the elements directly, while the rest first converts to α-AlH_3 and then decomposes. The release rates for hydrogen by the α, β, and γ polymorphs are higher in the less stable forms below 100°C, but all three have comparable release rates above 100°C [59, 60].

The regeneration of spent alane (AlH_3) from its used fuel—commonly aluminum metal and hydrogen—is a key challenge to the practical application of metastable hydrides as solid-state hydrogen carriers. Straightforward re-hydrogenation of aluminum is usually impossible at moderate pressure because it needs extremely high pressures (of the order of 7 kilobar at ambient temperature) to overcome thermodynamic requirements. This makes direct hydrogenation impractical for most purposes. As a result, other regeneration methods have been sought to facilitate the recycling of alane and other hydrides [61].

16.7.2 Recycling through Traditional Synthesis

One of the simple methods is to recycle the used aluminum, hydrogen, and other byproducts into the precursors of the conventional alane synthesis. Nevertheless, it is energy consuming, especially when lowering salt byproducts and reactivating the required reagents. The thermodynamic constraints of this process lead to a theoretical maximum regeneration efficiency of approximately 30%, which is not enough to meet the well-to-tank efficiency requirement of 60% and above.

16.7.3 Electrochemical Hydrogenation

Electrochemical processes provide a promising route. Low-pressure electrochemical hydrogenation directly onto aluminum can add hydrogen into aluminum but generally only in low concentrations (about 1000 ppm), without the formation of the targeted α-AlH_3 phase. More sophisticated methods have been devised, like employing an aluminum anode in a sodium aluminum hydride ($NaAlH_4$) dissolved in tetrahydrofuran (THF) electrolyte. Here, AlH_4^- ions serve as a source of hydrogen, oxidizing the aluminum at the anode to produce solvated alane (AlH_3•nTHF). This technique mirrors standard chemical synthesis but substitutes aluminum chloride with aluminum from the anode. The primary problems with this technique are sodium metal or sodium aluminum hydride (Na_3AlH_6) forming at the cathode and, at high current densities, causing issues such as dendrite growth. Further, effective recovery and reconversion of the cathodic product back to a usable hydride is required for a closed-cycle process [62].

16.7.4 Thermochemical Hydrogenation with Amines

Another regeneration pathway is thermochemical hydrogenation of aluminum in the presence of an amine to produce an alane-amine adduct (AlH_3•NR_3) under relatively moderate conditions. Several tertiary amines—trimethylamine (TMA), triethylenediamine (TEDA), and dimethylethylamine (DMEA)—can be employed for this end. This approach is appealing because it can be developed to treat aluminum powder from spent fuel directly. Yet the selection of amine influences both the ease with which adducts form and how easily alane [63] can be separated. More stable adducts, which form easily under mild conditions, are troublesome to separate, whereas less stable adducts are more easily separated and more challenging to create directly.

16.7.5 Adduct Separation and Ligand Exchange

The removal of alane from its adduct is one of the critical steps in regeneration. Generally, the ligand is stripped thermally under vacuum, but caution should be exercised to prevent hydride decomposition. A transamination or ligand exchange step can overcome the trade-off between formation and separation ease. For instance, DMEA hydrogenation of aluminum gives AlH_3•DMEA, which can be treated with triethylamine (TEA) to provide AlH_3•TEA. The latter adduct is more convenient for separation with recovery of pure AlH_3. The same approach might be extended to other metastable hydrides, e.g., lithium aluminum hydride ($LiAlH_4$) and magnesium aluminum hydride ($Mg(AlH_4)_2$) [64].

16.7.6 Outlook and Challenges

Metastable hydrides such as alane possess superior gravimetric and volumetric hydrogen densities, low temperatures for hydrogen release, and the capability of producing high pressures without external compressors. These characteristics make them desirable for portable power units, uncrewed devices, and large mobile platforms. The system-specific energy of integrating alane with a lightweight fuel cell can be over 800 Wh/kg, significantly surpassing existing lithium-ion batteries. Yet, these systems' commercialization depends upon the availability of low-cost and effective regeneration and synthesis strategies. Breakthroughs that avoid or minimize the use of solvents, allow for direct alane adduct formation from aluminum metal, and enhance alane/alane adduct separation will be critical to the eventual widespread use of these materials [65].

The effective and economical regeneration of ammonia borane was a priority of the CHSCoE since, without an effective and economical regeneration technology, the worldwide application of ammonia borane as a material for hydrogen storage is very improbable. The DOE target cost for delivered hydrogen is USD 2–3/kg at the pump (Table 16.1). This means that regeneration costs should be less than the USD 3/kg H_2 to realize the DOE target. Different methods were explored for regenerating spent ammonia borane (i.e., polyborazylene), which include the thiacatechol method, the alcohol method, the super acidic haloacid method, and the hydrazine method. The hydrazine method is the most refined among the regeneration methods in that it is a one-pot process where polyborazylene (spent AB) is directly converted into ammonia borane with the help of ammonia and hydrazine at 60 °C [65]. A general reaction pathway may be illustrated as follows [65]:

$$BNH_2 + N_2H_4 \rightarrow NH_3NH_3BH_3 + N_2 \tag{16.1}$$

- ΔHrxn, 25 °C=−6.95 kcal/mol
- ΔGrxn, 25 °C=−22.88 kcal/mol

The price of regenerating used ammonia borane with the hydrazine process is USD 45.73 per kilogram of hydrogen, with the main cost being hydrazine at USD 43.79 per kilogram of hydrogen.

Designing slurry-phase reactors with our methodology was primarily driven by the necessity to overcome widespread problems of reactor fouling and slugging, which have the potential to critically affect reactor performance and reliability. To mitigate fouling, we used a mechanically mediated reactive transport approach, coupling a motor-powered auger system with the design of a ship auger bit. This setup not only allowed solid material transport—involving ammonia borane (AB), spent AB, alane (AlH_3), and aluminum metal—through the reactor but also allowed for gas-phase transport along the centerline of the auger, as shown in Figure 16.5.

Mechanical mixing from the auger helped to improve gas separation of entrained hydrogen and to avoid agglomeration of solids, a typical reason for clogging. For slurry feed delivery, we chose

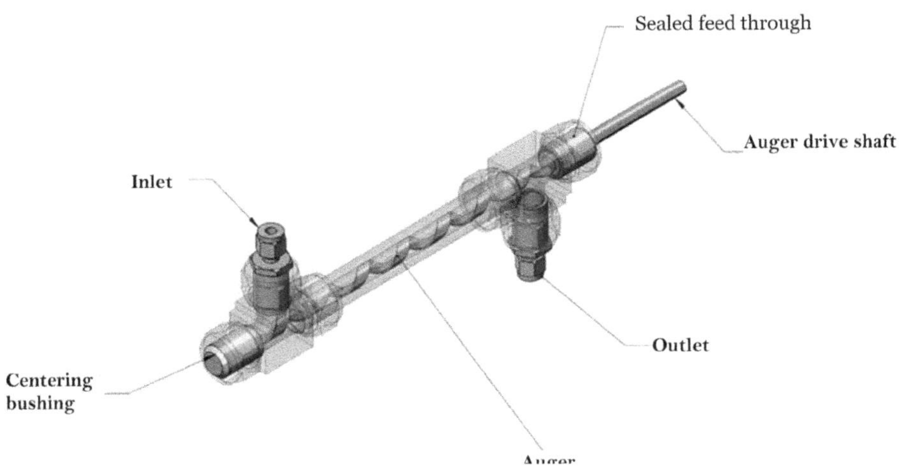

FIGURE 16.5 Schematic illustration of the slurry-based reactor system for hydrogen storage, consisting of the slurry feed system, auger reactor, heat exchanger, and gas-liquid separators [66].

a Watson–Marlow peristaltic pump, having first tried gear pumps that were found to be clogging and seizing. Syringe pumps were eschewed because they require homogeneous feed mixtures to ensure precise kinetic correlation. All slurry feed reservoirs were stirred continuously using a motor-driven impeller to ensure feed homogeneity, because phase separation may occur within 30 minutes, depending on slurry loading [67].

The testing system also experienced several design refinements to contend with unexpected difficulties, specifically ammonia borane slurries. The last refinement, illustrated in Figure 16.6, bore a close resemblance to that of the MPAB system (refer to Figure 16.6), with the incorporation of a feed reservoir mixer, flow conditioning, diborane conditioning, and particulate filter, all particularly designed to cope with the non-ideal nature of AB slurries. Analytical tools for gas-phase analysis comprised gas-phase IR spectroscopy and gas chromatography, which enabled the quantification of diborane, hydrogen, ammonia, and borazine. Slurries of alane were made with silicone oil, mechanical pump fluid, and tetraglyme with solids loadings between 20 and 65 wt%. Chemicals like lithium hydride and titanium t-butoxide were investigated to enhance alane dehydrogenation. The slurries of ammonia borane were prepared with silicone oil and Triton X-15 (an anti-foaming surfactant), attaining 50 wt% solids loading using particle size reduction and tip-sonication, which enhanced dispersion and flowability [68]. During experiments, strict safety and engineering controls were followed, mitigating flammability, dust explosion, pyrophoricity, nanoparticle toxicity, carcinogenic hazards, chemical compatibility, storage, shelf-life, and disposal issues. Despite these attempts, flow-through slurry-phase reactor experiments were difficult to obtain dehydrogenation kinetics and illustrate mechanically mediated reactive transport. Clogging and fouling of the reactors were noted for alane and AB slurries, along with ancillary downstream clogging of apparatus like valves and rotameters in AB slurries. Salient findings for alane slurries were sparse alane conversion data below 200°C owing to the total vaporization of tetraglyme and partial vaporization of silicone oil and mechanical pump fluid at elevated temperatures, and side reactions between alane and these carriers at temperatures above 200°C. Total conversion of alane was realized in a 20 wt% slurry of silicone oil at

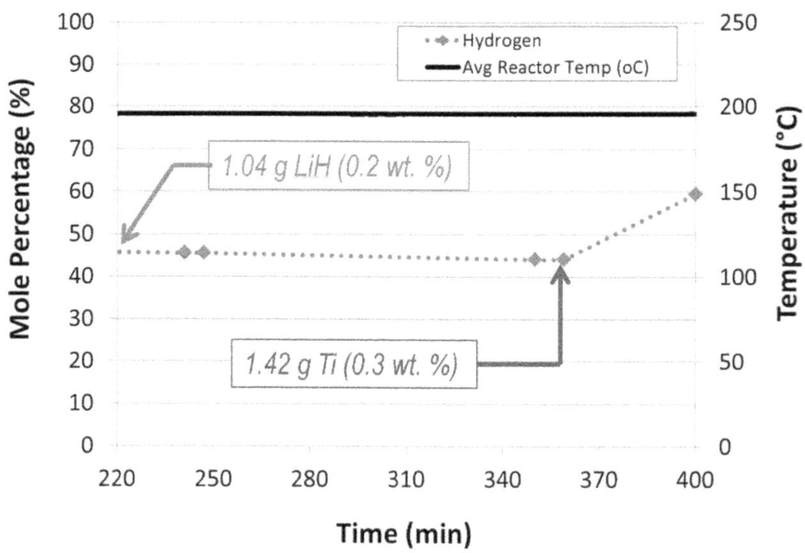

FIGURE 16.6 Effects of successive addition of lithium hydride (LiH) and titanium t-butoxide to a 60 wt% ATK alane slurry in mechanical pump fluid, in reactor conditions of auger speed = 12 rpm and feed flow rate = 1.02 mL/min, were studied to determine their influence on alane dehydrogenation kinetics in the auger-driven slurry-phase reactor [66].

(Open Reproduction Permission from American Chemical Society)

250°C, a space-time of 6.8 minutes, and an auger speed of 40 rpm. In a 50 wt% alane-silicone oil slurry, conversions of over 80% were obtained at an average temperature of 205°C, an auger speed of 12 rpm, and a space-time of 7.6 minutes. Lithium hydride added with titanium t-butoxide was found to almost double the measured hydrogen mole percent, but lithium hydride alone did not encourage alane dehydrogenation. These findings highlight the potential and the continued challenges of slurry-phase reactor systems for hydrogen release from advanced hydrides [69]. If LiH was added separately to the alane slurry, there was no notable enhancement in hydrogen release; the dehydrogenation rate remained the same as with the base slurry without additives. But adding LiH in sequence with titanium t-butoxide resulted in a remarkable catalytic effect [70].

16.8 CONCLUSION

The development and engineering of slurry-based hydrogen storage and release systems are a significant advancement over the shortcomings of solid-state hydrogen carriers such as alane and ammonia borane. Our chapter revolved around avoiding reactor fouling and slugging by mechanistically mediated reactive transport employing a motor-driven auger to enable motion of slurry solids while enabling gas-phase flow and efficient hydrogen separation. The addition of continuous mixing in the slurry feed reservoir and use of a peristaltic pump gave a stable and homogenous slurry supply with minimum phase separation and interruptions to operation. Upgrading the slurry feed system, auger reactor, heat exchanger, and gas–liquid separators formed a chain process that minimized high solid loadings and provided steady-state operations. Sequential addition of lithium hydride and titanium t-butoxide to alane slurries significantly enhanced dehydrogenation kinetics, nearly doubling the hydrogen yield over slurries with no such additives. Such a catalytic effect was most pronounced in high solids loading slurries (up to 60 wt%), which justifies the importance of additive selection and reactor design in optimizing hydrogen release efficiency. The mechanical agitation in the auger reactor avoided clogging and fouling and enabled gas-solid separation, achieving uniform reactor operation. Nevertheless, the challenges of reactor clogging, fouling of downstream equipment, and the requirement for strict safety control measures (flammability, dust explosion, and toxicity) show the complexity of slurry-phase reactor operation. A successful slurry-based hydrogen storage and release system demonstration opens new possibilities for portable and scalable hydrogen storage systems. The high energy density and rapid release rates of advanced hydrides, allied to the strength and versatility of slurry-phase reactor systems, make such systems ready for application in various applications, from portable power packs to unmanned vehicles and even large-scale hydrogen infrastructure. Future investigations should be focused on enhanced catalyst formulations, reactor materials, and integration of processes towards increased efficiency, safety, and cost-effectiveness. Ultimately, continued progress in slurry-based hydrogen storage technology has huge potential for enabling the widespread utilization of hydrogen as a clean, high-energy-density fuel.

ACKNOWLEDGMENTS

The author (S. Roy) would like to acknowledge Scheme for Transformational and Advanced Research in Sciences (STARS) (MoE-STARS/STARS-2/2023-0175) by the Ministry of Education, Government of India, for promoting translational India-centric research in sciences, implemented and managed by Indian Institute of Science (IISc), Bangalore, for their support.

REFERENCES

1. U.S. Department of Energy, Hanford Lifecycle Scope, Schedule and Cost Report, DOE/RL-2018-45, 2019.
2. Bhattacharjee, J., Roy, S., Welding Automation and Robotics, in Advanced Welding Technologies, Wiley, 2025, pp. 47–71. https://doi.org/10.1002/9781394331925.ch4.

3. Kelly, S. E., A Joule-Heated Melter Technology for the Treatment and Immobilization of Low-Activity Waste, RPP-48935, Hanford Site, 2011.
4. Silva, R. C., Experimental Characterization Techniques for Solid-Liquid Slurry Flows in Pipelines: A Review. Processes 10 (2022) 597. https://doi.org/10.3390/pr10030597.
5. Kocevska, S., Maggioni, G. M., Crouse, S. H., Prasad, R., Rousseau, R. W., Grover, M. A., Effect of Ion Interactions on the Raman Spectrum of NO_3^-: Toward Monitoring of Low-Activity Nuclear Waste at Hanford, Chem. Eng. Res. Des. **181** (2022) 173–194. https://doi.org/10.1016/j.cherd.2022.03.002.
6. Vienna, J. D., Skorski, D. C., Kim, D. S., Matyas, J., Glass Property Models and Constraints for Estimating the Glass to Be Produced at Hanford by Implementing Current Advanced Glass Formulation Efforts, United States, 2013.
7. Abdu, M. T., Khalifa, W., Abdelrahman, M. S., Investigation of Erosion–corrosion Failure of API X52 Carbon Steel Pipeline, Sci. Rep. **13** (2023) 20494. https://doi.org/10.1038/s41598-023-42556-6.
8. Westesen, A., Wells, B., Peterson, R., Identifying Challenge Sludges to Process at the Hanford Site, Part. Sci. Technol. **41**(4) (2022) 453–459. https://doi.org/10.1080/02726351.2022.2116372.
9. Siegfried, M. J., Stone, M. E., LAW Simulant Recipes for Evaluation of Real-Time, In-Line Monitoring Instruments, United States, 2020.
10. Févotte, G., In Situ Raman Spectroscopy for In-Line Control of Pharmaceutical Crystallization and Solids Elaboration Processes: A Review, Chem. Eng. Res. Des. **85** (2007) 906–920. https://doi.org/10.1205/cherd06229.
11. McCreery, R. L., Raman Spectroscopy for Chemical Analysis, John Wiley & Sons, Inc., New York, 2000.
12. Diprete, C. C., Howe, A. M., WTP Real-Time, In-Line Monitoring Program Task 2: Determine the Technical Basis for Process Control and Task 5: Process Control Challenges, United States, 2017.
13. Bhattacharjee, J., Roy, S., Shaikh, A. A., Artificial Intelligence and Machine Learning for the Optimization of Photocatalytic Performance, Energy **360** (2025) 100027.
14. Borissova, A., Khan, S., Mahmud, T., Roberts, K. J., Andrews, J., Dallin, P., Chen, Z. P., Morris, J., In Situ Measurement of Solution Concentration during the Batch Cooling Crystallization of L-Glutamic Acid Using ATR-FTIR Spectroscopy Coupled with Chemometrics, Cryst. Growth Des. **9**(2) (2009) 692–706. https://doi.org/10.1021/cg7010265.
15. Lin, M., Wu, Y., Rohani, S., Simultaneous Measurement of Solution Concentration and Slurry Density by Raman Spectroscopy with Artificial Neural Network, Cryst. Growth Des. **20**(3) (2020) 1752–1759. https://doi.org/10.1021/acs.cgd.9b01482.
16. Prasad, R., Crouse, S. H., Rousseau, R. W., Grover, M. A., Quantifying Dense Multicomponent Slurries with In-Line ATR-FTIR and Raman Spectroscopies: A Hanford Case Study, Ind. Eng. Chem. Res. **62**(39) (2023) 15962–15973. https://doi.org/10.1021/acs.iecr.3c01249.
17. Wong, S. W., Georgakis, C., Botsaris, G. D., Saranteas, K., Bakale, R., Online Estimation and Monitoring of Diastereomeric Resolution Using FBRM, ATR-FTIR, and Raman Spectroscopy, Ind. Eng. Chem. Res. **47**(15) (2008) 5576–5584. https://doi.org/10.1021/ie071179q.
18. Halder, S., Bhattacharjee, J., Roy, S., Das, B. C., Biswas, A. B., Surface Modification of Lignin by Cetyl Trimethyl Ammonium Bromide, in Jawaid, M., Ahmad, A., Meraj, A. (Eds.), Handbook of Lignin, Springer, Singapore, 2025.
19. Srivastava, R. R., Rajak, D. K., Ilyas, S., Kim, H., Pathak, P., Challenges, Regulations, and Case Studies on Sustainable Management of Industrial Waste, Minerals 13 (2023) 51. https://doi.org/10.3390/min13010051.
20. Awhangbo, L., Bendoula, R., Roger, J. M., Béline, F., Fault Detection with Moving Window PCA Using NIRS Spectra for Monitoring the Anaerobic Digestion Process, Water Sci. Technol. 81 (2020)367–382. https://doi.org/10.2166/wst.2020.117.
21. Wang, Q., Li, F., Jiang, X., Wu, S., Xu, M., On-Stream Mineral Identification of Tailing Slurries of Tungsten via NIR and XRF Data Fusion Measurement Techniques, Anal. Methods 12 (2020) 3296–3307. https://doi.org/10.1039/D0AY00322K.
22. Van Manen, H. J., Bloemenkamp, R., Van Den Brink, O. F., Focal Length Determination of Raman Immersion Ball Probes in Diverse Media, Appl. Spectrosc. 63 (2009) 378–380. https://doi.org/10.1366/000370209787598852.
23. Marquardt, B. J., Le, T., Burgess, L. W., Demonstration of a High-Precision Optical Probe for Effective Sampling of Solids by Raman Spectroscopy, in Raman Spectroscopy and Light Scattering Technologies in Materials Science, SPIE, 2001, pp. 62–69.

24. Cornel, J., Lindenberg, C., Mazzotti, M., Quantitative Application of in Situ ATR-FTIR and Raman Spectroscopy to a Crystallization Process, Ind. Eng. Chem. Res. 47 (2008) 4870–4882. https://doi.org/10.1021/ie800236v.
25. Kocevska, S., Maggioni, G. M., Rousseau, R. W., Grover, M. A., Spectroscopic Quantification of Target Species in a Complex Mixture Using Blind Source Separation and Partial Least-Squares Regression: A Case Study on Hanford Waste, Ind. Eng. Chem. Res. 60 (2021) 9885–9896. https://doi.org/10.1021/acs.iecr.1c01387.
26. Rinnan, Å., Pre-Processing in Vibrational Spectroscopy—When, Why and How, Anal. Methods 6 (2014) 7124–7129. https://doi.org/10.1039/C3AY42270D.
27. Rinnan, Å., Van Den Berg, F., Engelsen, S. B., Review of the Most Common Pre-Processing Techniques for Near-Infrared Spectra, TrAC, Trends Anal. Chem. 28 (2009) 1201–1222. https://doi.org/10.1016/j.trac.2009.07.007.
28. Strachan, C. J., Rades, T., Gordon, K. C., Rantanen, J., Raman Spectroscopy for Quantitative Analysis of Pharmaceutical Solids, J. Pharm. Pharmacol. 59 (2010) 179–192. https://doi.org/10.1211/jpp.59.2.0005.
29. Simone, E., Saleemi, A. N., Nagy, Z. K., Application of Quantitative Raman Spectroscopy for Monitoring Polymorphic Transformation in Crystallization Processes Using a Good Calibration Practice Procedure, Chem. Eng. Res. Des. **92**(4) (2014) 594–611. https://doi.org/10.1016/j.cherd.2013.11.004.
30. Schrader, B., Hoffmann, A., Keller, S., Near-Infrared Fourier Transform Raman Spectroscopy: Facing Absorption and Background, Spectrochim. Acta A 47 (1991) 1135–1148. https://doi.org/10.1016/0584-8539(91)80201-S.
31. Mining and Slurry Pipelines. Atmosi.com., June 16, 2025. https://www.atmosi.com/en/news-events/blogs/mining-and-slurry-pipelines/.
32. Mannar, P., Hydraulic Surge Analysis for Tata Steel's Slurry Pipeline. iFluids Engineering, March 28, 2023. https://ifluids.com/hydraulic-surge-analysis-for-tata-steels-slurry-pipeline/.
33. Wold, S., Sjöström, M., Eriksson, L., PLS-Regression: A Basic Tool of Chemometrics, Chemom. Intell. Lab. Syst. 58 (2001) 109–130. https://doi.org/10.1016/S0169-7439(01)00155-1.
34. Tse, P., Shafer, J., Bryan, S. A., Lines, A. M., Quantification of Raman-Interfering Polyoxoanions for Process Analysis: Comparison of Different Chemometric Models and a Demonstration on Real Hanford Waste, Environ. Sci. Technol. 55 (2021) 12943–12950. https://doi.org/10.1021/acs.est.1c02512.
35. Lines, A. M., Tse, P., Felmy, H. M., Wilson, J. M., Shafer, J., Denslow, K. M., Still, A. N., King, C., Bryan, S. A., Online, Real-Time Analysis of Highly Complex Processing Streams: Quantification of Analytes in Hanford Tank Sample, Ind. Eng. Chem. Res. 58 (2019) 21194–21200. https://doi.org/10.1021/acs.iecr.9b03636.
36. Mandal, D., Bhattacharjee, J., Mitra, D., Roy, S., Aspects of Energy Efficiency and Execution of Modern Technologies for Achieving Net Zero, Grad. J. Interdiscip. Res. Rep. Rev. 2 (2025) 150–163. https://doi.org/10.34256/gjir3.v2i03.39.
37. Bhattacharjee, J., Roy, S., Shaikh, A. A., Datta, P., Efficient Fabrication and Characterization of Doped Nanocomposites for Thermoelectric Materials, Nano Trends 10 (2025) 100109.
38. Shaikh, A. A., Bhattacharjee, J., Datta, P., Roy, S., A Comprehensive Review of the Oxidation States of Molybdenum Oxides and Their Diverse Applications, Sustain. Chem. Environ. 7 (2024) 100125.
39. Ricks, B., Li, Q., Long-Distance Slurry Pipeline Design in Cold Regions: Case Studies. OneMine, 2007. https://www.onemine.org/documents/long-distance-slurry-pipeline-design-in-cold-regions-case-studies.
40. Bhakta, D., Gayen, D., Bhattacharjee, J., Chatterjee, R., Dastidar, P., Roy, S., Repurposing Industrial and Agricultural Wastes as Sustainable Alternatives in Photovoltaic Cells and Dye Degradation, Chem. Select 9 (2024) 38.
41. Simon, L. L., Pataki, H., Marosi, G., Meemken, F., Hungerbühler, K., Baiker, A., Tummala, S., Glennon, B., Kuentz, M., Steele, G., Kramer, H. J. M., Rydzak, J. W., Chen, Z., Morris, J., Kjell, F., Singh, R., Gani, R., Gernaey, K. V., Louhi-Kultanen, M., . . . Chiu, M.-S., Assessment of Recent Process Analytical Technology (PAT) Trends: A Multiauthor Review. Organic Process Res. Develop. 19(1) (2015) 3–62. https://doi.org/10.1021/op500261y.
42. Nickolov, Z. S., Ozcan, O., Miller, J. D., FTIR Analysis of Water Structure and Its Significance in the Flotation of Sodium Carbonate and Sodium Bicarbonate Salts, Colloids Surf., A 224 (2003) 231–239. https://doi.org/10.1016/S0927-7757(03)00317-0.
43. Peak, D., Luther, G. W., Sparks, D. L., ATR-FTIR Spectroscopic Studies of Boric Acid Adsorption on Hydrous Ferric Oxide, Geochim. Cosmochim. Acta 67 (2003) 2551–2560. https://doi.org/10.1016/S0016-7037(03)00096-6.

44. Su, C., Suarez, D. L., Coordination of Adsorbed Boron: A FTIR Spectroscopic Study, Environ. Sci. Technol. 29 (1995) 302–311. https://doi.org/10.1021/es00002a005.
45. Kingma, K. J., Hemley, R. J., Raman Spectroscopic Study of Microcrystalline Silica, Am. Mineral. 79 (1994) 269–273.
46. Deo, G., Wachs, I. E., Predicting Molecular Structures of Surface Metal Oxide Species on Oxide Supports under Ambient Conditions, J. Phys. Chem. A 95 (1991) 5889–5895. https://doi.org/10.1021/j100168a033.
47. Junior, M. R. S., Costa, E. C., Ferreira, C. C., Bernar, L. P., da Silva, M. P., de Andrade Mâncio, A., Santos, M. C., da Mota, S. A. P., de Castro, D. A. R., Junior, S. D., Borges, L. E. P., Araújo, M. E., Machado, N. T., Simulation of Organic Liquid Product Deoxygenation through Multistage Countercurrent Absorber/Stripping Using CO_2 as Solvent with Aspen-HYSYS: Process Modeling and Simulation, Molecules 27 (2022) 2211. https://doi.org/10.3390/molecules27072211.
48. Griffith, W. P., Wickins, T. D., Raman Studies on Species in Aqueous Solutions: Part 1: The Vanadates, J. Chem. Soc. A (1966) 1087–1090. https://doi.org/10.1039/j19660001087.
49. Nielsen, J. R., Ward, N. E., Raman Spectrum and Structure of the Metaborate Ion, J. Chem. Phys. 5 (1937) 201. https://doi.org/10.1063/1.1750008.
50. Ma, Y., Yan, W., Sun, Q., Liu, X., Raman and Infrared Spectroscopic Quantification of the Carbonate Concentration in K_2CO_3 Aqueous Solutions with Water as an Internal Standard, Geosci. Front. 12 (2021) 1018–1030. https://doi.org/10.1016/j.gsf.2020.03.002.
51. Sharma, S. K., Mammone, J. F., Nicol, M. F., Raman Investigation of Ring Configurations in Vitreous Silica, Nature 292 (1981) 140–141. https://doi.org/10.1038/292140a0.
52. Kuebler, K. E., Jolliff, B. L., Wang, A., Haskin, L. A., Extracting Olivine (Fo-Fa) Compositions from Raman Spectral Peak Positions, Geochim. Cosmochim. Acta 70 (2006) 6201–6222. https://doi.org/10.1016/j.gca.2006.07.035.
53. Yang, Z., Peng, H., Wang, W., Liu, T., Crystallization Behavior of Poly(ε-caprolactone)/Layered Double Hydroxide Nanocomposites, J. Appl. Polym. Sci. 116 (2010) 2658–2667. https://doi.org/10.1002/app.31787.
54. Marshall, C. P., Dufresne, W. J. B., Rufledt, C. J., Polarized Raman Spectra of Hematite and Assignment of External Modes, J. Raman Spectrosc. 51 (2020) 1522–1529. https://doi.org/10.1002/jrs.5824.
55. Chamritski, I., Burns, G., Infrared- and Raman-Active Phonons of Magnetite, Maghemite, and Hematite: A Computer Simulation and Spectroscopic Study, J. Phys. Chem. B 109 (2005) 4965–4968. https://doi.org/10.1021/jp048748h.
56. Birnbaum, H. K., Zeides, F., Sirois, E., Rozenaka, P., Spooner, S., Lin, J. S., Hydrogen in Aluminum, J. Alloys Compd. 254 (1997) 260–264.
57. Zidan, R., Garcia, B. L., Fewox, C. S., Stowe, A. C., Gray, J. R., Harter, A. G., Aluminium Hydride: A Reversible Material for Hydrogen Storage, Chem. Commun. (2009) 3717–3719.
58. Martínez-Rodríguez, M. J., Joseph, B. L. G., Teprovich, J. A., Jr., Knight, D. A., Zidan, R., Advances in the Electrochemical Regeneration of Aluminum Hydride, Appl. Phys. A 4 (2012) 545–550.
59. Ni, C., Wegrzyn, J. E., Zhou, W., Celebi, Y., Graetz, J., N-Alkylpiperidine α-Alane Compounds and Their Applications in Alane Regeneration, Int. J. Hydrog. Energy 38 (2013) 9779–9785.
60. Sutton, A. D., Burrell, A. K., Dixon, D. A., Garner, E. B., Gordon, J. C., Nakagawa, T., Ott, K. C., Robinson, J. P., Vasilliu, M., Regeneration of Ammonia Borane Spent Fuel by Direct Reaction with Hydrazine and Liquid Ammonia, Science 18 (2011) 1426–1429.
61. Summerscales, O. T., Gordon, J. C., Regeneration of Ammonia Borane from Spent Fuel Materials, Dalt. Trans. 42 (2013) 10075–10084.
62. Semelsberger, T. A., Davis, B., Rekken, B., Paik, B., Tafoya, J., Chemical Hydrogen Rate Modeling, Validation, and System Demonstration LANL Project Overview, 2013 DOE Annual Merit Review, Unpublished Work, 2013.
63. Bhattacharjee, J., Roy, S., Advanced Techniques Using Ionic Liquids for Efficient Separation of Sugar Alcohols, Innov. Discov. 2 (2025) 3.
64. Bhattacharjee, J., Roy, S., A Comprehensive Review on Integrated Photo Rechargeable Batteries-Supercapacitors, and Their Techno-Economic Feasibility, J. Photochem. Photobiol. 25 (2025) 100257. https://doi.org/10.1016/j.jpap.2025.100257.
65. Butlak, A. V., Kondrat, Y. V., Mazur, A. S., Timoshkin, A. Y., Thermal Decomposition of Ammonia Borane at 357 K, Russ. J. Gen. Chem. 85 (2015) 2505–2508.

66. Semelsberger, T., Graetz, J., Sutton, A., Rönnebro, E. C. E., Engineering Challenges of Solution and Slurry-Phase Chemical Hydrogen Storage Materials for Automotive Fuel Cell Applications, Molecules 26 (2021) 1722. https://doi.org/10.3390/molecules26061722.
67. Semelsberger, T. A., Brooks, K. P., Chemical Hydrogen Storage Material Property Guidelines for Automotive Applications, J. Power Sources 279 (2015) 593–609. https://doi.org/10.1016/j.jpowsour.2014.10.090.
68. Biswas, B., Bhattacharjee, J., Roy, S., Biswas, A. B., Lignin Degradation of Water Hyacinth for Bioethanol Production Using Yeast, in Jawaid, M., Ahmad, A., Meraj, A. (Eds.), Handbook of Lignin, Springer, Singapore, 2025. https://doi.org/10.1007/978-981-97-2664-6_53-1.
69. Bhattacharjee, J., Roy, S., A Review on Photocatalysis and Nanocatalysts for Advanced Organic Synthesis, Hybrid Adv. 6 (2024) 100268. https://doi.org/10.1016/j.hybadv.2024.100268.
70. Shore, S. G., Parry, P. W., The Crystalline Compound Ammonia-Norane, H3NBH3, J. Am. Chem. Soc. 77 (1955) 6084–6085.

Index

A

abrasion, 24
abrasive contact parameters, 36
acidity and alkalinity, 6
additive manufacturing, 126, 207
adduct separation, 271
advanced materials, 114
advanced surface engineering approaches, 187
advantages (FBE coating), 169
AI in slurry erosion, *see* artificial intelligence (slurry erosion)
alloys, 43, 122
alane (aluminum hydride), 269
analysis
  flow-induced erosion, 231
  slurry flow, 73
ANSYS (CFD modeling/meshing), 235
antagonistic effect, 39
antifouling/biofouling, 193
applications, 130
  coatings/pipelines, 166
  industrial, 160
  slurry erosion research, 217
  spiral concentrators, 71
assessment (slurry erosion), 80
arc PVD, 158
artificial intelligence (slurry erosion), 211
assessment (performance testing), 206

B

bends (pipe geometry/erosion), 235
beneficiation, 68
Bitter's model, 30
boundary conditions (solver scheme), 236

C

carburizing, 124
case study/studies
  coating materials, 178
  erosive wear, 243
cavitation testing, 130
ceramics, 43
cermets, 189
CFD/computational fluid dynamics, 68, 73, 84, 251
  erosion mapping, 251
  modeling, 235, 262
closed-loop test systems, 33
clustering (wear patterns), 216
coating(s), 152
  design considerations, 192
  materials (slurry resistant), 167
  methods, 152
  techniques (industrial), 169
  technologies (AI trends), 222
cold spray, 157, 187
  technique, 176
component fracture, 44
composite coatings (slurry erosion), 178
composite materials, 43, 105, 108, 114
computation(al)
  methods, 59
  modelling (slurry erosion), 215
  study, 85
  techniques, 71
concentration, 27
Coriolis tester, 32, 40
corrosion, 24
  protection (internal coatings), 166
curvature ratio (effect of r/D), 238
cutting, 26, 35
CVD technique, 159

D

data acquisition (AI models), 218
data preprocessing (feature engineering), 219
deformation, 26
density of particles, 9, 27, 28
design parameters, 68
detonation spray coatings, 171
diameter (pipeline diameter), 264
digital twins (predictive maintenance), 223
disperse phase model, 234
ductility, 28
dynamics (fluid flow), 73

E

EBPVD, 158
elastic modulus, 28
electrochemical, 39
  hydrogenation, 270
  measurement techniques, 39
electrochemical impedance spectroscopy (EIS), 40
electrochemical noise analysis (ENA), 40
electroless plating, 159
empirical equations for erosion rates, 30
epoxy coatings (pipelines), 167
erosion, 3, 24, 262
  analysis, 84
  analysis (velocity effect), 238
  fundamentals, 106
  intensification, 141
  mechanism, 153, 212
  model (Oka et al.), 234
  in pipe bends, 240
  prediction (ML methods), 215
  protection need, 155
  rate, 238
  rate prediction, 217
  resistance, 83, 114
  resistant coatings, 166
  slurry & solid particle, 111
  testing protocols, 250
  wear evaluation, 113
erosion–abrasion, 34, 36, 41

erosion–corrosion, 37–39, 41, 182
erosive wear, 85, 243

F

fabrication (metallic), 120
fatigue, 26
FBE powder (production process), 168
feature engineering, 219
field-scale validation, 252
Finnie's model, 30
flow
  conditions, 28
  dynamics, 141
  geometry, 29
  induced erosion, 231
  phenomenon, 3
  regime, 29
  velocity, 13, 27
fluid flow field, 73
food slurry (processing), 267
functionally graded coatings, 149
functionally graded materials, 128
fundamentals (slurry erosion), 199
fusion-bonded epoxy (FBE), 168
future prospects, 131
future trends (AI in erosion control), 222

G

galvanic current measurements, 40
geometric parameters (influence), 76
geometry (pipeline domain), 235
governing principles, 71
grain size, 128
grooving, 44

H

hardfacing, 143
  techniques, 160
hardness
  of particles, 11, 27
  target, 28
head loss, 49
heat treatment, 205
high-velocity air fuel (HVAF) coatings, 175
high-velocity oxygen fuel (HVOF) coatings, 173
  sprayed coatings, 188
HVOF spraying, 156
hydraulic simulation (surge analysis), 266
hydraulic turbines, 80, 88
hydrostatic pressure (effects), 193

I

impact angle, 29, 213
impaction/direct impact, 26
impact wear, 139
implementation/outcomes, 265
indentation, 26
industrial applications, 115, 160
industrial deployment (AI models), 220
industrial optimization, 261
industry context (slurry systems), 245
industry-scale coatings, 168
influence of microstructure, 128
inline viscosity sensors, 268
internal coatings, 166
  pipelines, 166
IoT sensors (erosion control), 222

J

jet erosion tester, 32, 41
jet impingement testing, 129

K

kinetics (of alane), 269

L

laser cladding, 159, 187
laser hardening, 125
laser surface
  modification, 144
  treatment, 202
ligand exchange, 271
liquid phase model, 233
localized corrosion, 38
localized material loss, 44

M

machine learning (slurry erosion), 215
marine (applications), 131
marine protection, 182
material(s)
  behavior (under slurry erosion), 200
  characterization (erosion-resistant alloys), 249
  considerations, 147
  HVOF coatings, 189
  properties, 27, 34
  response (ductile vs. brittle), 141
mathematical modeling, 233
mechanism(s)
  erosion-abrasion, 34
  erosion-corrosion, 37
  slurry erosion, 17, 25, 138, 153, 199
meshing (fluid domain), 235
metallic materials, 120
metals, 43
micro-fatigue, 35
micro-/nano-structured coatings, 149
microstructure, 28, 128, 188
mineral slurry, 243
mining (applications), 131
MMC reinforcements, 110
models (erosion rate), 30
model training & validation, 219
monitoring, 263
multisized slurry (diameter optimization), 264

N

nanostructured coatings, 189
Ni-based (nickel superalloys), 189
numerical analysis, 68
numerical modelling, 71

O

open-loop test systems, 33
optimization (cost/diameter), 264

P

particle
  behavior (turbulence effects), 253
  characteristics, 26, 35, 140
  kinetics, 139
  segregation, 76
  size, 4, 10, 26, 34, 214
  velocity, 213
performance
  assessment, 206
  evaluation, 129
  evaluation methods, 148
  HVOF coatings, 191
pH, 38
phase distribution, 128
physics-informed AI, 222
pipelines, 42, 53, 231
pitting, 44
plasma/HVOF (thermal spray), 142
plasma spray coatings, 170
plasma spraying, 156
plasma transferred arc (PTA), 187
plowing, 26, 35
polymer coatings (pipelines), 167
polymers, 43
polyurethane coatings, 167
post-processing (slurry erosion resistance), 197
potentiodynamic polarization, 39
powder flame spraying, 157
precipitation hardening, 129
prediction (slurry flow), 73
problem identification, 245
properties
  hardness/toughness, 188
  of slurries, 27, 35
protection (surface), 136
protective coatings, 179
pump(s), 80
  impellers, 42
  surface, 84
PVD coatings, 157
  methods, 115

Q

quantifying synergism, 42

R

recycling (traditional synthesis), 270
reliability, 265
rheological properties, 50

S

segregation mechanism, 76
SEM/experimental methods, 112
separation mechanisms, 68
settling properties, 5
shape of particles, 7, 27, 34
shot peening, 203
SLM (selective laser melting), 126
slurry
  flow characteristics, 73
  general, 3, 261
  parameters, 107
  particles, 82
  pot test, 31, 40
  pumps, 4
  resistant coatings, 166
slurry erosion, 3, 16, 24, 26, 38, 41, 49, 80, 88, 111, 136, 152, 199
  coatings, 166
  mechanism, 212
  testing, 25, 130
small-scale coatings (laboratory), 167
smart surfaces/AI & ML design trends, 150
sol-gel coatings, 187
solid concentration of slurry, 12
solid particle erosion, 111
solution approach, 264
solution strategy (erosion control), 255
spiral concentrators (description), 68
sputter technology, 158
submarine applications, 182
surface alloying & cladding, 204
surface characteristics, 214
surface coatings
  challenges, 161
  need, 155
surface engineering, 136
  techniques, 142
surface hardening, 124
surface post-processing techniques, 201
surface texturing and patterning, 145
surface treatment, 182
  heat treatments, 146
surge mitigation, 266
synergistic effects, 214

T

temperature
  control (zoned heating), 268
  produced during flow, 15, 27
testing (slurry), 24
test loops, 33
test methods of slurry erosion, 31
test setups (jet/slurry/rotating disc), 112
thermal spray coatings, 123, 142, 201
thermal spraying (types), 170
thermodynamics (of alane), 269
toughness, 28
turbines, 80
turbulence, 29
  intensity, 239
  modelling, 72
types of coatings (slurry pipelines), 166

U

unsupervised learning, 216

V

validation (flow model), 237
valves, 42
velocity effect (erosion rate), 238
viscosity, 27
    control/sensors, 267–268

W

wall thinning, 44
wear, 243, 263
wear zones (slurry systems), 252
wire arc spraying, 157, 177
working principles (separation), 70

For Product Safety Concerns and Information please contact our EU
representative GPSR@taylorandfrancis.com
Taylor & Francis Verlag GmbH, Kaufingerstraße 24, 80331 München, Germany

www.ingramcontent.com/pod-product-compliance
Ingram Content Group UK Ltd.
Pitfield, Milton Keynes, MK11 3LW, UK
UKHW052051230926
475702UK00025B/1108
9781032910604